全国高等医药院校药学类实验教材

基础物理学实验

（第三版）

主　编　赵　喆

编　者　（以姓氏笔画为序）

马　骄　　王晓飞　　亓　霞

支壮志　　邓岩浩　　孙　言

孙宝良　　苏婷婷　　李玉娟

李百芳　　张　翼　　张红艳

赵　喆　　赵　耀

中国健康传媒集团

中国医药科技出版社

内 容 提 要

《基础物理学实验》采用双语体系编写，注重教学实践与教学改革的结合，遵循由浅入深、循序渐进的教学原则，在内容上精简了原有的基础性实验，增设了提高性、综合性与设计性实验，引进了计算机仿真实验，融入与医药学科相关的新技术与当代科学技术相关的新成果。本教材适合高等医药院校本科、专科师生使用。

图书在版编目（CIP）数据

基础物理学实验/赵喆主编. —3 版. —北京：中国医药科技出版社，2019.10
（2024.8 重印）
全国高等医药院校药学类实验教材
ISBN 978-7-5214-1339-7

Ⅰ.①基… Ⅱ.①赵… Ⅲ.①物理学-实验-医学院校-教材 Ⅳ.①O4-33

中国版本图书馆 CIP 数据核字（2019）第 202332 号

美术编辑　陈君杞
版式设计　友全图文

出版	中国健康传媒集团｜中国医药科技出版社
地址	北京市海淀区文慧园北路甲 22 号
邮编	100082
电话	发行：010-62227427　邮购：010-62236938
网址	www.cmstp.com
规格	787×1092mm $^1/_{16}$
印张	22
字数	507 千字
初版	2006 年 3 月第 1 版
版次	2019 年 10 月第 3 版
印次	2024 年 8 月第 3 次印刷
印刷	北京京华铭诚工贸有限公司
经销	全国各地新华书店
书号	ISBN 978-7-5214-1339-7
定价	65.00 元

获取新书信息、投稿、
为图书纠错，请扫码
联系我们。

版权所有　盗版必究
举报电话：010-62228771
本社图书如存在印装质量问题请与本社联系调换

前　言

物理学是一门重要的基础学科，是整个自然科学的基础。物理实验是物理课程的重要组成部分，是学生步入高校接触的一门实践性基础课程。实验教学是高等药学院校最基本的教学形式之一，在培养学生科学的思维与方法、创新的意识与能力，全面推进素质教育等诸多方面具有重要作用。

物理学是药学各专业的重要基础课，药学研究和生产中处处融有物理学的基本理论、基本方法和研究思路，物理量的变化对药物的性质和药物的运动规律有较大影响。随着现代科学与实验技术的迅速发展，对高等医药院校的物理实验课教学提出了更高的要求，大量先进的物理学实验进入教学课程体系成为必然趋势。因此，在新的形势下，为培养适应21世纪的高素质的药学专业人才，加大对实验项目、实验内容、实验手段、实验方法的改革势在必行。

本教材由工作在教学一线、具有丰富的理论与实验教学经验的教师编写而成。在编写过程中，力求做到将教学实践与教学改革相结合，将物理理论与医药专业相结合，将经典理论与现代技术相结合，将基本能力训练与创新能力培养相结合，遵循由浅入深、循序渐进的教学原则，在结构和内容上都进行了重大改革。在结构上，将实验课题进行了具有层次化的板块设置；在内容上，精简了原有的基础性实验，增设了提高性、综合性与设计性实验，引进了计算机仿真实验，融入与药学学科相关的新技术与当代科学技术相关的新成果。同时，为提高学生的科技英语水平，本教材采用双语体系编写，为实验课程改革构建数字化、信息化的教学平台。

本教材可供高等医药院校本科生使用，也可供药学专业专科和函授学生使用。

参加本书编写人员：赵喆、孙宝良、李玉娟、张翼、支壮志、马骄、邓岩浩、张红艳、李百芳、亓霞、赵耀、苏婷婷、孙言、王晓飞。

插图绘制：赵喆、支壮志。

本教材在编写过程中得到了编者所在院的大力支持，外语部战中亮老师为本书的翻译工作提供了许多帮助，在此一并表示感谢。

由于编者水平有限，书中难免存在疏漏谬误之处，希望读者提出宝贵意见，以便今后加以修正完善。

编　者
2019年8月

目录

第一章 绪论 …………………………………………………………………… (1)
Chapter One　Exordium …………………………………………………… (1)
　第一节　物理实验课程的基本要求和程序 …………………………………… (1)
　Section 1　Basic Requirements and Procedures in Physics Experiment …… (2)
　第二节　误差、有效数字和数据处理 ………………………………………… (4)
　Section 2　Error，Significant Digits and Data Processing ………………… (21)

第二章 基础性实验 …………………………………………………………… (43)
Chapter Two　Basic Experiment …………………………………………… (43)
　实验一　基本长度测量 ………………………………………………………… (43)
　Experiment 1　Measurement of Length ……………………………………… (49)
　实验二　物体密度的测定 ……………………………………………………… (56)
　Experiment 2　Determination of Density ……………………………………… (60)
　实验三　液体变温黏滞系数的测定 …………………………………………… (65)
　Experiment 3　Determination of the Coefficient of Viscosity for Liquid of Varied
　　　　　　　　Temperature …………………………………………………… (69)
　实验四　液体表面张力系数的测量 …………………………………………… (73)
　Experiment 4　Determination of Surface Tension Coefficient of Liquid …… (80)
　实验五　万用电表的使用 ……………………………………………………… (88)
　Experiment 5　Manipulation of Multimeter …………………………………… (94)
　实验六　示波器的使用 ………………………………………………………… (102)
　Experiment 6　Usage of Oscilloscope ………………………………………… (107)
　实验七　惠斯通电桥测电阻 …………………………………………………… (115)
　Experiment 7　Measurement of Resistance with Wheatstone Bridge ……… (122)
　实验八　RLC 交流电路 ……………………………………………………… (127)
　Experiment 8　RLC Alternating Current Circuit …………………………… (130)
　实验九　用线式电位差计测电池的电动势 …………………………………… (133)
　Experiment 9　Measurement of the Battery EMF with Wire Potentiometer … (136)
　实验十　透镜曲率半径的测量 ………………………………………………… (140)
　Experiment 10　Measurement of Radius of the Lens Curvature …………… (145)
　实验十一　利用劈尖干涉测量厚度 …………………………………………… (150)
　Experiment 11　Determine Thickness by Interference of a Wedge – sharped
　　　　　　　　　Film …………………………………………………………… (152)

第三章 提高性实验 …………………………………………………………… (155)
Chapter Three　Improving Experiment …………………………………… (155)
　实验十二　分光计的调节和使用 ……………………………………………… (155)
　Experiment 12　Adjustment and Use of Spectrometer ……………………… (163)

· 1 ·

实验十三 用衍射光栅测定光波波长 ………………………………………… (170)
Experiment 13 Determination of the Wavelength of Light Using a Diffraction
　　　　　　　Grating ……………………………………………………… (174)
实验十四 用阿贝折射计测定液体的折射率 ……………………………… (178)
Experiment 14 Determination of Liquid Index of Refraction by Abbe
　　　　　　　Refractometer ……………………………………………… (182)
实验十五 利用旋光现象测定液体的旋光率和浓度 ……………………… (186)
Experiment 15 Determine Specific Rotation and Concentration of Liquid Utilizing
　　　　　　　Optical Phenomena ………………………………………… (191)
实验十六 用模拟法描绘静电场 …………………………………………… (196)
Experiment 16 Simulation Method for Depicting Electrostatic Field …………… (202)
实验十七 核磁共振 ………………………………………………………… (210)
Experiment 17 Nuclear Magnetic Resonance …………………………………… (214)
实验十八 激光全息照相 …………………………………………………… (219)
Experiment 18 Laser Hologram Photography …………………………………… (224)

第四章 综合性与设计性实验 ………………………………………………… (230)
Chapter Four　Comprehensive and Designing Experiment ……………………… (230)
实验十九 指端光电容积法测量脉搏波 …………………………………… (230)
Experiment 19 Fingertip Pulse Wave Measurement Using Photoelectric Plethysmography
　　　　　　　……………………………………………………………… (233)
实验二十 血液流变学参数测定 …………………………………………… (237)
Experiment 20 Determination of Hemorheology Parameter ……………………… (241)
实验二十一 利用压力传感器测定人体血压 ……………………………… (247)
Experiment 21 Human Body Blood Pressure Measurement Using a Pressure
　　　　　　　Sensor ……………………………………………………… (252)
实验二十二 用箱式电位差计测量温差电动势 …………………………… (258)
Experiment 22 Measurement of Thermal Electromotive Force with the Box – type
　　　　　　　Potentiometer ……………………………………………… (263)
实验二十三 人耳听觉听阈的测量 ………………………………………… (269)
Experiment 23 Determination of the Human Ear Auditory Threshold Curve …… (272)
实验二十四 热敏电阻温度计的制作 ……………………………………… (276)
Experiment 24 The Assembly Make of Thermistor Thermometer ………………… (279)

第五章 计算机仿真实验 ……………………………………………………… (282)
Chapter Five　The Computer Simulation Experiment …………………………… (282)
实验二十五 虚拟仪器基础 ………………………………………………… (282)
Experiment 25 The Basic of Virtual Instrument ………………………………… (293)
实验二十六 测试二极管并确定其极性 …………………………………… (307)
Experiment 26 Testing Diodes and Determining Their Polarities ……………… (310)
实验二十七 电路仿真实验 ………………………………………………… (313)
Experiment 27 Simulating Circuits ……………………………………………… (320)
实验二十八 使用myDAQ捕获声音信号 ………………………………… (327)
Experiment 28 Capturing Sound with myDAQ ………………………………… (332)

附表 ………………………………………………………………………………… (338)
参考文献 …………………………………………………………………………… (344)

第一章 绪 论

Chapter One　Exordium

第一节　物理实验课程的基本要求和程序

物理实验是一门独立设置的实践性极强的课程。物理学概念的建立、原理和定律的发现都是以严格的物理实验为基础并通过实验加以证明。做好物理实验对学生深入理解物理学基本原理、掌握物理学的实际应用、探寻物质运动规律、培养学生的实验动手能力具有重要意义。

在现代药学中，物理学的理论和实验方法得到了广泛的应用。物理实验是学生进入大学后接受系统实验方法和实验技能的开端，通过物理实验的基本训练，必将为学生今后的学习和工作奠定良好的基础。

一、教学要求

开设物理实验这门课程的主要目的是培养学生具有良好的实验素养、基本实验技能、理论联系实际和独立工作能力。本课程的基本教学要求如下。

（1）使学生在物理实验基础知识、基本方法和实验技能诸方面受到系统且严格的训练。

（2）学习如何根据实验要求确定合理的实验方案，正确选择和使用基本仪器，掌握各种测量技术和实验方法，对实验数据进行处理（包括有效数字计算、误差分析及作图等），判断和分析实验结果等等。

（3）学习物理实验方法，探求物理规律，观察和分析物理现象，通过实验加深对一些重要物理规律的认识和理解，并用学过的理论及实验知识指导实验、分析实验中存在的问题。

（4）通过实验培养学生严肃认真、实事求是的科学态度和工作作风。

二、教学环节

物理实验是学生在教师指导下独立进行实验的一种实践活动，因此在实验过程中应当发挥学生的主观能动性，有意识地培养学生独立工作的能力和严谨的工作作风。每个实验都有三个教学环节：预习实验、进行实验和撰写实验报告。

1. 预习实验

课前应仔细阅读实验教材，了解实验的原理和方法，并基本了解实验仪器的使用方法，特别是每个实验后所附的仪器介绍部分（最好到实验室看一下实验设备），在此基础上写出实验预习报告。要明确哪些物理量是直接测量量、哪些是间接测量量，用什么方法和测量仪器来测定等等。正确写出实验步骤、画出实验电路图（如电磁学实验）及填写数据记录表格。

2. 进行实验

实验时应遵守实验室规章制度，仔细阅读有关的仪器使用说明书和仪器的使用注意事项，在教师的指导下正确使用仪器。对电磁学实验，必须由教师检查电路的连接正确无误后，方可接通电源进行实验。实验进行时，应合理操作，认真思考，仔细观察，把实验数据细心地记录在预习报告的表格内。记录时用钢笔或圆珠笔，不要用铅笔。如需要删去已记录的数据，可用笔划掉，同时注明原因。此外，还要记下所用仪器的型号、编号、规格，便于以后核对数据时查用。

注：在实验过程中需特别注意安全。一旦发现异常现象，应立即切断电源、火源、气源，并报告实验指导教师，有必要请拨打"119"火警电话。

3. 撰写实验报告

先对数据进行整理计算，然后用简洁的文字撰写实验报告。报告应字迹清晰、文理通顺、图表正确。实验中的原理图、电路图应用尺规画出，对实验结果的图解表示必须仔细认真、力求准确，并利用尺规或曲线板画在坐标纸上。作图一律用铅笔。

实验报告一般应包括以下内容。

（1）实验名称、实验者姓名、实验日期。

（2）实验目的。

（3）实验仪器（型号、编号、规格）及装置。

（4）实验原理和方法。要用自己的语言简要地叙述，不要盲目抄书。

（5）实验步骤，数据记录（将预习报告上所记录的数据仔细地转记于实验报告上）及计算（包括误差计算）。

（6）实验结果及讨论（如实验中观察到的现象的分析、改进实验的建议、实验后的体会、实验中存在的问题、回答思考题等等）。

通常，实验报告按上述要求撰写即可，对有特殊要求的实验报告，可参见各实验的具体说明。

Section 1　Basic Requirements and Procedures in Physics Experiment

Physics experiment is a practical course. The establishment of physics concept, the discoveries of the principle and the law are all based on physical experiments and in turn it serves the experiment. It is vital for students to do physics experiments so that they can understand the physics basic principle thoroughly, as well as to develop their abilities to utilize the physics.

In modern pharmacy, the physics theory and the experimental technique are widely used. Physical experiment is the first step for college students to accept the systematic experimental techniques and the skills. Through the basic training of physics experiment, it will lay a good foundation for students' study in the future.

Teaching requirements

The main purpose of physics experiment is to develop a good habit for the students in experiments, the basic experimental skills and the ability to apply theory to reality. The requests are as follows.

(1) To give the students a strict training in elementary knowledge, methods and skills.

(2) To learn how to determine a correct experiment plan, choose the basic instruments, grasp different kinds of survey technology and the experimental techniques, process experiment data (including significant digits computation, error analysis and mapping, etc), judge and analyze the experiment results and so on.

(3) To master physics experiments' techniques, observe and analyze physical phenomenon with physics experimental technique, deepen the understanding on the law of important physics, and use the experimental knowledge to instruct and analyze the problems in the experiments.

(4) To develop the students' scientific attitude towards work.

Teaching process

The physical experiment is a practical activity for the students under the teachers' instruction. Thus the teacher should encourage the students' subjective initiative in the experimental process, and consciously develop their abilities to work independently. Each experiment has three steps: preparation, experimenting and experiment report.

1. Preparation

Students should read the experimental teaching materials carefully, and understand the principle of experiment and the basic measuring method of equipment, especially the introduction attached to the instruments (you'd better have a look in the laboratory), as well as the experiment report based on the preparation. You must be clear about which physical quantities are directly measured and which are indirectly measured, and know which one is the best method and equipment and so on. Write down the procedure correctly and draw the experimental circuit diagram (for example electromagnetics experiment) and the data form.

2. Experimenting

You should observe the laboratory rules and regulations while experimenting. Read the instruction carefully and use the equipments under the teachers' instruction. As for the electromagnetics experiment, you can do the experiment after the teacher's check on the circuit connection. When doing the experiment, you should operate correctly, observe carefully and record data in details. When recording, you should use pens or ball pens, do not use pencils. If you want to get rid of the recorded data, you can scratch it with pens and write down the reason. In addition, you must take down the type, serial number, specification of the equipment so that you can check the data later.

Note: In the process of experiment, please keep safety always in your mind. If there is anything abnormal, cut off the electric power, fire and gas supply, and then report to the experiment instructors. Please dial "119" if it is emergent.

3. Experiment report

Your experiment report should contain correct graphs which are clear and logical. The experimental schematic diagram and the circuit diagram should be drawn with the ruler and compasses, and the graphic solution to the experiment must be drawn carefully with ruler, compasses or French curve on the coordinate paper. Pencils are needed in all the mappings.

Generally, the experiment report should include:

(1) The experiment title, name and date.

(2) Experiment objectives.

(3) Equipments (type, serial number, specification).

(4) Experimental principle. To narrate briefly in your own words, do not copy the book.

(5) Procedure, data (which has been recorded on the preparation should be rewritten on experiment report carefully) and the computation (including error).

(6) Experiment results and the discussion (for example the analysis of the phenomenon observed, some suggestions about the improvement of the experiment, the problems in the experiment and the thought on the experiment, etc).

Usually, the experiment report should be written according to the above requests, if there are any other special requirements, your report should base on the request.

第二节　误差、有效数字和数据处理

一、测量与误差

（一）物理量的测量

测量是人类认识世界和改造世界的基本手段，通过测量，人对客观事物可以获得定量的概念，总结出它们的规律性，从而建立起相应的物理定律和物理定理。

测量是将待测的物理量与选定的同类单位量进行比较的过程。测量分为直接测量和间接测量。直接测量是用仪器直接将待测量与选定的同类单位量进行比较，即直接在仪器上读出待测量的数值。例如，用秒表测量时间，用米尺测量长度等等。间接测量是由几个直接测量出的物理量通过已知的公式或定律进行计算，间接求出待测量。例如，直接测量出圆柱体的直径 D、高度 H 和质量 m，由公式 $\rho = \dfrac{m}{\frac{1}{4}\pi D^2 H}$，即可求出圆柱体的密度，大多数物理量都是通过间接测量得到的。

（二）误差的定义及分类

由于测量仪器、实验条件及观察者感官等种种因素的限制，测量值与客观存在的真值之间总有一定的差异，这就是我们所说的测量误差。误差存在于一切测量之中，贯穿测量过程的始终。测量误差的大小反映出我们的测量与客观真值的接近程度，讨论误差的来源，消除或减小测量的误差是提高测量的准确程度、使测量结果更为可信的关键。

根据误差产生的原因及对实验的影响性质，可将其分为系统误差和偶然误差。

1. 系统误差

系统误差的主要来源为仪器本身制造上的不完善，仪器未经很好地校准，测量的外部条件与仪器要求的环境条件不符，测量方法和技术不完善，以及测量者不正确的习惯等。系统误差通常在每次测量中都有一定大小，一定符号或按一定规律变化，可以通过校正仪器，改进测量方法、修正公式和定律等手段加以消除或减小。

2. 偶然误差

（1）偶然误差的定义：这种误差是由许多不稳定的偶然因素引起的。即使在相同条件下，对同一量进行多次重复测量时（甚至在极力消除或改正一切明显的系统误差之后），每次测量结果仍然不同，呈现出无规则的随机变化，由此产生的误差称为偶然误差（或随机误差）。

（2）产生偶然误差的原因

① 由于观察技术及个人感官的限制，使观察者在调节仪器和估计读数时带来误差，使结果时大时小。

② 在实验过程中周围环境和状态难以控制的不规则的变化。

偶然误差是一种无规则的涨落，看不出它们的规律性，如图1－2－1（a）所示。当测量的次数足够多时，偶然误差服从一定的统计规律，测量结果总是在真值（A）附近涨落，由于这种误差的偶然性，因此它是不可消除的，但是增加测量的次数可以减小测量的偶然误差。系统误差的存在使测量值按一定的大小及方向偏离真值，不能用多次重复测量的办法来发现及克服它的影响。通常情况下，系统误差和偶然误

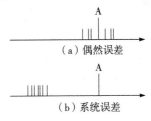

图1－2－1 误差分布

差是同时存在的，如图1－2－1（b）所示。系统误差小则测量的准确度高，偶然误差小表明测量的可重复性好，或者说是精密度高。精确度高是指偶然误差和系统误差都小。

如上所述，误差与错误是两个完全不同的概念，错误是实验者仪器使用不正确，或实验方法不合理，或违反操作规程，或粗心大意读错数据等，学生应加强责任感，避免错误的发生，设法减小误差。

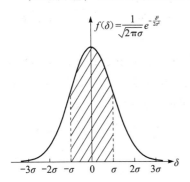

图1－2－2 偶然误差的正态分布

（3）偶然误差的统计分布规律：大量的实验发现，重复测量次数趋近无穷多时，测量误差δ趋近正态分布（图1－2－2），$f(\delta)$表示误差δ出现的概率。该分布具有如下特点：正负误差出现的概率相等，绝对值较小的误差出现的次数较多，很大的误差通常不出现，随机误差的算术平均值趋于零；若无系

统误差，测量的平均值趋于真值。当测量次数较少时将偏离正态分布，增加测量次数可以减少测量误差。

二、有效数字

（一）仪器的精密度

仪器的精密度（又称精度）是指在正确使用测量仪器时，所能测得的最小的准确值，它一般由仪器的分度（仪器所标示的最小分划单位）决定。例如，用 mm 分度尺测量物体的长度，其精密度就是 1mm。

（二）有效数字

1. 有效数字

由于仪器精密度的限制，测得的任何一个物理量的数值的位数只能是有限的。正确而有效地表示测量和计算结果的数字，称为有效数字。它是由全部准确数字和一位欠准确数字构成的。如图 1-2-3 所示，用最小分度是毫米的米尺测量木块的长度，使木块的一端

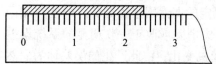

图 1-2-3

与米尺的零刻度线对齐，在另一端从米尺上读取木块长度的数据，由图中可以看到 A 比 23mm 长约半个刻度，则测量结果可以记为 23.6mm。在这三位数字中，前两位数 23 是从尺上最小分度以上准确读得的，因此是可靠的，称之为"准确数字"；而从尺子的最小刻度中估读出的 0.6mm，其 6 这一位是从毫米刻度以下估计得来的，若换另一人来读数，也可能估计成 5 或 7，故 6 这一位是欠准确数字，称这样的数字为"可疑数字"。显然，可疑数字是包含有误差的数字，但它还是在一定程度上反映了被测量实际大小的信息，因此也是有效的，一般来说，可疑数字只取一位。在测量中，全部可靠数字和估读的一位可疑数字一起构成有效数字。

2. 用有效数字表示测量值时应注意的问题

（1）测量值有效数字的位数不能随便增减，有效数字的位数与测量的相对误差有关，有效数字的位数愈多，测量值的相对误差愈小。

（2）测量值中的非零数字都是有效数字（1，2，3，4，5，6，7，8，9），"0"不总是有效数字。例如，测得一物体的长度为 0.0050800m，它的有效数字位数为五位。"5"前面的 3 个"0"只表示小数点的位置，不是有效数字，其他三个"0"是有效数字。最后的"0"表示测量的估计值，因此是有效数字，其他两个"0"均处于有效数字之间，故均是有效数字。

（3）有效数字的位数与单位的换算无关。为了突出有效数字位数，我们采用科学记数法表示有效数字。这一方法规定了小数点前一律取一位非零数字，其他照写，所引起的不同数位由 10 的幂次来补充。例如，0.0010300kg 可表示为 1.0300×10^{-3} kg；125.20ms 可表示为 1.2520×10^{2} ms 或 1.2520×10^{-1} s。这些例子说明了单位的改变不影响有效数字的位数。

（4）有效数字的进位法。有效数字通常采用四舍五入法，例如：将 1.625 取为三位有效数字，应写作 1.63；取两位有效数字，应记为 1.6。还有另一种经常采用的

"四舍六入五凑偶"方法,即尾数小于5则舍,大于5则入,等于5则将其尾数凑成偶数。例如:将1.625取为三位有效数字,可记为1.62;将1.615取为三位有效数字,仍记为1.62;将1.626取为三位有效数字,则记为1.63。本教材采用的是四舍五入法。

(5) 常数(如π、$\sqrt{2}$、e、$\frac{1}{4}$…)的有效数字位数是无限的,可根据具体问题适当选取,一般情况下,至少比测量值多保留一位。

(三) 有效数字的运算规则

1. 加减法运算规则

在有效数字的加减法运算中,计算结果的有效数字的最后一位与各分量中可疑位的最高位置保持一致。

例 1-2-1 普通算术式

$$X = 71.3 - 0.753 + 6.26 + 271$$

有效数字计算:

$$X = 71.3 - 0.8 + 6.3 + 271 = 348$$

数字下面的"●"表示可疑位的最高位置。

2. 乘除法运算规则

在有效数字的乘除法运算过程中,计算结果的有效数字位数与各分量中有效数字位数最少的分量保持一致。

例 1-2-2

$$X = \frac{39.50 \times 4.08 \times 0.0013}{868.01} = \frac{39.5 \times 4.08 \times 0.0013}{868} = 2.4 \times 10^{-4}$$

可见,此式分量中有效数字位数最少的为0.0013,为两位有效数字,故将其作为标准。

3. 乘方、开方等运算中有效数字的取法

在乘方、开方、三角函数等运算中,结果的有效数字位数与底数的有效数字位数保持一致。举例如下。

例 1-2-3

$$\sqrt{625} = 2.50 \times 10$$
$$25^2 = 6.3 \times 10^2$$

例 1-2-4

$$\cos 32.7° = 0.842$$

例 1-2-5

$$\lg 19.28 = 1.285$$

注意:在较复杂的多步计算过程中,各中间结果在运算过程中按四舍五入处理后应多保留一位。

另外,对物理公式中不是由实验测出的某些数值,例如,测量圆柱体的直径D和长度H,求其体积V的公式$V = \frac{1}{4}\pi D^2 H$中的$\frac{1}{4}$不是测量值,则在确定V的有效数字位

数时不必考虑$\frac{1}{4}$的有效数字位数。

由有效数字的运算规律可见，对不同准确度的数字进行计算时，其结果的有效数字位数应取得恰当。取少了，会带来附加的计算误差，降低结果的精度；取多了，从表面上看似乎精度很高，实际上毫无意义，反而给人以错误的印象和带来不必要的繁复。

在实验中正确运用有效数字，不仅能如实反映测量结果而且可以简化运算、节省时间，即使用计算器也应养成合理取有效数字的习惯。

三、直接测量值误差的估计

由于误差的存在，在直接测量中不可能准确地测量出待测量的真值，为使测量更加准确，往往需要进行反复多次的测量，然而各次测量的结果不同，我们该如何表示其测量结果呢？

（一）单次直接测量的误差估计

有些实验，由于在动态中测量，不容许对被测量做重复测量；也有些实验的精密度要求不高；或在间接测量中，其中某一物理量的误差对最后的结果影响较小。在这些情况下可以对被测量只测一次，对于单次测量的误差，一般是根据所用仪器、测量对象、实验方法和实验者的经验来估计误差。在一般情况下，对于偶然误差很小的测量值，可将仪器出厂说明书或仪器标牌上注明的仪器误差作为单次测量的误差。如果查不到仪器误差，也可以取用仪器最小分度的一半作为单次测量的绝对误差。对于偶然误差较大的，应根据情况采用仪器最小刻度或甚至更大的合理数值。

例 1-2-6　用千分尺测某物体长度（一次）为 3.647mm（尾数 7 是估计值），而千分尺的最小刻度为 0.01mm，那么我们可取 0.005mm 作为一次测量结果的绝对误差。

例 1-2-7　用米尺测一摆线长，如果米尺使用得正确，则读数误差是测量误差的主要成分，摆的上下两端读数误差各取 0.5mm，则长度测量误差可取为 1mm。

例 1-2-8　用秒表测量一物体运动的时间间隔，如果秒表的系统误差不必考虑，则测量的误差主要是由启动和制动秒表时，手的动作和目测协调的情况决定的。一般可估计启动、制动时各有 0.1s 误差，总的误差为 0.2s。

例 1-2-9　用天平称量物体质量时，空载时天平指针的停点和加砝码后天平指针的停点一般是不一致的，其差异将引起测量误差，当此二停点之差不超过一个分度时，可取测量误差为天平指针停点移动一分度时两侧的质量差额。天平指针停点移动一个分度，天平两侧质量差称为天平的感量（或称感度）。

（二）多次直接测量的误差估计

为了减少偶然误差，在可能的情况下，总是采用多次测量，将多次测量的算术平均值作为测量的结果。如果对某物理量进行重复测量，每次的测量值分别为：x_1，x_2，x_3，$\cdots x_i$，$\cdots x_n$，则其算术平均值为

$$\bar{x} = \frac{x_1 + x_2 + x_3 + \cdots + x_n}{n} = \frac{1}{n}\sum_{i=1}^{n} x_i \qquad (1-2-1)$$

根据误差理论，多次测量值的算术平均值比各个测量值更可能接近真值。我们可将 $\bar{x}$ 看作该物理量的近似真值。

根据误差的定义式可知，误差不能确定，只能估计。估计偶然误差的方法有平均绝对误差、标准差及相对误差。

1. 平均绝对误差

我们把第 i 次测量值与平均值之差称为偏差（或残差）

$$d_i = x_i - \bar{x} \tag{1-2-2}$$

n 次测量的偏差的绝对值的平均值称为平均绝对偏差或平均绝对误差，用 Δx 表示，即

$$\Delta x = \frac{1}{n}[|x_1 - \bar{x}| + |x_2 - \bar{x}| + \cdots + |x_n - \bar{x}|] = \frac{1}{n}\sum_{i=1}^{n}|x_i - \bar{x}| \tag{1-2-3}$$

这样定义的误差不能说明平均值是大于还是小于真值。通常将真值表示为

$$x = \bar{x} \pm \Delta x$$

意思是，真值在 $\bar{x} + \Delta x$ 与 $\bar{x} - \Delta x$ 之间。此误差是真值与平均值相差的绝对量，它反映测量值偏离真值的范围，故常称为绝对误差，它的单位与测量值的单位相同。绝对误差可以表示测量的优劣，但并不全面，通常还需引进相对误差加以补充。

2. 相对误差

相对误差是绝对误差与最佳值（算术平均值）的比，它是没有单位的。相对误差是用来比较不同测量对象可靠性程度的指标，常用百分数来表示，即

$$E_r = \frac{\Delta x}{\bar{x}} \times 100\% \tag{1-2-4}$$

故又称为百分误差。为了说明相对误差的物理意义，下面举例加以说明。例如现测得两个物体的长度为

$$L_1 = (23.50 \pm 0.03)\text{cm}, L_2 = (2.35 \pm 0.03)\text{cm}$$

则其相对误差分别为

$$E_{r_1} = \frac{0.03}{23.50} \times 100\% = 0.13\%$$

$$E_{r_2} = \frac{0.03}{2.35} \times 100\% = 1.3\%$$

从绝对误差来看，两者相等，但从相对误差来看，后者是前者的 10 倍。显然，第一个测量比第二个测量更准确些。因此，实验的测量结果，应同时用绝对误差和相对误差表示。用相对误差表示的测量结果表示式为

$$x = \bar{x}(1 \pm E_r)$$

绝对误差和相对误差的关系为

$$\Delta x = \bar{x} \cdot E_r$$

例 1-2-10 将某一物体的长度测量 5 次，得到的测量值分别为：$L_1 = 3.41\text{cm}$, $L_2 = 3.43\text{cm}$, $L_3 = 3.45\text{cm}$, $L_4 = 3.44\text{cm}$, $L_5 = 3.42\text{cm}$，求最后的测量结果。

解：算术平均值为

$$\bar{L} = (3.41 + 3.43 + 3.45 + 3.44 + 3.42) = 3.43(\text{cm})$$

各次偏差的绝对值（即各次绝对误差）为

$$\Delta L_1 = |3.41 - 3.43| = 0.02\text{cm}$$
$$\Delta L_2 = |3.43 - 3.43| = 0.00\text{cm}$$
$$\Delta L_3 = |3.45 - 3.43| = 0.02\text{cm}$$
$$\Delta L_4 = |3.44 - 3.43| = 0.01\text{cm}$$
$$\Delta L_5 = |3.42 - 3.43| = 0.01\text{cm}$$

平均绝对误差为

$$\Delta L = \frac{1}{5}(0.02 + 0.00 + 0.02 + 0.01 + 0.01) = 0.012 \approx 0.02(\text{cm})$$

相对误差为

$$E_r = \frac{\Delta L}{\bar{L}} \times 100\% = \frac{0.02}{3.43} \times 100\% = 0.00583 \approx 0.59\%$$

最后的测量结果表示为

$$L = (3.43 \pm 0.02)\text{cm}$$

或

$$L = 3.43(1 \pm 0.59\%)(\text{cm})$$

从上面几个例子可以看出：

（1）绝对误差一般只取一位有效数字，且采取只进不舍进位法，即其后的数字不分大小（零除外）一律进位。例如：$\Delta M = 0.013\text{g}$ 应写成 $\Delta M = 0.02\text{g}$。

相对误差一般只取两位有效数字，进位方式同上。例如：$E_r = 1.21\%$ 写成 1.3%；0.8% 写成 0.80%；10.4% 写成 11%。

（2）无论直接或间接测量的结果，其主值（单次测量值、平均值或计算结果）位数取舍的最后依据是：它的末位必须与绝对误差所在的位对齐，即测量值的最后一位应与绝对误差同数量级。测量值末位以后的数字，则采取四舍五入。例如：计算得到某物理量的测量值为 18.659 单位，而绝对误差是 0.04 单位，则最后结果写为 (18.66 ± 0.04) 单位。

（3）当误差参加运算时，为了避免产生附加误差，在运算中应多保留一位有效数字。

3. 标准误差、标准偏差和平均值的标准偏差

（1）标准误差：在图 1-2-2 中，σ 是一个与实验条件有关的常数，称之为正态分布的标准误差。

$$\sigma = \lim_{n \to \infty} \sqrt{\frac{\sum_{i=1}^{n}(x_i - \bar{x})^2}{n}} \tag{1-2-5}$$

（2）标准偏差：在实际测量中，n 不可能 $\to \infty$，对一测量列，只要满足 $n \geq 5$（实验中一般取 $5 \leq n \leq 10$），则有一标准误差的代替者标准偏差（偏差 = 测量值 - 算术平均值）。有限次测量中某一次测量的结果的实验标准偏差为

$$S = \sqrt{\frac{\sum_{i=1}^{n}(x_i - \bar{x})^2}{n-1}} \quad (\text{贝塞尔公式}) \qquad (1-2-6)$$

（3）n 次测量的算术平均值的标准偏差为

$$S_m = \frac{S}{\sqrt{n}} = \sqrt{\frac{\sum_{i=1}^{n}(x_i - \bar{x})^2}{n(n-1)}} \qquad (1-2-7)$$

平均值的标准偏差比任何一次测量的实验标准差小，增加测量次数，可以减少平均值的标准偏差，提高测量的准确度。但是，$n>10$ 以后，n 再增加，平均值的标准偏差减小缓慢，因此，在物理实验教学中一般取 n 为 6~10 次。

根据统计理论，测量值 x_i 中的 68.3% 落在 $\bar{x} \pm S$ 区域内，x_i 中的 95.4% 落在 $\bar{x} \pm 2S$ 内。真值在 $\bar{x} \pm S_m$ 区域内的概率为 68.3%，而在 $\bar{x} \pm 2S$ 区域内的概率为 95.4%。这些概率值叫作置信度。

实验结果可以表示为 $\bar{x} \pm S_{mx}$，其中 S_{mx} 表示被测量 x 的 n 次测量的平均值的标准差。这时相对误差可表示为

$$E_r = \frac{S_{mx}}{\bar{x}} \times 100\% \qquad (1-2-8)$$

例 1-2-11 测量电阻 R 8 次，求最后的测量结果。

R（Ω）	d_i（mΩ）	d_i^2（mΩ)^2
4.615	−10	100
4.638	13	169
4.597	−28	784
4.634	9	81
4.613	−12	144
4.623	−2	4
4.659	34	1156
4.623	−2	4

解： $\bar{R} = 4.625\,\Omega$，$S_{um} = 2442\,(\text{mΩ})^2$

因此 $S = \sqrt{\dfrac{2442}{7}} = 19\,(\text{mΩ}) = 0.019\,(\Omega)$

$$S_m = \frac{0.019}{\sqrt{8}} = 0.007\,(\Omega)$$

结果表示为 $R = \bar{R} \pm S_m = (4.625 \pm 0.007)\,\Omega$

在一般情况下，即测量次数不多，如在 50 次以下时，标准偏差取两位是适当的；如果只取一位，这仅仅在测量次数很少，标准偏差的第一位的数值是较大的数时才适当，这是根据统计理论做出的结论。

4. 测量不确定度

（1）测量不确定度是指由于测量误差的存在而对被测量值不能确定的程度，是对

被测量真值在某个量值范围内的评定。

不确定度通过"量值范围"和"置信概率"来表达。不确定度给出了被测量的真值在某一范围 $[X-U, X+U]$ 内的可能性有多大，表示出测量结果的可信程度有多大，明确指出了被测量的真值处于某一范围中的概率。在 σ、2σ、3σ 三种范围中的概率分别是：$P_\sigma = 68.3\%$，$P_{2\sigma} = 95.4\%$；$P_{3\sigma} = 99.7\%$。

不确定度和误差是两个不同的概念。误差表示测量结果对真值的偏离，是一个确定的值；不确定度表明测量值的分散性，表示一个区间。由于真值是不知道的，测量误差只是理想的概念；不确定度则可以根据实验、资料、经验等信息进行定量确定。不确定度大，不一定误差的绝对值也大，两者不应混淆。

（2）不确定度的评定及合成：测量不确定度一般包含几个分量，按其数值评定的方法，可分为两大类：采用统计方法评定的 A 类标准不确定度分量和采用其他非统计方法评定的 B 类标准不确定度分量。A、B 两类不确定度只是评定方法的不同，不是不确定度性质的不同。有些情况下只需进行 A 类或 B 类评定，更多情况下要综合 A、B 两类评定的结果。

A 类标准不确定度：在相同条件下多次重复测量用统计方法计算（主要涉及随机误差）求得的不确定度，用 u_A 表示。用算术平均值的实验标准偏差 $s(\bar{x}_i)$ 来表征，它与单侧测量结果 x_i 的实验标准差 $s(x_i)$ 的关系为

$$u_A = s(\bar{x}_i) = \frac{s(x_i)}{\sqrt{n}} = \sqrt{\frac{\sum_{i=1}^{n}(x_i - \bar{x})^2}{n(n-1)}} \qquad (1-2-9)$$

B 类标准不确定度：是指用其他非统计方法评定的不确定度分量，例如仪器误差、未定系统误差的估值等，用 u_B 表示。

B 类标准不确定度的信息来源一般有：以前的观数据；对有关技术资料和测量仪器特性的了解和经验；生产部门提供的技术说明文件；校准证书、鉴定证书或其他文件提供的数据、准确度等级等；手册或某些资料给出的参考数据及其不确定度。

已知信息表明被测量之测量值分散区间的半宽为 Δ，且落在 $x_i - \Delta$ 至 $x_i + \Delta$ 区间的概率为 100%。通过对其分布的估计可得出 B 类标准不确定度 u_B 为：

$$u_B = \frac{\Delta}{k_i} \qquad (1-2-10)$$

式中包含因子 k_i（也可表示为置信系数 C）取决于测量值的分布规律。常用分布 $u(x_i)$ 与 k_i 的关系见表（1-2-1）。

表 1-2-1 常用分布 $u(x_i)$ 与 k_i 的关系

分布类型	P (%)	k_i	$u(x_i)$
矩形	100	$\sqrt{3}$	$a/\sqrt{3}$
正态	99.73	3	$a/3$
三角	100	$\sqrt{6}$	$a/\sqrt{6}$
梯形 $\beta = 0.71$	100	2	$a/2$

在实验教学情况下，为简化起见，一般估计为矩形（均匀）分布。其中，Δ 取仪器的误差限或实际测量估计的误差极限值，这样 B 类不确定度 u_B 为

$$u_B = \frac{\Delta}{\sqrt{3}}$$

合成标准不确定度的评定：

当测量结果是由若干个其他量的值求得时，按其他各量的方差和协方差算得的标准不确定度，称为合成标准不确定度。它是测量结果标准偏差的估计值，用符号 u_c 表示。

若测量结果中含统计不确定度分量 u_{a_1}，u_{a_2}，…，非统计不确定度分量 u_{b_1}，u_{b_2}，…，且它们互相独立，合成标准不确定度表征为：

$$u_c = \sqrt{\sum_{i=1}^{m} u_{a_i}^2 + \sum_{i=1}^{n} u_{b_i}^2} \quad (1-2-11)$$

m，n 分别表示 A、B 两类不确定度分量的个数。如果 $m=n=1$ 时，则

$$u_c = \sqrt{u_a^2 + u_b^2}$$

教学实验中，A 类不确定度只有一个分量

$$u_A = s(\bar{x}_i) = \frac{s(x_i)}{\sqrt{n}} = \sqrt{\frac{\sum_{i=1}^{n}(x_i - \bar{x})^2}{n(n-1)}} \quad (测量次数 n \geq 6 时)$$

B 类不确定度有多个分量。

由仪器产生的不确定度：$u_{b_1} = \Delta_{仪}/\sqrt{3}$；根据实际情况估计误差极限值 Δ，$u_{b_2} = \Delta/\sqrt{3}$

B 类不确定度：$u_B = \sqrt{u_{b_1}^2 + u_{b_2}^2}$

（3）总不确定度（扩展不确定度）的评定：将合成不确定度 $u_c(x)$ 乘以一个包含因子 k（或记为 C）（也称为置信因子），即得扩展不确定度 $U = ku_c(x)$。

一般来说，实验测量值 x 落在区间 $\bar{x} - u_c(x)$ 至 $\bar{x} + u_c(x)$ 的概率大约只有 68%。扩展置信区间，可以提高置信概率。k 的取值有两种：不需要准确给出置信概率时，k 值可取 2～3。k 值取 2，则 $U = 2u_c(x)$ 的置信概率约为 95%；k 值取 3，$U = 3u_c(x)$ 的置信概率约为 99%。在物理实验中，扩展不确定度按此方法评定。在实验报告中，简化起见，k 值统一取 2。

（4）测量结果的表示：计算完不确定度之后的结果表示如下：$X = \bar{x} \pm U$（单位）。此表示式的意义说明计算真值以一定的概率（≥95%）落在 $(\bar{x} - U, \bar{x} + U)$ 区间之内。其中，$U = 2u$。$k = 2$，u 为合成不确定度，一般置信概率大约只有 68.3%。

例 1-2-12 用分度值为 0.02mm 的游标卡尺测得圆柱直径 d 分别为：2.594cm、2.592cm、2.596cm、2.592cm、2.590cm、2.592cm。计算直径的测量结果。

解：算术平均值 $\bar{d} = (2.594 + 2.592 + 2.596 + 2.592 + 2.590 + 2.592)/6 = 2.593$（cm）

测量的平均值标准差

$$s(\bar{d}) = \sqrt{\frac{\sum_{i=1}^{6}(d_i - \bar{d})^2}{n(n-1)}} = 0.0009(\text{cm})$$

不确定度 B 类分量 $u_B = \Delta_仪/\sqrt{3} = 0.002/\sqrt{3}$ （cm）

合成标准不确定度 $u_c = \sqrt{u_A^2 + u_B^2} = \sqrt{0.0009^2 + (0.002/\sqrt{3})^2} = 0.0015$ （cm）

扩展不确定度　　　$U = 2u_c = 2 \times 0.0015 = 0.003$ （cm）

测量结果表示　　　$d = (2.593 \pm 0.003)$ cm

四、间接测量结果的误差估计

多数情况下，实验结果都是经过间接测量得到的，即先对诸多有关量进行直接测量，再通过函数关系计算出间接测量的结果。例如测定固体物质的密度，如果物体的形状为长方体，直接测量出其质量 m 及长 x、宽 y、高 z，即可用公式 $\rho = \dfrac{m}{xyz}$ 计算出固体物质的密度。间接测量的结果来源于直接测量，由于直接测量有误差，间接测量也必然有误差。根据直接测量的误差来计算间接测量的误差叫作误差的传递，或误差的合成。

误差合成的方式通常有误差的算术合成及误差的方和根（误差的平方和的根）合成。

1. 最大误差法估计间接测量的误差

设间接测量值 $N = f(x, y, z\cdots)$，其中 N 为间接测量量，$x, y, z\cdots$ 为直接测量量，且 $x = \bar{x} \pm \Delta x$，$y = \bar{y} \pm \Delta y$，$z = \bar{z} \pm \Delta z\cdots$，且 $x, y, z\cdots$ 为相互独立的变量。间接测量量的误差传递指的是当直接测量量 $x, y, z\cdots$ 各量存在误差 $\Delta x, \Delta y, \Delta z\cdots$ 时，间接测量量 ΔN 为多少，这恰好是求 N 的全微分的问题。据全微分的表示式有

$$dN = \frac{\partial f}{\partial x}dx + \frac{\partial f}{\partial y}dy + \frac{\partial f}{\partial z}dz + \cdots$$

考虑到误差限宁大勿小，最坏的情况取其绝对值并用"Δ"代替"d"，则有

$$\Delta N = \left|\frac{\partial f}{\partial x}\Delta x\right| + \left|\frac{\partial f}{\partial y}\Delta y\right| + \left|\frac{\partial f}{\partial z}\Delta z\right| + \cdots \qquad (1-2-12)$$

称上式为算术平均误差限的传递公式。

由上式可推出对于加减乘除的算术平均误差限的传递公式有如下简单的表示式：

加减法：$\Delta N = \Delta x + \Delta y + \Delta z + \cdots$

乘除法：$E_r = E_x + E_y + E_z + \cdots$

为了便于学生使用时查阅，以供参考，现将常用函数关系的误差计算公式列表如下（表 1-2-2）。

表 1-2-2　常用函数关系的误差计算公式

函数关系式 $N = f(x, y, z\cdots)$	误差	
	绝对误差 ΔN	相对误差 $E_r = \dfrac{\Delta N}{N}$
$N = x + y + z$	$\Delta x + \Delta y + \Delta z$	$\dfrac{\Delta x + \Delta y + \Delta z + \cdots}{\bar{x} + \bar{y} + \bar{z}}$
$N = x - y$	$\Delta x + \Delta y$	$\dfrac{\Delta x + \Delta y}{\bar{x} - \bar{y}}$
$N = x \cdot y$	$\bar{x} \cdot \Delta y + \bar{y} \cdot \Delta x$	$\dfrac{\Delta x}{\bar{x}} + \dfrac{\Delta y}{\bar{y}}$

续表

函数关系式 $N=f(x, y, z\cdots)$	误差	
	绝对误差 ΔN	相对误差 $E_r = \dfrac{\Delta N}{N}$
$N = x \cdot y \cdot z$	$\bar{y} \cdot \bar{z} \cdot \Delta x + \bar{x} \cdot \bar{z} \cdot \Delta y + \bar{x} \cdot \bar{y} \cdot \Delta z$	$\dfrac{\Delta x}{\bar{x}} + \dfrac{\Delta y}{\bar{y}} + \dfrac{\Delta z}{\bar{z}}$
$N = \dfrac{x}{y}$	$\dfrac{\bar{y} \cdot \Delta x + \bar{x} \cdot \Delta y}{\bar{y}^2}$	$\dfrac{\Delta x}{\bar{x}} + \dfrac{\Delta y}{\bar{y}}$
$N = x^n$	$n \cdot \bar{x}^{n-1} \cdot \Delta x$	$n \cdot \dfrac{\Delta x}{\bar{x}}$
$N = \sqrt[n]{x}$	$\dfrac{1}{n} \bar{x}^{(1/n)-1} \cdot \Delta x$	$\dfrac{1}{n} \cdot \dfrac{\Delta x}{\bar{x}}$

从上表可以看出，在计算间接测量误差时，在加减运算中，先算绝对误差是比较方便的。而在乘、除、乘方、开方等运算中，先算相对误差比较方便，然后通过 $\Delta N = E_r \cdot \bar{N}$，求出 ΔN。

另外，间接测量量的平均值 $\bar{N} = f(\bar{x}, \bar{y}, \bar{z}, \cdots)$，结果表示为 $N = \bar{N} \pm \Delta N$，相对误差 $E_r = \dfrac{\Delta N}{\bar{N}} \times 100\%$。

2. 标准偏差传递的基本公式

用最大误差估计间接测量误差时，在计算中，由于误差本身或正或负是不可知的，因此，式中 Δx，Δy，$\Delta z \cdots$ 前的系数均取绝对值。这样估计的误差将有些偏大，更精确的估计方法是将各误差平方后相加再开方，称之为标准偏差传递的基本公式，即

$$\sigma = \sqrt{\left(\frac{\partial f}{\partial x}\right)^2 \sigma_x^2 + \left(\frac{\partial f}{\partial y}\right)^2 \sigma_y^2 + \left(\frac{\partial f}{\partial z}\right)^2 \sigma_z^2 + \cdots} \qquad (1-2-13)$$

表（1-2-3）为一些常用函数关系的标准偏差传递公式。

表 1-2-3　常用函数关系的标准偏差传递公式

数学运算关系式 $N=f(x, y, z\cdots)$	误差	
	绝对误差 σ_N	相对误差 $\dfrac{\sigma_N}{N}$
$x \pm y$	$\sqrt{\sigma_x^2 + \sigma_y^2}$	$\dfrac{\sqrt{\sigma_x^2 + \sigma_y^2}}{\bar{x} \pm \bar{y}}$
$x \cdot y$	$\sqrt{\bar{y}^2 \sigma_A^2 + \bar{x}^2 \sigma_B^2}$	$\sqrt{\left(\dfrac{\sigma_x}{\bar{x}}\right)^2 + \left(\dfrac{\sigma_y}{\bar{y}}\right)^2}$
$\dfrac{x}{y}$	$\dfrac{\sqrt{\bar{y}^2 \sigma_x^2 + \bar{x}^2 \sigma_y^2}}{\bar{y}^2}$	$\sqrt{\left(\dfrac{\sigma_x}{\bar{x}}\right)^2 + \left(\dfrac{\sigma_y}{\bar{y}}\right)^2}$
x^n	$n \cdot \bar{x}^{n-1} \cdot \sigma_x$	$n \cdot \dfrac{\sigma_x}{\bar{x}}$
$k \cdot x$	$k \sigma_x$	$\dfrac{\sigma_x}{\bar{x}}$

例 1-2-13　求圆环面积 $S = \dfrac{\pi}{4}(D^2 - d^2)$ 的误差公式。

解：上式可改写为 $S = \dfrac{\pi}{4} D^2 - \dfrac{\pi}{4} d^2$，故由差的绝对误差公式，将各量的绝对误差相

加可得出绝对误差公式

$$\Delta S = \frac{\pi}{4} \cdot 2\bar{D} \cdot \Delta D + \frac{\pi}{4} \cdot 2\bar{d} \cdot \Delta d = \frac{\pi}{2}(\bar{D} \cdot \Delta D + \bar{d} \cdot \Delta d)$$

其相对误差公式可以除以 $\bar{S} = \frac{\pi}{4}(\bar{D}^2 - \bar{d}^2)$ 得出，即

$$\frac{\Delta S}{\bar{S}} = \frac{2(\bar{D} \cdot \Delta D + \bar{d} \cdot \Delta d)}{(\bar{D}^2 - \bar{d}^2)}$$

综上所述，实验数据的处理步骤与测量结果的正确表达在实际测量过程中，估计测量误差一般按下列程序进行。

（1）对被测量进行多次测量，获得一组数据，将它们列成表格。

（2）计算被测值的算术平均值。

（3）按要求求绝对误差或标准误差及相对误差。

（4）结果应表示为 $\begin{cases} x = \bar{x} \pm \Delta x \\ E_r = \frac{\Delta x}{\bar{x}} \cdot 100\% \end{cases}$

（5）最后，在直接测量值及其误差已知的条件下，利用误差传递公式先算出间接测量值的误差，然后用误差决定间接测量值的有效数字位数，最后把间接测量结果正确地表示出来。

例 1-14 用单摆测定重力加速度，使用仪器有米尺和停表。直接测量结果用平均值及平均绝对误差表示。

现测得 20 个全振动时间为 28.4s，测量误差为 0.1s；
摆长为 $L = \bar{L} \pm \Delta L = (0.500 \pm 0.001)$ m。

解： 由题意，得全振动的周期

$$\bar{T} = \frac{28.4}{20} = 1.42\text{s}, \Delta T = \frac{1}{20} \times 0.1 = 0.005\text{s}$$

则，重力加速度

$$\bar{g} = \frac{4\pi^2 \bar{L}}{\bar{T}^2} = \frac{4\pi^2 \times 0.500}{1.42^2} = 9.779\text{m/s}^2$$

长度测量的相对误差

$$\frac{\Delta L}{\bar{L}} = \frac{0.001}{0.500} = 0.20\%$$

周期测量的相对误差

$$\frac{\Delta T}{\bar{T}} = \frac{0.1}{28.4} = 0.35\%$$

重力加速度的相对误差，按算术合成

$$\frac{\Delta g}{\bar{g}} = 2\frac{\Delta T}{\bar{T}} + \frac{\Delta L}{\bar{L}} = 2 \times 0.35\% + 0.20\% = 0.90\%$$

于是

$$\Delta g = 9.779 \times 0.009 = 0.088011 = 0.09\text{m/s}^2$$

故结果为

$$g = \bar{g} \pm \Delta g = (9.78 \pm 0.09) \text{m/s}^2$$

例 1-15 用单摆测重力加速度。实验中采用精度较高的仪器,直接测量结果用平均值及平均值的标准差表示为:摆长 $L = (92.9 \pm 0.1)$ cm ($S_{mL} = 0.1$ cm),周期 $T = (1.936 \pm 0.004)$ s ($S_{mT} = 0.004$ s)。

解:由题意,$\bar{g} = \dfrac{4\pi^2 \bar{L}}{\bar{T}^2} = \dfrac{4\pi^2 \times 92.9}{1.936^2} = 979 \text{cm/s}^2$

利用公式

$$\frac{S_{mg}}{\bar{g}} = \sqrt{\left(\frac{1}{\bar{L}}\right)^2 S_{mL}^2 + \left(\frac{2}{\bar{T}}\right)^2 S_{mT}^2}$$

有

$$\frac{S_{mg}}{\bar{g}} = \sqrt{\left(\frac{1}{92.9}\right)^2 \times 0.1^2 + \left(\frac{1}{1.936}\right)^2 \times 0.004^2} = 4.27 \times 10^{-3} = 0.43\%$$

则

$$S_{mg} = 979 \times 0.43 = 4 \text{cm/s}^2$$

测量结果为

$$g = \bar{g} \pm S_{mg} = (979 \pm 4) \text{cm/s}^2$$

五、数据处理方法

(一) 列表法

在记录实验数据时经常需要列表,因为数据表可以简单明了地反映出有关物理量之间的对应关系,便于检查测量结果是否正确合理,有助于分析物理量之间的规律性。列表力求简单明了,反映出有关物理量间的对应关系;表中的数据要正确反映出测量结果的有效数字,以表明测量的准确程度;表中各符号代表的物理意义要交待清楚并标明单位,单位应写在标题栏内,不要重复地记在表内各个数字上;表中不能说明的问题,可在表下附加说明。

(二) 作图法

处理实验数据也常采用作图的方法。将测得的数据描点作图,并分析图像,是找到测量值间对应函数关系的最常用方法之一。

1. 作图法的优点

(1) 可以方便地求出所需要的某些实验结果。

(2) 易于发现实验中的测量错误。

(3) 由于图线是依据许多数据点描出的平滑曲线,因此对测量的数据有修正作用,具有多次测量取平均值的意义。

(4) 在图线上能够直接读出没有测量的点,而且在一定条件下,可以从图线的延伸部分读到测量范围以外的点。

2. 作图规则

(1) 将测量的数据按一定规律列成相应的表格。

(2) 根据情况选用合适的坐标纸,如直角坐标纸、对数坐标纸或极坐标纸等。确定坐标纸的大小及坐标轴的比例。图纸的大小应根据测量数据的有效数字来选择，使测量数据中的可靠数字在图上也是可靠的，即图中的一个小格对应数据中可靠数字的后一位。数据中的一位可疑数字在图中应是估计的。使整个图线比较匀称地充满整个图纸，横轴与纵轴的比例可以不同，坐标轴的起点也不一定是零值。

（3）图纸与坐标轴的比例选定后，标明坐标轴。一般情况下，函数值取 Y 轴，自变量取 X 轴。注明其代表的物理量（或符号）及单位；在坐标轴上每隔一定间距标出该物理量的数值，在图纸上适当位置写明图的名称及必要的说明。

（4）标点与连线：根据测量的数据，用削尖的铅笔或钢笔，以"＋"或"×"等符号在图上标出各实验点的准确位置（同一图纸上不同的曲线应使用不同的符号），用直尺或曲线尺将各点连成光滑曲线（图线画好后，符号也不应擦去，以便复核及保留数据的记录）。由于误差的影响，曲线不一定通过所有的点，但要求曲线两边的偏差点有比较均匀的分布，个别偏离较大的点应舍去或重新测量。

图 1-2-4 测量误差

（5）用误差短线标出测量误差，如图 1-2-4 所示。

综上所述，作图法有许多优点，但也有不足，如求得的值准确性不太高，有效数字位数不能太多等。

3. 逐差法

所谓逐差法，就是把测量数据中的因变量进行逐项相减或按顺序分为高、低两组进行对应项相减，然后将所得差值作为因变量的多次测量值进行数据处理的方法。逐差法是针对自变量等量变化，因变量也做等量变化时，所测得有序数据等间隔相减后取其逐差平均值得到的结果。其优点是充分利用了测量数据，具有对数据取平均的效果，可及时发现差错或数据的分布规律，及时纠正或及时总结数据规律。它也是物理实验中处理数据常用的一种方法，常用于处理自变量等间距变化的数据。

例 1-16 以测弹簧的倔强系数为例：在弹簧下端依次加 1g，2g，…，8g 的砝码，记下弹簧端点在标尺上的位置 n_1，n_2，…，n_8。对应于每增加 1g 砝码弹簧相应伸长为

$$\Delta n_1 = n_2 - n_1$$
$$\Delta n_2 = n_3 - n_2$$
$$\cdots$$
$$\Delta n_7 = n_8 - n_7$$

伸长量平均值：

$$\overline{\Delta n} = \frac{\Delta n_1 + \Delta n_2 + \cdots + \Delta n_7}{7}$$
$$= \frac{(n_2 - n_1) + (n_3 - n_2) + \cdots + (n_8 - n_7)}{7}$$
$$= \frac{(n_8 - n_1)}{7}$$

中间测量值没起作用，为了发挥多次测量的优势，应采用以下逐差法。

先将数据分为两组：n_1，n_2，n_3，n_4；n_5，n_6，n_7，n_8

再取对应差值项的平均值

$$\overline{\Delta n} = \frac{(n_5 - n_1) + (n_6 - n_2) + \cdots + (n_8 - n_4)}{4}$$

即得每增加4g砝码对应的伸长平均值。

六、统计最佳直线

对于较复杂的函数关系，由于它们是非线性的，所以图形都是曲线，而由曲线上求解实验方程参数很不方便，且难以从图中判断结果是否正确，因此常选用不同的变量来代替原来的变量（称为变量置换法），将曲线改成直线加以处理，只要确定了斜率和截距即可确定该直线。例如，对 $xy = k$，可以将 $x-y$ 曲线改为 $y-1/x$ 为轴的 $y-1/x$ 图线，则曲线变为直线。线性回归（线性拟合）就是由实验数据组 (x_i, y_i) 确定斜率和截距的过程。最小二乘法认为：若最佳拟合的直线为 $y = f(x)$，则所测各 y_i 值与拟合直线上相应的点 $y_i = f(x_i)$ 之间的偏离的平方和为最小。处理过程如下：

假设我们根据描点情况可判定函数的关系为

$$y = bx + a \quad (1-2-14)$$

式中，b 和 a 为分别为该直线的斜率和截距。由于描出的点 (x_i, y_i) 总是不可能全部落在式（1-2-14）所表示的直线上，于是每个 x_i 对应的 y_i 总有偏差

$$\varepsilon_i = y_i - y$$

求各点的偏差的平方和

$$E = \sum_{i=1}^{n} \varepsilon_i^2 = \sum_{i=1}^{n} (y_i - y)^2$$

$$= \sum_{i=1}^{n} [y_i - (a + bx_i)]^2$$

$$= \sum_{i=1}^{n} [y_i^2 - 2y_i(a + bx_i) + a^2 + 2abx_i + b^2x_i^2] \quad (1-2-15)$$

当 E 为极小时，则我们选取的直线（1-2-14）即为最佳统计直线。为确定这条直线，我们现在需要的就是确定 a 和 b。

E 为极小时的必要条件为 $\frac{\partial E}{\partial a} = 0$，$\frac{\partial E}{\partial b} = 0$

于是由（1-2-15）式得

$$\frac{\partial E}{\partial a} = 0, \sum_{i=1}^{n} (-2y_i + 2a + 2bx_i) = 0$$

即

$$\sum_{i=1}^{n} y_i = na + b \sum_{i=1}^{n} x_i \quad (1-2-16)$$

及

$$\frac{\partial E}{\partial b} = 0, \sum_{i=1}^{n} (-2y_i x_i + 2ax_i + 2bx_i^2) = 0$$

即
$$\sum_{i=1}^{n} x_i y_i = b \sum_{i=1}^{n} x_i^2 + a \sum_{i=1}^{n} x_i \quad (1-2-17)$$

由式（1-2-16）解得
$$a = \frac{\sum_{i=1}^{n} y_i}{n} - b \frac{\sum_{i=1}^{n} x_i}{n}$$

或
$$a = \bar{y} - b\bar{x} \quad (1-2-18)$$

即所求直线通过的平均坐标。

将式（1-2-18）代入式（1-2-17），得
$$\sum_{i=1}^{n} x_i y_i = b \sum_{i=1}^{n} x_i^2 + b\bar{x} \sum_{i=1}^{n} x_i + \bar{y} \sum_{i=1}^{n} x_i$$

等式两边除以 n，得
$$\overline{xy} = b\overline{x^2} - b\bar{x}^2 + \bar{x} \cdot \bar{y}$$

因此
$$b = \frac{\overline{xy} - \bar{x} \cdot \bar{y}}{\overline{x^2} - \bar{x}^2} \quad (1-2-19)$$

可见，只要我们求出实验数据的 $\bar{x}$，$\bar{y}$，$\overline{xy}$，$\overline{x^2}$，$\bar{x}^2$，即可计算出 a 及 b。

我们可以利用相关系数 r 来检查所得的方程是否合理，r 定义为
$$r = \frac{\overline{xy} - \bar{x} \cdot \bar{y}}{\sqrt{(\overline{x^2} - \bar{x}^2)(\overline{y^2} - \bar{y}^2)}} \quad (1-2-20)$$

r 的值通常在 -1 与 1 之间，r 值越接近 1，数据越密集于直线近旁。反之，r 值小于 1 而接近 0 时，则表示数据很离散。

习　题

1. 一电流表的最小分度为 0.1A，现用它测量一回路中的电流，6 次所测得的读数都是 2.5A，问这电流一定是稳定的吗？

2. 测量一激光器的输出功率，7 次所测得的数值分别为：2.2、12.7、18.9、0.6、7.9、22.4、0.1，单位为 mW。计算出其平均值为 9.3mW，平均绝对误差为 7.7mW，平均值的标准差为 3.5mW，据此可以得出什么结论？

3. 指出下列各量是几位有效数字：8.640g，200.04g，0.005cm，1.00×10^{-5}cm，12.00cm。

4. 改正下列各表达式中的错误：

$L = (1.54 \pm 0.02)$ m $= (1540 \pm 20)$ mm；

$m = 1.23$kg $= 1230$g；

$t = (8.50 \pm 0.1)$ s。

5. 请计算出下列各式的结果：

(1) $y = \dfrac{ab}{2a - b}$

$a = 2.0517 \quad b = 4.1032$

(2) $y = \dfrac{p(a + c)}{ab - cd}$

$a = 3.1416 \quad b = 10.002 \quad c = 31.416 \quad d = 1.0001 \quad p = 31.4287$

(3) $y = a + 5b - 3c - 4d$

$a = 382.02 \quad b = 1.03754 \quad c = 56 \quad d = 0.001036$

(4) $y = \dfrac{8abc}{\pi^2 dlm}$

$a = 800.0 \quad b = 105.35 \quad c = 112.6 \quad d = 0.03005 \quad l = 12.98 \quad m = 0.97$

6. 现测量一铜线的长度 L，6 次的测量值分别为：60.499m，60.498m，60.500m，60.503m，60.497m，60.505m。试计算其算术平均值、平均绝对误差、平均值的标准差及相对误差。

7. 计算物体的重力势能 E_p ($E_p = mgh$)。各已知量用平均值及平均绝对误差分别表示为：$m = (0.1000 \pm 0.0005)$ kg；$g = (9.80 \pm 0.04)$ m/s²；$h = (0.689 \pm 0.002)$ m。

8. 计算以下长方形的面积 A ($A = ab$)。已知量用平均值及平均值的标准差表示为：$\bar{a} = 24.345$mm，$S_{ma} = 0.006$mm；$\bar{b} = 50.368$mm，$S_{mb} = 0.008$mm。

9. 改变温度 t，测某铜线的电阻 R（共 5 次），数据如下：

t (℃)	R (Ω)
17.8	3.554
20.9	3.687
37.7	3.827
48.2	3.969
58.8	4.105

设最佳统计直线为 $R = R_0(1 + \alpha t)$，试确定常数 R_0 及 α。其中 R_0 为 0℃时的电阻，α 为铜线的电阻温度系数。

（赵　喆　张红艳）

Section 2　Error, Significant Digits and Data Processing

1. Measurement and error

(1) Measurement: Measurement is a essential method for human being to perceive and transform the world. Through measuring, people may obtain concept of quantity, summarize regularity of it, and thus establish the corresponding law and theorem of physics.

Measurement is a process comparing unknown object with selected unit, includs direct measurement and indirect measurement. Direct measurement is a process comparing unknown object directly with selected similar unitage of the instrument. It means we can read out the value from the instrument. For example, survey time with a stopwatch, survey length with a metre rule and so on. Indirect measurement can be calculated by surveying some direct measurements and the known formula or the law. For example, after directly measuring the diameter, height and mass of a cylinder, its density can be obtained by formula $\rho = \dfrac{m}{\frac{1}{4}\pi D^2 H}$. And the majority of various kinds of physical quantity are obtained through the indirect measurement.

(2) Error definition and error classification: Because of the limits of measuring instruments, experiment condition and the sense of observers, etc, the observed value always has a certain amount of difference with the true value, and this is what we call the measuring error. Measuring error consists in process of all measurements. The measuring error reflects how the result of the measurement closes to the "true" value, so discovering the reason of error, minimizing or even eliminating error are proved to be the key of increasing measuring accuracy and its reliability. On the basis of its causes and influences, measuring errors are classified into systematic error and random error.

1) Systematic error: Systematic error is caused by disadvantages of the instrument structure, inaccurate and unsuitable adjustment, imperfection of measuring method and technique, and improper habit of experimenter, etc. Systematic error usually has a certain magnitude, a certain symbol and changes in a regular way for every measurements. And it can be minimized or eliminated by means of accurate adjustment instrument, improvement of measuring method, and amendment of formula or law, etc.

2) Random error

①Random error: This kind of error is caused by many unstable accidental factors. Even if under the same condition, several measurements of the same quantity on the same subject will not in general be the same (even if all obvious system errors have been eliminated or corrected), and it shows irregular and stochastic changes. And the error caused in this way is defined as random error.

②Causes of random error

a. Due to the limits ability of human sense, such as seeing, hearing, feeling, and so on, the reading error may be caused when the viewer adjusts the instrument and estimates the reading, which will possibly make the results fluctuate.

b. Irregular change of the environment and the condition during the process of experiment.

The random error is a kind of ruleless fluctuation without any regularity, as shown in Fig. 1-2-1 (a). When the measuring times are enough, the random error obeys the certain statistical rules. And the measuring result always fluctuates nearby the true value (A). As a result of the random error, it cannot be entirely eliminated, but if we increase the measuring times, the random error may be reduced. The observed value may deviate a certain size

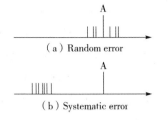

(a) Random error

(b) Systematic error

Fig. 1-2-1　Error distribution

and in the direction comparing with the true value according to the existence of systematic error. It cannot be discovered and overcome by measuring many times. In the usual situation, the systematic error and the random error exist simultaneously, as shown in Fig. 1-2-1 (b). A small systematic error means the accuracy is high, and a small random error means the measuring repeatability is well, or the accuracy is high. The high precision means the accidental error and the system error are all small.

Mentioned above, the error and mistake are two different conceptions completely. The mistake is the experimenter's incorrect operation of instrument, unreasonable experimental technique, out-of sequence operation or misreading the data and so on. The student should strengthen the responsibility, avoid occurrence of the mistake and try to reduce the error.

c. The statistical distribution rule of random error: By a large number of experiments, it has been shown that when the repeated measurement times tend to be infinity larger, the measurement error is approaching normal distribution (Fig. 1-2-2). $f(\delta)$ represents the probability of error δ. This distribution has the following features: there is equal probability of error margins; the times of small absolute errors occurs more; the larger errors rarely occurs; the arithmetic mean of the random error tend to reach zero; without the systematic errors, the measurement tends to reach the mean value. When the measured frequency will deviate from the normal and increase the number of measurements, the measurement error can be reduced.

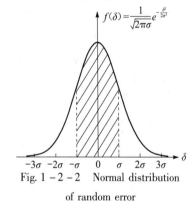

Fig. 1-2-2　Normal distribution of random error

2. Significant digits

(1) Precision of instrument: When a measuring instrument is used correctly, precision of the instrument (called precision) is determined by the smallest unit which it can measure. It depends on the instrument graduation (the smallest graduation of the instrument). For example, we measure the length of an object with a meter ruler (divided into a thousand divisions called millimeter), and its precision is 1mm.

(2) Significant digits

1) Significant digits: Significant digits which are also called significant figures. As a result of the limit of the instrument precision, the digits of each quantity to be measured is only limited. The value which can express the measurement and the computed result correctly and effectively is called significant digits. It consists of all accurate digits and an estimated digit. As shown in Fig. 1-2-3, to measure the length of a block of wood with a metre ruler which the least scale is

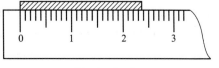

Fig. 1-2-3

millimeter, align at one end of the block with the zero mark, and read the number from another end of the block, from Fig. 1 – 2 – 3 we may see the block is longer about a half graduation than 23 millimeters, Then the block is measured to be 23.6 millimeters. The value 23.6cm would have three significant digits, the first two numbers 2 and 3 are accurately read from the ruler, so they are significant and reliable, called them "accurate digits". But the number 6, the last digit of the value is an estimated value found between two adjacent millimeter marks, if another person reads it, 6 is possibly estimated 5 or 7, therefore the 6 is an estimated value, we call such a digit "suspicious digit". Obviously, the estimated value involves error, but it still reflects the measuring information in some degree, therefore it is also effective. Generally speaking, the suspicious digit should be taken only one. So in the measurement, the significant digits is constituted by all the reliable digits and one suspicious digit.

2) Problems worthy of attention with significant digits

①Significant digits of the measured quantity is not allowed to increase or decrease casually, and it has something to do with the relative error. The more the number of the significant digits is, the smaller its relative error of the measured quantity is.

②All non zero digits in a measurement are significant (1, 2, 3, 4, 5, 6, 7, 8, 9). Not all zeroes are significant. For example, if you reported the measured length of an object to be 0.0050800m, it would have only five significant digits. The first three zeroes only serve to locate the decimal point and are not significant. The other three zeroes are significant. The last zero is the estimated digit and it is significant. The other two are between two significant digits and will always be significant in this position.

③The number of the significant digits has nothing to do with unit conversion. In order to stand out the significant digits, we may express the significant digits with the scientific notation. It means any quantity may be expressed as an integral power of 10, or as the product of two numbers, one of which is an integral power of 10 and the other a non – zero digit locates left to the decimal. For example, 0.0010300 (kg) may be written as 1.0300×10^{-3} (kg); 125.20ms may be written as 1.2520×10^{2} (ms) or 1.2520×10^{-1} (s). These examples shows unit conversion does not affect the number of the significant digits.

④Rounding off. It is often be used for significance digits, the basic rules for rounding is, when rounding to a place value, look at the value to the right of that position. If this value is 5 or bigger, round up. When the first digit dropped is less than five, the last digit retained should remain unchanged. For example, 1.625 should be written as 1.63 when taking three significant digits, and it should be recorded as 1.6 when taking two significant digits. Moreover there is another kind method to be used, it is "four – drop six – add and five – collected even" method, it rounds up if the next digit is greater than five and rounds down if the next digit is less than five, specially with the next digit is five it rounds up and down half the time based on whether the digit remained is even, and this ensures the last digit remained to become even. For example: 1.625 will be written as 1.62 when taking three significant digits, and 1.615 still be recorded as 1.62 when taking three significant digits, 1.626 is recorded as 1.63 when taking three significant digits. This teaching material uses the round off method.

⑤The number of significant digits for constant (for example π, $\sqrt{2}$, e, $\frac{1}{4}$ and so on) is infinite, it should be taken suitably according to the measured quantity. In the ordinary condition, at least it should retain one more than the measured value.

3) Arithmetic operations with significant digits

①Significant digits in addition and subtraction: In a calculation involving addition and subtraction, the number of decimal places (not number of significant digits) in the answer should be the same as the least number of decimal places in any of the numbers being added or subtracted.

Example 1-2-1 Add these values

$$X = 71.3 - 0.753 + 6.26 + 271$$

In Addition and Subtraction, we get

$$X = 71.3 - 0.8 + 6.3 + 27\underline{1} = 34\underline{8}$$

The point under the number expresses the least precisely position of significant digits.

②Significant digits in multiplication and division: In a calculation involving multiplication and division, the number of significant digits in the answer should equal the least number of significant digits in any one of the numbers being multiplied and divided.

Example 1-2-2

$$X = \frac{39.50 \times 4.08 \times 0.0013}{868.01} = \frac{39.5 \times 4.08 \times 0.0013}{868}$$
$$= 2.4 \times 10^{-4}$$

Obviously, in this equation the component 0.0013 has the least significant digits (two significant digits), therefore we take it as a standard.

③Other operations: In the operation of exponents, logarithms and trigonometric function, as a rule, the number of the result should have the same number of significant digits as their arguments. Some examples are as following

Example 1-2-3

$$\sqrt{62\underline{5}} = 2.5\underline{0} \times 10$$
$$25^2 = 6.\underline{3} \times 10^2$$

Example 1-2-4

cos32.7° = 0.84$\underline{2}$

Example 1-2-5

lg19.2$\underline{8}$ = 1.28$\underline{5}$

Note: When doing multi-step calculations, keep at least one more significant digit in intermediate results than needed in your final answer.

Moreover, some quantities which are not determined by the experiment, for example, to calculate volume of a circular cylinder after measuring its diameter and length, then use the formula $V = \frac{1}{4}\pi D^2 H$, where $\frac{1}{4}$ is not a measured quantity, then we do not need to consider its number of significant digits.

Obviously, by the operation rules of significant digits, the number of significant digits of the result should be appropriate when we carry on the computation for quantities with different accuracy. If we take less, an additional calculation error will be produced, and the precision of the experiment will be reduced; and if we take more, as if the experiment precision is very high, in fact it is insignificance, instead of it there will bring us mistake impression and needless trouble.

Correctly utilizing significant digits in experiment, it can not only truthfully reflect the measuring result, but also simplify the operation and save time. Good habit should be educated to take reasonable number of the significant digits even if a calculator is used.

3. Evaluating error of the direct measured value

As a result of the existence of error, it is impossible for a measured value to obtain its true value accurately for a direct measurement, so we often need measure it repeatedly many times to make the experiment more accurate. However each measuring result is still different, how should we express the result?

(1) Evaluating the error of a single direct measured value: Some experiments, since the measurement is in a dynamic situation, it is not allowed to be measured the value repeatedly; sometimes the accuracy is not required strictly in some experiments; or in a indirect measurement, the error of a physical quantity is too small to influence the final result, under these conditions the measured value may be determined only once. Regarding the single measurement, the error generally is evaluated according to the instrument used, the measured object, the experimental method and experimenter's experience. In ordinary cases, when the random error is very small, we estimate the single measurement error according to the instruction book or the instrument error indicated in the label. If the instrumental error does not be looked up, we may take one-half of least graduation of the instrument as the absolute error for a single measurement. We may take the least graduation of the instrument or even a greater reasonable value as the error for which the random error is larger.

Example 1-2-6 A length of a certain object measured with a micrometer (one time) is 3.647mm (mantissa 7 is an estimated value), the least scale of the micrometer is 0.01mm, then we may take 0.005mm as its absolute error for the result.

Example 1-2-7 Measure the length of a cycloid with a metre ruler, if you use it correctly, then the reading error is main constituent for the measuring error, take the reading error of each ends respectively as 0.5mm, then the error of the length may be taken as 1mm.

Example 1-2-8 Determine interval of time for a motive object with a stopwatch, if the stopwatch's systematic error does not need to be considered, then the measuring error is mainly decided by starting and braking the stopwatch. Generally we estimate both starting and braking respectively existing an error of 0.1s, and then the total error is 0.2s.

Example 1-2-9 Weigh an object using a balance, the stop-points between the balance idling and after adding the weights generally are inconsistent, the difference will cause measuring error, when it does not greater than one graduation, we may take the difference between two sides qualities as the measured error when the balance pointer stop-spot moves one graduation. At the moment the quality difference is called the balance sensitivity.

(2) Estimating error of repeated direct measurements: In order to reduce the random error, if possible, we always measure the value for many times, and take their arithmetic mean value as the measuring result. Now measure a physical quantity repeatedly, and each measured value is respectively x_1, x_2, x_3, $\cdots x_i$, $\cdots x_n$, so its mean value is

$$\bar{x} = \frac{x_1 + x_2 + x_3 + \cdots + x_n}{n} = \frac{1}{n}\sum_{i=1}^{n} x_i \qquad (1-2-1)$$

According to the error theory, the mean value the best approximate of the true value. We may take the $\bar{x}$ as the approximate true value of this physical quantity.

We know the error cannot be determined, it only can be estimated according to the definition of error. There are many methods for evaluating the random error, such as, the mean absolute error, the standard deviation and the relative error.

①The mean absolute error: We define the deviation as the difference of the th measured value and the mean value

$$d_i = x_i - \bar{x} \qquad (1-2-2)$$

the mean value of absolute value of the deviation for n times is called the mean absolute deviation, or the mean absolute error, which we shall denote by Δx. and

$$\Delta x = \frac{1}{n}[\,|x_1 - \bar{x}| + |x_2 - \bar{x}| + \cdots + |x_n - \bar{x}|\,]$$

$$= \frac{1}{n}\sum_{i=1}^{n} |x_i - \bar{x}| \qquad (1-2-3)$$

Define the error like this cannot show the mean value is larger or smaller than the true value. The true value is usually expressed as

$$x = \bar{x} \pm \Delta x$$

The meaning is, the tolerance interval in the measurement is from $\bar{x} - \Delta x$ to $\bar{x} + \Delta x$. The error is an absolute value of the difference between the true value and the mean value, it reflected the range of the measured value deviating the true value, so generally it is called the absolute error, its unit is the same as the measured value unit. The measurement may be reflected from the absolute error, but it is not certainly comprehensive, usually also the relative error must be introduced to the supplement.

②The relative error: The relative error is the ratio of the absolute error and the true value (arithmetic mean value), it is does not have unit. The relative error is a target used for expressing the reliable degree of the different measured object, it is commonly expressed with percent, namely

$$E_r = \frac{\Delta x}{\bar{x}} \times 100\% \qquad (1-2-4)$$

Therefore it is called the percent error. In order to explain the physics significance of the relative error, an example is given as following. For example, measure lengths of two objects, they are

$$L_1 = (23.50 \pm 0.03) \text{cm}, \quad L_2 = (2.35 \pm 0.03) \text{cm}$$

Then their relative errors respectively are

$$E_{r_1} = \frac{0.03}{23.50} \times 100\% = 0.13\%$$

$$E_{r_2} = \frac{0.03}{2.35} \times 100\% = 1.3\%$$

Their absolute errors are equal (0.03cm), but their relative errors are different, the first is 10 times bigger than the second. Obviously, the first measurement is more accurate than the second one. Therefore, the result of the experiment should be expressed using simultaneously the absolute error and the relative error. The result of the measurement expressed using the relative error is

$$x = \bar{x}(1 \pm E_r)$$

The relation of the absolute error and the relative error is

$$\Delta x = \bar{x} \cdot E_r$$

Example 1-2-10 Measure the length of an object for five times, the measured values are respectively
$L_1 = 3.41 \text{cm}, L_2 = 3.43 \text{cm}, L_3 = 3.45 \text{cm}, L_4 = 3.44 \text{cm}, L_5 = 3.42 \text{cm}$, find the final measuring result.

Solution: First, the mean value is

$$\bar{L} = (3.41 + 3.43 + 3.45 + 3.44 + 3.42) = 3.43 \text{ (cm)}$$

The absolute value of each deviation is:

$\Delta L_1 = |3.41 - 3.43| = 0.02 \text{cm}$

$\Delta L_2 = |3.43 - 3.43| = 0.00 \text{cm}$

$\Delta L_3 = |3.44 - 3.43| = 0.01 \text{cm}$

$\Delta L_5 = |3.42 - 3.43| = 0.01 \text{cm}$

The mean absolute error is

$$\Delta L = \frac{1}{5}(0.02 + 0.00 + 0.02 + 0.01 + 0.01) = 0.012 \approx 0.02 \text{ (cm)}$$

The relative error is

$$E_r = \frac{\Delta L}{L} \times 100\% = \frac{0.02}{3.43} \times 100\% = 0.00583 \approx 0.59\%$$

The expression of the final result is

$$L = (3.43 \pm 0.02) \text{ cm}$$

or

$$L = 3.43\ (1 \pm 0.59\%) \text{ cm}$$

From the above example, we know

1) The absolute error generally only takes one significance digit, and generally it only rounds up no matter what its next is, one is added to the significance digits (except zero). For example: The $\Delta M = 0.013$ gram, it should be written as $\Delta M = 0.02$ gram.

The relative error generally only takes two significant digits, it also rounds up no matter what its next is, one is added to the significance digits (except zero), for example: $E_r = 1.21\%$ should be written as 1.3%, and 0.8% should be written as 0.80% and 10.4% should be written as 11%.

2) Regardless of the result of direct measurement or indirect measurement, its main value (single measured value, mean value or computing result) the final basis for significant digits is: its last digit must be aligned with position of the digit of absolute error. Namely the last digit of the measured value should have the same order of magnitude with the absolute error. For the last digit of the measured value, it adopts the rounding off method. For example: by computing to obtain a measured value of a physical quantity, it is 18.659 unit, its absolute error is 0.04 unit, then the final result is written as (18.66 ± 0.04) unit.

3) When the error participates in the operation of calculating, in order to avoid the additive error, it should retain more significant digits in the operation.

③Standard deviation, standard error and mean value of standard deviation

1) Standard error: In the Figure 1-2-2, the formula δ is a constant related to the experimental conditions, which is called as the normal distribution of the standard error.

$$\sigma = \lim_{n \to \infty} \sqrt{\frac{\sum_{i=1}^{n}(x_i - \bar{x})^2}{n}} \quad (1-2-5)$$

2) The standard deviation: In actual measurement, n is impossible $\to \infty$, for a measuring line, as long as $n \geq 5$ (generally the experiment $5 \leq n \leq 10$), is placed with a standard error by the standard deviation (the error = the measured value − the arithmetic mean value). In a limited measurement, the experimental standard deviation of results of measurements is as follows:

$$S = \sqrt{\frac{\sum_{i=1}^{n}(x_i - \bar{x})^2}{n-1}} \quad \text{(Bessel equation)} \quad (1-2-6)$$

3) n times standard deviation of the arithmetic mean of measurements:

$$S_m = \frac{S}{\sqrt{n}} = \sqrt{\frac{\sum_{i=1}^{n}(x_i - \bar{x})^2}{n(n-1)}} \qquad (1-2-7)$$

The standard deviation of mean deviation is smaller than that of any standard measurement test. By increasing the number of measurements, the mean standard deviation can be reduced to improve the accuracy. However, $n > 10$ later, n adds, the mean of standard deviation is little slow, so the general admission n is 6 to 10 times in the physics experiment teaching.

According to the statistical theory, 68.3% of the measured value x_i locates in the region of $\bar{x} \pm S$, 95.4% of x_i falls in the region of $\bar{x} \pm 2S$. The probability, which the true value is in the region of $\bar{x} \pm S_m$, is 68.3%, but the probability which the true value is in the region of $\bar{x} \pm 2S$ is 95.4%. These probability values are called the confidence levels.

The experiment result may be expressed as $\bar{x} \pm S_{mx}$, where S_{mx} is the standard deviation of mean value for the measurement of n times. By now the relative error may be represented as

$$E_r = \frac{S_{mx}}{\bar{x}} \times 100\% \qquad (1-2-8)$$

Example 1-2-11 Determine the resistance eight times.

R (Ω)	d_i (mΩ)	d_i^2 (mΩ)^2
4.615	-10	100
4.638	13	169
4.597	-28	784
4.634	9	81
4.613	-12	144
4.623	-2	4
4.659	34	1156
4.623	-2	4

First we get $\bar{R} = 4.625\Omega \qquad S_{um} = 2442$ (mΩ)^2

Therefore

$$S = \sqrt{\frac{2442}{7}} = 19\text{m}\Omega = 0.019\Omega$$

$$S_m = \frac{0.019}{\sqrt{8}} = 0.007\Omega$$

The expression of the result is

$$R = \bar{R} \pm S_m = (4.625 \pm 0.007)\Omega$$

In ordinary circumstances, when the times of measurement are not less than 50 times, the standard deviation takes two digits is suitable; it is suitable to take one digit merely when the times of measurement is very few and the first digit of standard deviation S is a larger number, this is a conclusion made according to the statistical theory.

④The measurement uncertainty

1) The measurement uncertainty refers to that due to the presence of measurement error and the extent of measured value can not be determined, it is assessed by measuring the true value within a range of values.

The extent of measurement uncertainty can be expressed by the "magnitude range" and "confidence probability". It gives the true value which is measured at a range of $[X - U, X + U]$, shows the possibility and credibility of the measurement results. It is clear that the true value is measured in a range of probability. In the three ranges of probability of σ、2σ、3σ are: $P_\sigma = 68.3\%$, $P_{2\sigma} = 95.4\%$; $P_{3\sigma} = 99.7\%$.

The measurement uncertainty and the error are two different concepts. The error represents the value of deviation of the true value. It is a definite value. However, the measurement uncertainty shows the dispersion measurements which indicate a range. Since the true value is not known, the measurement error can exist the ideal concept. The degree of uncertainty can be qualified by the experiment, data and experience and other information. In case of the large of degree of uncertainty, the absolute value of error may not be large, so they should not be confused.

2) The assessment and synthetic of uncertainty degree: The measurement uncertainty typically contains several components. The numerical evaluation methods can be divided into two categories: adopt the statistical methods assessed Class A standard uncertainty components and the other non – statistical methods assessed Class B standard uncertainty components. The uncertainty degree of Class A and Class B are the two types of different assessment methods while they are not different in nature. In some cases it can be rated only by Class A and Class B. However, in most cases it can be integrated the results of Class A and Class B.

The uncertainty of Class A standard: under the same conditions, the statistical calculation methods can be repeated used to obtain the uncertainty degree with u_A (mainly related to random error). With the experimental standard deviation of the arithmetic mean $s(\bar{x}_i)$, it characterizes the relationship between the unilateral experimental result x_i and the experimental standard measurements $s(x_i)$

$$u_A = s(\bar{x}_i) = \frac{s(x_i)}{\sqrt{n}} = \sqrt{\frac{\sum_{i=1}^{n}(x_i - \bar{x})^2}{n(n-1)}} \qquad (1-2-9)$$

The uncertainty of Class B standard: with other non – statistical evaluation methods, such as the instrument error, the undecided system error value can be shown with u_B.

The information source of Class B standard uncertainty can be from in general: the previous data, the relevant technical information and characteristics of measuring instrument knowledge and experience, the data calibration certificate provided by the productive sector, the identification certificate or other document and accuracy provided etc. as well as the manual or some reference data and uncertainty information given.

The known information indicates that the measured value measured by the range of the dispersion of the half width Δ, and the probability falls to $x_i - \Delta$ to $x_i + \Delta$ the interval is 100%. Through its estimated distribution, it can be drawn from the Class B standard uncertainty u_B is:

$$u_B = \frac{\Delta}{k_i} \qquad (1-2-10)$$

The formula contains factor k_i (can also be expressed as a ratio C) depends on the distribution of measured values. The relationship of the common distribution $u(x_i)$ and k_i is in the following table (Tab. 1-2-1):

Tab. 1-2-1   The relationship between the common distribution $u(x_i)$ and k_i

Distribution type	$P(\%)$	k_i	$u(x_i)$
Rectangular distribution	100	$\sqrt{3}$	$a/\sqrt{3}$
Normal distribution	99.73	3	$a/3$
Triangular distribution	100	$\sqrt{6}$	$a/\sqrt{6}$
Trapezium distribution $\beta = 0.71$	100	2	$a/2$

Under the situations of experimental teaching, for the simplicity, the general estimate is rectangular distribution. Among them, take the instrument error Δ limit or estimate the actual measurement error limits, such as the Class B uncertainty u_B:

$$u_B = \frac{\Delta}{\sqrt{3}}$$

The combined standard uncertainty assessment:

When the measurement result is the amount determined by a number of other values, the standard uncertainty of each other by the variance and covariance calculated called as the combined standard uncertainty. It is an estimate of the standard deviation of the measurement results, represented by the symbol u_c.

If the measurement results with statistical uncertainty components: u_{a_1}, u_{a_2}, $\cdots$; non-statistical uncertainty components: u_{b_1}, u_{b_2}, $\cdots$, and they are independent of each other, the combined standard uncertainty characterized as:

$$u_c = \sqrt{\sum_{i=1}^{m} u_{a_i}^2 + \sum_{i=1}^{n} u_{b_i}^2} \qquad (1-2-11)$$

m, n represent the number of A, B component of the two types of uncertainty. If $m = n = 1$, then

$$u_c = \sqrt{u_a^2 + u_b^2}$$

In the teaching experiment, Class A is only one component of uncertainty

$$u_A = s(\bar{x}_i) = \frac{s(x_i)}{\sqrt{n}} = \sqrt{\frac{\sum_{i=1}^{n}(x_i - \bar{x})^2}{n(n-1)}} \text{ (number of measurements } n \geq 6\text{)}$$

Class B uncertainty has multiple components:

Uncertainty generated by the instrument: $u_{b_1} = \Delta_{\text{instrument}}/\sqrt{3}$; Estimated error limit according to the actual situation Δ: $u_{b_2} = \Delta/\sqrt{3}$

So Class B uncertainty is as follows: $u_B = \sqrt{u_{b_1}^2 + u_{b_2}^2}$

3) The assessment of total uncertainty (expanded uncertainty): The combined uncertainty $u_c(x)$ multiplied by a coverage factor k (or denoted C) (also called a factor), which was expanded uncertainty $U = ku_c(x)$.

In general, the experimental measurements x falls to the interval $\bar{x} - u_c(x)$ to $\bar{x} + u_c(x)$, the probability is only about 68%. In the extended confidential interval, the probability can be increased. There are two k values: no need to give the accurate confidential probability, k values advisable 2 – 3. When k takes the value 2, the confidential probability of $U = 2u_c(x)$ is about 95%. When k takes the value 3, the probability of $U = 3u_c(x)$ is approximate 99%. In the physics experiments, the expanded uncertainty method can be evaluated. In the lab report, k value unified 2.

4) The measurement results: After the uncertainty is calculated, the results can be expressed as follows: $x = \bar{x} \pm U$ (unit). The meaning of this expression is explained to calculate the certain probability of true value ($\geqslant 95\%$) falls within the range of ($\bar{x} - U$, $\bar{x} + U$). $U = 2u$, $k = 2$, u is the combined uncertainty, in general, the probability is about 68%.

Example 1 – 12: The cylinder diameters d measured by vernier caliper 0.02mm are respectively: 2.594cm, 2.592cm, 2.596cm, 2.592cm, 2.590cm, 2.592cm. Please calculate the measurement results of the diameter.

Solution: The arithmetic mean
$$\bar{d} = (2.594 + 2.592 + 2.596 + 2.592 + 2.590 + 2.592)/6 = 2.593(\text{cm})$$

The mean standard deviation measurements
$$s(\bar{d}) = \sqrt{\frac{\sum_{i=1}^{6}(d_i - \bar{d})^2}{n(n-1)}} = 0.0009(\text{cm})$$

Class B component of uncertainty
$$u_B = \Delta_{\text{instrument}}/\sqrt{3} = 0.002/\sqrt{3}(\text{cm})$$

Combined standard uncertainty
$$u_c = \sqrt{u_A^2 + u_B^2} = \sqrt{0.0009^2 + (0.002/\sqrt{3})^2} = 0.0015(\text{cm})$$

Expanded uncertainty: $U = 2u_c = 2 \times 0.0015 = 0.003$ (cm)

The measurement result is expressed as: $d = (2.593 \pm 0.003)$ (cm)

4. Estimating error of indirect measurement result

In the most cases, the experiment result is generally obtained through indirect measurement. It means that the values of direct quantities are measured first, then calculate the result of indirect quantities through some relation. For example, to determine density of an object, if the object is a cuboid, its qualities m, x, y, and z are measured directly, and then calculates the density of the object by formula $\rho = \dfrac{m}{xyz}$. The result of the indirect measurement originates from the direct measurement, and because the direct measurement has error, the indirect measurement also inevitably has error. The error of the indirect measurement calculated according to the error of direct measurement is called the error propagation, or error synthesis.

The ways of error synthesis usually have the error arithmetic synthesis and square and root of error (root of error square sum) synthesizes.

(1) To Estimate the error of indirect measurement by maximum error method: For an indirect measurement, assume $N = f(x, y, z\cdots)$, where N is the value of indirect measurement, $x, y, z\cdots$ are values of direct measurement, and also $x = \bar{x} \pm \Delta x$, $y = \bar{y} \pm \Delta y$, $z = \bar{z} \pm \Delta z\cdots$, where $x, y, z\cdots$ are mutual independent variables. The error propagation of indirect measurement means that when direct measurement values $x, y, z\cdots$ have errors $\Delta x, \Delta y, \Delta z\cdots$, ΔN is right the indirect measurement error, and to calculate ΔN is exactly a question of calculating the entire differential calculus of N. The entire differential expression is as follows

$$dN = \frac{\partial f}{\partial x}dx + \frac{\partial f}{\partial y}dy + \frac{\partial f}{\partial z}dz + \cdots$$

Considering the rule that the limit of error is preferred bigger to smaller, in the worst situation we take their absolute values and replace "d" by "Δ", then has:

$$\Delta N = \left|\frac{\partial f}{\partial x}\Delta x\right| + \left|\frac{\partial f}{\partial y}\Delta y\right| + \left|\frac{\partial f}{\partial z}\Delta z\right| + \cdots \qquad (1-2-12)$$

(1-2-12) is called the propagation formula of arithmetic mean error.

On the basis of the above formula, a simplified propagation formula of the arithmetic operations mean error can be induced as follows:

In addition and subtraction

$$\Delta N = \Delta x + \Delta y + \Delta z + \cdots$$

In multiplication and division

$$E_r = E_x + E_y + E_z + \cdots$$

For the sake of convenient consulting and reference, a table of error formula of functional relation, Tab. 1-2-2, is listed below

Tab. 1-2-2 Error formula of functional relation

functional relational expression $N = f(x, y, z\cdots)$	error	
	absolute error ΔN	relative error $E_r = \dfrac{\Delta N}{N}$
$N = x + y + z$	$\Delta x + \Delta y + \Delta z$	$\dfrac{\Delta x + \Delta y + \Delta z + \cdots}{\bar{x} + \bar{y} + \bar{z}}$
$N = x - y$	$\Delta x + \Delta y$	$\dfrac{\Delta x + \Delta y}{\bar{x} - \bar{y}}$
$N = x \cdot y$	$\bar{x} \cdot \Delta y + \bar{y} \cdot \Delta x$	$\dfrac{\Delta x}{\bar{x}} + \dfrac{\Delta y}{\bar{y}}$

续表

functional relational expression $N=f(x, y, z\cdots)$	error	
	absolute error ΔN	relative error $E_r = \dfrac{\Delta N}{N}$
$N = x \cdot y \cdot z$	$\bar{y} \cdot \bar{z} \cdot \Delta x + \bar{x} \cdot \bar{z} \cdot \Delta y + \bar{x} \cdot \bar{y} \cdot \Delta z$	$\dfrac{\Delta x}{\bar{x}} + \dfrac{\Delta y}{\bar{y}} + \dfrac{\Delta z}{\bar{z}}$
$N = \dfrac{x}{y}$	$\dfrac{\bar{y} \cdot \Delta x + \bar{x} \cdot \Delta y}{\bar{y}^2}$	$\dfrac{\Delta x}{\bar{x}} + \dfrac{\Delta y}{\bar{y}}$
$N = x^n$	$n \cdot \bar{x}^{n-1} \cdot \Delta x$	$n \cdot \dfrac{\Delta x}{\bar{x}}$
$N = \sqrt[n]{x}$	$\dfrac{1}{n} \bar{x}^{(1/n)-1} \cdot \Delta x$	$\dfrac{1}{n} \cdot \dfrac{\Delta x}{\bar{x}}$

Concluded from the above table, it is convenient to first calculate the absolute error when computing indirect measurement error in addition and subtraction or first calculate the relative error in the operation of multiply, divide, mathematical power, evolution, etc, then by $\Delta N = E_r \cdot \bar{N}$, extracts ΔN.

Besides, the mean value of indirectly measurement is $\bar{N} = f(\bar{x}, \bar{y}, \bar{z}, \cdots)$, and the final result is expressed as $N = \bar{N} \pm \Delta N$, its relative error is $E_r = \dfrac{\Delta N}{\bar{N}} \times 100\%$.

(2) The basic formula of standard deviation propagation: To evaluate the error of indirect measurement by maximum error method, when estimating indirect measurement error by maximum error method, in this process, it is unknown yet whether the error itself is positive or negative, therefore, all coefficients of Δx, Δy, $\Delta z\cdots$ in the formula, are in the form of absolute value. The error estimated in this way is definitely a bit larger, to be more precise, it is better to take the sum of the squares of those error values, and then take its root, the fundamental formula out of this process is called standard deviation propagation, namely

$$\sigma = \sqrt{\left(\dfrac{\partial f}{\partial x}\right)^2 \sigma_x^2 + \left(\dfrac{\partial f}{\partial y}\right)^2 \sigma_y^2 + \left(\dfrac{\partial f}{\partial z}\right)^2 \sigma_z^2 + \cdots} \qquad (1-2-13)$$

The table below lists some practical formulas of standard deviationpropagation (Tab. 1 -2 -3):

Tab. 1 -2 -3 Common formulas of standard deviation propagation

functional relation $N=f(x, y, z\cdots)$	error	
	absolute error σ_N	relative error $\dfrac{\sigma_N}{\bar{N}}$
$x \pm y$	$\sqrt{\sigma_x^2 + \sigma_y^2}$	$\dfrac{\sqrt{\sigma_x^2 + \sigma_y^2}}{\bar{x} \pm \bar{y}}$
$x \cdot y$	$\sqrt{\bar{y}^2 \sigma_A^2 + \bar{x}^2 \sigma_B^2}$	$\sqrt{\left(\dfrac{\sigma_x}{\bar{x}}\right)^2 + \left(\dfrac{\sigma_y}{\bar{y}}\right)^2}$
$\dfrac{x}{y}$	$\sqrt{\dfrac{\bar{y}^2 \sigma_x^2 + \bar{x}^2 \sigma_y^2}{\bar{y}^2}}$	$\sqrt{\left(\dfrac{\sigma_x}{\bar{x}}\right)^2 + \left(\dfrac{\sigma_y}{\bar{y}}\right)^2}$
x^n	$n \cdot \bar{x}^{n-1} \cdot \sigma_x$	$n \cdot \dfrac{\sigma_x}{\bar{x}}$
$k \cdot x$	$k\sigma_x$	$\dfrac{\sigma_x}{\bar{x}}$

Example 1-2-13 Find the error formula of a cirque area.

Solution: The formula of a cirque area is $S = \frac{1}{4}(D^2 - d^2)$, and it may be rewritten as

$$S = \frac{\pi}{4}D^2 - \frac{\pi}{4}d^2$$

From above absolute error formula, we get

$$\Delta S = \frac{\pi}{4} \cdot 2\overline{D} \cdot \Delta D + \frac{\pi}{4} \cdot 2\overline{d} \cdot \Delta d = \frac{\pi}{2}(\overline{D} \cdot \Delta D + \overline{d} \cdot \Delta d)$$

Its relative error formula may be obtained divided both sides by $\overline{S} = \frac{\pi}{4}(\overline{D}^2 - \overline{d}^2)$, it is

$$\frac{\Delta S}{\overline{S}} = \frac{2(\overline{D} \cdot \Delta D + \overline{d} \cdot \Delta d)}{(\overline{D}^2 - \overline{d}^2)}$$

In summary, in process of actual measurement, estimating the measuring error generally carries on according to the following procedure to process data and express the result correctly.

After measuring the intended object several times and the set of data of it being recorded, a table of the data is hence listed.

To calculate the mean value of these values you measured.

To calculate the absolute error or the standard error and the relative error according to the requirement.

The result should be expressed as $\begin{cases} x = \bar{x} \pm \Delta x \\ E_r = \frac{\Delta x}{\bar{x}} \cdot 100\% \end{cases}$

Finally, under the condition of direct measuring value and its error are known, to calculate the error of indirect measuring value by the transmission formula first, then confirm the significant figures of the indirect measurement value by the error, and finally express the result of the indirect measurement correctly.

Example 1-2-14 Determining acceleration of gravity with a simple pendulum, the instrument, a metre ruler and a stopwatch are used. Please express the direct measurement result with the mean value and the mean absolute error.

At present, the measured time of 20 entire vibration is 28.4 seconds, and its measurement error is 0.1 seconds. The length of the pendulum is $L = \overline{L} \pm \Delta L = (0.500 \pm 0.001)$ m. So the cycle is

$$\overline{T} = \frac{28.4}{20} = 1.42(\text{s})$$

$$\Delta T = \frac{1}{20} \times 0.1 = 0.005(\text{s})$$

And the acceleration of gravity is

$$\bar{g} = \frac{4\pi^2 \overline{L}}{\overline{T}^2} = \frac{4\pi^2 \times 0.500}{1.42^2} = 9.779(\text{m/s}^2)$$

The relative error of the length is

$$\frac{\Delta L}{\bar{L}} = \frac{0.001}{0.500} = 0.20\%$$

The relative error of the cycle is

$$\frac{\Delta T}{\bar{T}} = \frac{0.1}{28.4} = 0.35\%$$

And according to arithmetic synthesis, the relative error of the acceleration of gravity is

$$\frac{\Delta g}{\bar{g}} = 2\frac{\Delta T}{\bar{T}} + \frac{\Delta L}{\bar{L}} = 2 \times 0.35\% + 0.20\% = 0.90\%$$

And then

$$\Delta g = 9.779 \times 0.009 = 0.088011 = 0.09 (\text{m/s}^2)$$

Hence the result is

$$g = \bar{g} \pm \Delta g = (9.78 \pm 0.09)(\text{m/s}^2)$$

Example 1-2-15 Determine acceleration of gravity with a simple pendulum, using a higher precision instrument. Please express the direct measurement result with the mean value and the standard error of mean value. Measure that

length of the pendulum is $L = (92.9 \pm 0.1)$ cm ($S_{mL} = 0.1$ cm)

the cycle is $T = (1.936 \pm 0.004)$ s ($S_{mT} = 0.004$ s)

Acceleration of gravity is

$$\bar{g} = \frac{4\pi^2 \bar{L}}{\bar{T}^2} = \frac{4\pi^2 \times 92.9}{1.936^2} = 979 \text{cm/s}^2$$

Using formula

$$\frac{S_{mg}}{\bar{g}} = \sqrt{\left(\frac{1}{\bar{L}}\right)^2 S_{mL}^2 + \left(\frac{2}{\bar{T}}\right)^2 S_{mT}^2}$$

We have

$$\frac{S_{mg}}{\bar{g}} = \sqrt{\left(\frac{1}{92.9}\right)^2 \times 0.1^2 + \left(\frac{1}{1.936}\right)^2 \times 0.004^2}$$

$$= 4.27 \times 10^{-3} = 0.43\%$$

$$S_{mg} = 979 \times 0.43 = 4(\text{cm/s}^2)$$

So the result is

$$g = \bar{g} \pm S_{mg} = (979 \pm 4)(\text{cm/s}^2)$$

5. Data processing method

(1) Tabulation method: Tabulation is often used while recording the experiment data, because the data-sheet can reflect the corresponding relation of these related physical quantities simply and clearly, it is convenient for checking whether it is correct or reasonable for the measuring result, and it contributes to analyse the regularity among the physical quantities. Tabulation is required to make everything simple and perspicuous, and it can reflect the corresponding relations among these related physical quantity; the data in the table must reflect the significant digit for a measurement result correctly, so it can indicate the accurate degree of the measurement; In the table, the physics significance of various marks must be explained clear-

ly and marked with the unit, the unit should be written in the title bar, not in each numeral of the table repeatedly; if there is something that cannot be explained clearly in the table, a suffix may be added under the table.

(2) Graphic method: Graphic method often is used in processing experiment data, make a graph with the data measured and analyze the graph. It is one of common methods to find the corresponding functional relation among the measured values.

1) Advantages of graphic method

①It is convenient to find some experiment results needed.

②It is easy to discover the measuring mistake in the experiment.

③Because the graph is a smooth curve tracing many data points, therefore it has a correcting action to the measured data, and it has the significance of mean value for repeated measurements.

④The points, which are not determined in the measurement, can be read on the graph. Moreover under the certain condition, the points beyond the scope of the measurement may be read from the extending graph.

2) Rules for plotting a graph

①To list a table corresponding to the measured data according to the certain rules.

②To select appropriate coordinate – paper according to the situation, such as right angle axes paper, log paper or polar coordinates paper and so on. Confirm the size of the coordinate paper and its proportion of coordinate axes. The size of the paper should be choosed according to the significant figures, and make the reliable number credible on the chart for the measured data, namely the least lattice in a chart is corresponding to the last digit of the reliable number of the data. A suspicious digit of the data in the chart should be estimated. The entire graph should suffuse symmetrically in the entire paper, the proportion of horizontal axis and vertical axis may be different, and the starting points of the coordinate axis are not sure necessarily as zero value.

③After confirming the proportion of the paper and the coordinate axis, mark the coordinate axis. In the ordinary circumstances, take function Y as the vertical axis, and take the independent variable X as the horizontal axis. Indicate the physical quantity that they represent (or symble) and their unit; Mark the value of this physical quantity for each the certain interval in the coordinate axis, write clearly the name of the chart and give essential explanation.

④Mark points and draw line: According to the measured data, with a sharpened pencil or a pen, mark the accurate position of various experiment points using the symble of " + " or " × " and so on (the different curves on the identical paper should use different mark), link each points as a smooth curve with a straight edge or a curve ruler (After having been drawn the graph, the marks should not be scraped off, in order to check and the reserve the record). As a result of the influence of error, the curve is not necessarily through all points, but it is requested the deviation has a quite even distribution on both sides, points with lager individual deviation should be deleted or remesured.

⑤Mark the measured error with the error short line as shown in Fig. 1 − 2 − 4.

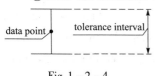

Fig. 1 − 2 − 4

In summary, graphic method has many advantages, but it also has its insufficiency, for instance, the accuracy of the obtained value is not enough, the number of significant figures cannot be too many and so on.

3) Successive minus method: Successive minus method is the data processing method as follows: subtracting the dependent variable data item by item or dividing the data into high and low groups, and corresponding subtraction, then the successive minus result is as a result of repeated measurement variables. As for the same amount of change for the independent variable, the dependent variable also changes and the measured data and other intervals ordered whichever is subtracted by the average difference between the results obtained. The advantage is to make use of the measured data fully and can detect the errors or data distribution and promptly correct or timely summary the data rule. It is also commonly used as a method of data processing physics experiments, commonly used in data processing variables.

Example 1 − 2 − 16 Take measuring the spring stiffness for example: increase weight at the lower end of the spring 1g, 2g, ⋯, 8g, noting the position of the spring endpoint ruler n_1、n_2、⋯、n_8. The elongations of the spring corresponding to increasing 1g weight respectively are:

$$\Delta n_1 = n_2 - n_1$$
$$\Delta n_2 = n_3 - n_2$$
$$\cdots$$
$$\Delta n_7 = n_8 - n_7$$

Elongation average value:

$$\overline{\Delta n} = \frac{\Delta n_1 + \Delta n_2 + \cdots + \Delta n_7}{7}$$
$$= \frac{(n_2 - n_1) + (n_3 - n_2) + \cdots + (n_8 - n_7)}{7}$$
$$= \frac{(n_8 - n_1)}{7}$$

The intermediate measurements did not work. In order to play the advantages of multiple measurements, the method of successive minus can be used.

Firstly, the data can be divided into two groups

$$n_1, \ n_2, \ n_3, \ n_4$$
$$n_5, \ n_6, \ n_7, \ n_8$$

Then take the average of the average value for successive minus

$$\overline{\Delta n} = \frac{(n_5 - n_1) + (n_6 - n_2) + \cdots + (n_8 - n_4)}{4}$$

That was the elongation average corresponding to increasing 4g weights each.

6. The best statistical straight line

Regarding a complex functional relation, because of their nonlinear, their graph all is a curve. It is very inconvenient to solve the experiment equation parameter by the curve, and it is also very difficult to judge whether the result is correct. Therefore, some different variables often be selected to replace the original variables (to be called variables substitution method), alter a curve to a straight line, then determine this straight line as long as conform its slope and intercept. For example, to $xy = k$, we may change the $x - y$ curve to a graph of $y - 1/x$ taking $y - 1/x$ as axis, then the curve becomes a straight line. The linear regression (linear fitting) is a process of determining slope and intercept according to the experiment data. Least square method is that, if the best-fit straight line is $y = f(x)$, then the sum of square is the smallest for the deviation between the measured various values y_i and the corresponding values in the fitting straight line.

Suppose we may confirm the function relation according to the situation of tracing points is

$$y = bx + a \qquad (1-2-14)$$

Where b and a are the slope and the intercept of the line. It is not possible for the points all to fall in the straight line which traces $(1-2-14)$, thereupon for each y_i corresponding to x_i there always is a deviation

$$\varepsilon_i = y_i - y$$

Find the sum of squares for each deviation

$$E = \sum_{i=1}^{n} \varepsilon_i^2 = \sum_{i=1}^{n} (y_i - y)^2$$

$$= \sum_{i=1}^{n} [y_i - (a + bx_i)]^2$$

$$= \sum_{i=1}^{n} [y_i^2 - 2y_i(a + bx_i) + a^2 + 2abx_i + b^2 x_i^2] \qquad (1-2-15)$$

When E is minimum, we select the straight line $(1-2-14)$ as the best statistical straight line. In order to determine this straight line, we now need to calculate a and b.

The essential condition for minimum E is

$$\frac{\partial E}{\partial a} = 0, \frac{\partial E}{\partial b} = 0$$

And then by formula $(1-2-15)$

$$\frac{\partial E}{\partial a} = 0, \sum_{i=1}^{n} (-2y_i + 2a + 2bx_i) = 0$$

Namely

$$\sum_{i=1}^{n} y_i = na + b \sum_{i=1}^{n} x_i \qquad (1-2-16)$$

And
$$\frac{\partial E}{\partial b} = 0, \sum_{i=1}^{n}(-2y_i x_i + 2a x_i + 2b x_i^2) = 0$$

So
$$\sum_{i=1}^{n} x_i y_i = b\sum_{i=1}^{n} x_i^2 + a\sum_{i=1}^{n} x_i \quad (1-2-17)$$

From (1-16) we get
$$a = \frac{\sum_{i=1}^{n} y_i}{n} - b\frac{\sum_{i=1}^{n} x_i}{n}$$

Or
$$a = \bar{y} - b\bar{x} \quad (1-2-18)$$

Namely it is the average coordinate which the straight line we request passes through. Substitute (1-18) to (1-17), we have

$$\sum_{i=1}^{n} x_i y_i = b\sum_{i=1}^{n} x_i^2 + b\bar{x}\sum_{i=1}^{n} x_i + \bar{y}\sum_{i=1}^{n} x_i$$

Divided both sides of the equation by n, we get

$$\overline{xy} = b\,\overline{x^2} - b\bar{x}^2 + \bar{x} \cdot \bar{y}$$

Hence
$$b = \frac{\overline{x \cdot y} - \bar{x} \cdot \bar{y}}{\overline{x^2} - \bar{x}^2} \quad (1-2-19)$$

Obviously, so long as we get the value of $\bar{x}$, $\bar{y}$, $\overline{xy}$, $\overline{x^2}$, $\bar{x}^2$, according to the data we have determined, then we can calculate a and b.

We may check whether the obtained equation is reasonable using the correlation coefficient r, which is defined as

$$r = \frac{\overline{x \cdot y} - \bar{x} \cdot \bar{y}}{\sqrt{(\overline{x^2} - \bar{x}^2)(\overline{y^2} - \bar{y}^2)}} \quad (1-2-20)$$

The value of r is usually between -1 and 1, and while the value of r is closer to 1, the data are concentrated to the line. Otherwise, when r is smaller than 1 and near to 0, the data is discrete.

Exercises

1. The least graduation of an ampere meter is 0.1A, and we determine the electric current with it, readings for six times are all 2.5A, is this electric current certainly stable?

2. Determine output power of a laser for 7 times, the readings are respectively 2.2, 12.7, 18.9, 0.6, 7.9, 22.4, 0.1. Its unit is milliwatt. Figure out its mean value is 9.3 mW, the mean absolute error is 7.7 mW, standard deviation of the mean value is 3.5 mW, and what conclusion can you make?

3. How many significant digits have the following quantities: 8.640 g, 200.04 g, 0.005 cm, 1.00×10^{-5} cm, 12.00 cm.

4. Correct the mistakes for the following expressions.

$L = (1.54 \pm 0.02)\text{m} = (1540 \pm 20)\text{mm}$

$m = 1.23\text{kg} = 1230\text{g}$

$t = (8.50 \pm 0.1)\text{s}$

5. Try these exercises.

(1) $y = \dfrac{ab}{2a - b}$

Where $a = 2.0517$, $b = 4.1032$.

(2) $y = \dfrac{p(a + c)}{ab - cd}$

Where $a = 3.1416$, $b = 10.002$, $c = 31.416$, $d = 1.0001$, $p = 31.4287$.

(3) $y = a + 5b - 3c - 4d$

Where $a = 382.02$, $b = 1.03754$, $c = 56$, $d = 0.001036$.

(4) $y = \dfrac{8abc}{\pi^2 dlm}$

Where $a = 800.0$, $b = 105.35$, $c = 112.6$, $d = 0.03005$, $l = 12.98$, $m = 0.97$.

6. Measure the length of a copper wire for six times, the values respectively are: 60.499 m, 60.498 m, 60.500 m, 60.503 m, 60.497 m, 60.505 m, we try to calculate its arithmetic mean value, the mean absolute error, the standard deviation of the mean value and the relative error.

7. Compute the gravitational potential energy of an object, various known quantities all indicates using the mean value and the mean absolute error ($E_p = mgh$), they are

Mass: $m = (0.1000 \pm 0.0005)$ kg

Gravity acceleration: $g = (9.80 \pm 0.04)$ m/s^2

Height: $h = (0.689 \pm 0.002)$ m

8. Compute the area of a rectangle $A = ab$, the known quantity indicated with the mean values and the standard deviations, they are

$\bar{a} = 24.345$ mm $S_{ma} = 0.006$ mm

$\bar{b} = 50.368$ mm $S_{mb} = 0.008$ mm

9. Measure the resistance of a copper wire (altogether for five times) by changing temperature t,

t (℃)	17.8	20.9	37.7	48.2	58.8
R (Ω)	3.554	3.687	3.827	3.969	4.105

Suppose the best statistical straight line is $R = R_0(1 + \alpha t)$, we try to definite the constant R_0 and α, R_0 is the resistance when temperature is at 0 ℃, α is resistance temperature coefficient for the copper wire.

(Zhao Zhe, Zhao Yao)

第二章 基础性实验

Chapter Two Basic Experiment

实验一 基本长度测量

【实验目的】

(1) 掌握一般游标原理,学会正确使用游标卡尺和螺旋测微计。
(2) 掌握有效数字的计算,巩固误差理论与数据处理方法。

【实验器材】

游标卡尺、螺旋测微计、金属圆环、钢球。

【实验原理】

1. 游标卡尺及读数

游标卡尺是比常用的米尺(分度值1mm)更加精密的长度测量仪器。为了提高测量的精确度,在米尺(即主尺)上附加一个能够滑动的有刻度的副尺(即游标),从而构成了游标尺,如图2-1-1所示。

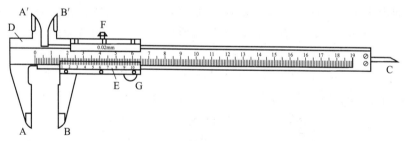

A、B. 外径量爪;A'、B'. 内径量爪;C. 深度尺;D. 主尺;
E. 游标;F. 固定螺钉;G. 微调推轮

图2-1-1 游标卡尺

下面以十分游标为例,如图2-1-2所示,给出游标卡尺测量长度 l 的普遍表达式:

$$l = x + k\delta_x \tag{2-1-1}$$

式中,x 为游标"0"线所在处主尺刻度的整毫米数;k 代表游标上的第 k 条线与主尺上的某刻度线对得最齐;δ_x 为游标的准确度。则

$$\delta_x = \frac{y}{n} \tag{2-1-2}$$

式中,n 为游标上刻有的总的最小格数;y 为主尺上的最小分度值(对十分游标 $n=10$,$y=1$mm)。因此,对图2-1-2所示的情况,有:$x = 41$mm,$k = 5$,$\delta_x = 0.1$mm,由式

· 43 ·

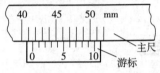

图 2-1-2 十分游标

（2-1-1），所测长度为

$$l = 41\text{mm} + 5 \times 0.1\text{mm} = 41.5\text{mm} = 4.15\text{cm}$$

注意：当游标"0"线恰与某分度刻线重合时，估计位仍不能略去，因为数值的有效数位直接反映测量的准确程度，所以读数的最后一位务必写上"0"，这个数字"0"也是有效数字。

2. 螺旋测微计（千分尺）及读数

对于螺距为 x 的螺旋，每转一周，螺旋将前进（或后退）一个螺距，如果转 $\frac{1}{n}$ 周，则螺旋就移动 $\frac{x}{n}$。设螺旋的螺距为 0.5mm，当它转动 $\frac{1}{50}$ 圆周时，螺旋将转动 $\frac{0.5}{50}=0.01$（mm）。如果转动 3 圈又 $\frac{24}{50}$ 圆周时，螺旋就移动 $3 \times 0.5 + \frac{24}{50} \times 0.5 = 1.5 + 0.24 = 1.74$（mm）。因此借助螺旋的转动将螺旋的角位移转变为直线位移，可进行长度的精密测量。

螺旋测微计的构造如图 2-1-3 所示，它由主尺和一个套在主尺上的能够旋转的副尺构成。副尺上沿圆周分为 50 分度，螺旋的螺距为 0.5mm，副尺上每转过一个格，螺杆 2 将前进（或后退）0.01mm，即为其测量精度。主尺刻在固定管 4 上，横线上方是毫米刻度，下方是半毫米刻度。读数也分为两步：①从副尺的前沿（活动套管的前沿）在固定套管 4（主尺）上的位置读出整圈数。②从固定套管上的横线所对的活动套管（副尺）上的分格数读出不足一圈的小数，二者相加即为测量值。如图 2-1-4 所示，其中图 2-1-4（b）的读数应为 5.382mm，图 2-1-4（c）所示读数为 5.882mm，二者的差别就在于副尺边缘的位置不同，前者没超过 5.5mm，而后者超过了 5.5mm。

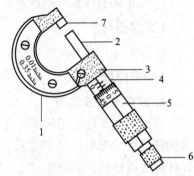

1. 尺架；2. 微动螺杆；3. 锁紧装置；
4. 固定套管；5. 微分筒；
6. 棘轮旋柄；7. 测砧

图 2-1-3 螺旋测微计的构造

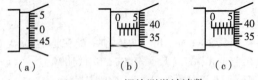

(a)　　　(b)　　　(c)

图 2-1-4 螺旋测微计读数

【实验步骤】

1. 用游标卡尺测量金属圆环的体积

（1）用拇指推动游标，使量爪 A、B 贴合，检查游标"0"线是否和主尺"0"线对齐，如不对齐，应记下零点读数，以校正测量结果。

（2）用游标卡尺测量圆环的内径 d、外径 D 和高 h，取不同部位各测量五次，将所测数据记录在表 2-1-1 内，并计算出测量的平均值、绝对误差和相对误差，写出测量的结果表达式。

(3) 计算圆环的体积及相对误差和绝对误差，写出圆环体积的正确表达式。

2. 用螺旋测微计测量钢球的体积

(1) 测量钢球的直径 d 五次，将所测数据记录在表 2-1-2 内，计算测量平均值、绝对误差和相对误差，写出测量的结果表达式。

(2) 计算出钢球的体积、相对误差及绝对误差，写出其结果表达式。

【数据记录及处理】

表 2-1-1 用游标卡尺测量柱形圆环的体积

项目 次数	外直径 D （cm）	内直径 d （cm）	深度 h （cm）	环体积 V （cm³）
1				
2				
3				
4				
5				$\overline{V} = \frac{\pi}{4}(\overline{D}^2 - \overline{d}^2) \cdot \overline{h}$
平均值				
绝对误差				
相对误差				
结果表示				

表 2-1-2 用螺旋测微计测钢球的体积 单位：_____

次数 项目	1	2	3	4	5	平均值	绝对误差	相对误差
直径								
体积	$\overline{V} = \frac{\pi}{6}\overline{d}^3 =$							
ΔV	$\Delta V = \frac{\pi}{2}\overline{d}^2 \cdot \Delta d =$							
E_V	$E_V = \frac{\Delta V}{\overline{V}} \times 100\% =$							
结果表示	$V = \overline{V} \pm \Delta V =$							

【注意事项】

(1) 游标卡尺和螺旋测微计是最常用的精密量具，使用时应注意维护。用完后应立即放回盒内，不许随便放在桌上，更不许放在潮湿的地方，只有这样才能保持它的准确度，延长其使用寿命。

(2) 每测一次，需要注意改变测量位置，使之具有平均的意义。

(3) 记录与计算均应按有效数字的规则进行处理。

【预习要点】

(1) 使用螺旋测微器应注意哪些问题？

(2) 使用游标卡尺应注意哪些问题？

【思考题】

(1) 由实验室的气压计上观察游标的准确度是多少？同时准确读出当天的气压是

多少？并正确地记录在实验报告上。

（2）试确定下列几种游标卡尺的准确度，并将它填入表格的空白处（表2-1-3）。

表2-1-3 测量游标卡尺准确度

游标分度数（格数）	10	10	20	20	50
与游标分度数对应的主尺度数（mm）	9	19	19	39	49
测量准确度（mm）					

【仪器介绍】

1. 游标卡尺

见图2-1-1，主尺D（按米尺刻度）与量爪A、A'相联，游标E与量爪B、B'及深度尺C相联。用拇指推动小轮G，游标即可紧贴着主尺滑动，量爪A、B用来测量厚度和外径，量爪A'、B'用来测量内径，深度尺C用来测量槽的深度，它们的读数值都是由游标0线和主尺的0线之间的距离来表示的，F为固定螺钉。

游标尺在构造上的主要特点：在游标上n个等分格子的总长与主尺上$(n-1)$个等分格子的总长相等。设y代表主尺上一个格子的长度，x代表游标上一个格子的长度，则有

$$nx = (n-1)y$$

那么，主尺与游标上每个格子的差值（即游标的准确度）为

$$\delta_x = y - x = \frac{1}{n}y$$

米尺的最小分度是1mm，即$y=1$mm。n取10时，即十分游标，对应的$\delta_x = \frac{1}{10}$mm $=0.1$mm。

下面以十分游标为例，说明游标卡尺的读数方法。当量爪A、B合拢时，游标上的"0"线与主尺上的"0"线重合。如图2-1-5所示，这时游标上第一条刻线在主尺第一条刻线的左边0.1mm处，第二条刻线在主尺第二条刻线左边0.2mm处……依此类推。这就是利用游标进行测量的依据。

在量爪A、B间放进一张厚度为0.1mm的纸片，这时与量爪B相联的游标就要向右移动0.1mm，如图2-1-6所示。游标上的第一条刻线与主尺上的第一条刻线重合，而游标上的其他任一刻线不与主尺上的任一条刻线重合；如果纸片的厚度为0.2mm，那么游标上的第二条刻线与主尺上的第二条刻线重合……依此类推。反过来讲，如果游标上第二条刻线与主尺上的第二条刻线重合，那么纸片的厚度就是0.2mm……；当游标"0"线与主尺上的第一条刻线重合时，则量爪A、B间的物体长度为1mm。

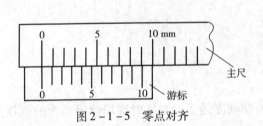

图2-1-5 零点对齐

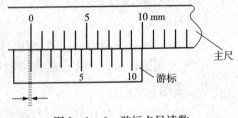

图2-1-6 游标卡尺读数

由上述可知，利用游标可以读出游标"0"线与主尺上前面最靠近的一条刻线之间的长度，即1mm之内的长度。如图2-1-2所示，游标"0"线处在41mm和42mm刻线之间，并且游标上第五条线与主尺上的刻线对齐，因此，量爪A、B间物体的长度$l=41.5$mm。

如前所述，对于十分游标，$\delta_x=0.1$mm，这是由主尺的刻度值y和游标尺的刻度值x之差给出的，不是估读的，它是十分游标能读出的最小数值，即十分游标的分度值。在测量A、B量爪间物体长度l时，由于用了游标，毫米以下一位，即$k\delta_x$是精确的。因此，根据有效数字的一般规则，读数的最后一位是读数的误差所在一位，前述长度l应该写为：$l=41.50$mm$=4.150$cm。最后加上"0"，表示读数误差出现在最后一位。

由此可见，使用游标可以提高读数的准确程度，游标卡尺的估读误差不大于$\frac{1}{2}\delta_x$。

还有一种常见的游标是"二十分游标"（$n=20$），即将主尺上的19mm等分成游标上的20格，如图2-1-7所示，或者将主尺上的39mm等分成游标上的20格，如图2-1-8所示，这样它们的准确度δ_x为

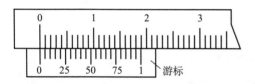

图2-1-7 游标卡尺准确度

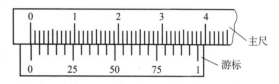

图2-1-8 游标卡尺准确度

$$\delta_x = 1.00 - \frac{19}{20} = 0.05\text{mm} \quad (2-1-3)$$

或

$$\delta_x = 2.00 - \frac{39}{20} = 0.05\text{mm} \quad (2-1-4)$$

在后一种情况下，δ_x是主尺上两格（2mm）与游标上的一格之差。

二十分游标上常刻有0、25、50、75、1等标度，以便于直接读数。如游标上第5根线（标25）与主尺上某刻线对齐，则读数的尾数为$5\times\delta_x=0.25$mm，即可直接读出。其估读误差为$\frac{1}{2}\delta_x$，在百分之一毫米位数上，因此，读数的尾数之后不再加"0"。如$l=0.25$mm之后不再加"0"。

另一种常用的游标是"五十分游标"（$n=50$），即主尺上49mm分成游标上的50格，见图2-1-9。五十分游标的分度值$\delta_x=1.00-\frac{49}{50}=0.02$mm，游标上刻有0、1、2…9标度，以便于读数。五十分游标的读数也写到百分之一毫米这一位上。

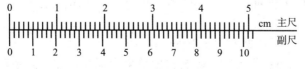

图2-1-9 五十分游标

综上所述，游标尺的分度值是由主尺与游标上的一格的长度之差值决定的，各种常用的游标卡尺的读数都写到百分之一毫米这一位上。使用游标尺时，可一手拿物体，另一手持尺，如图 2-1-10 所示。

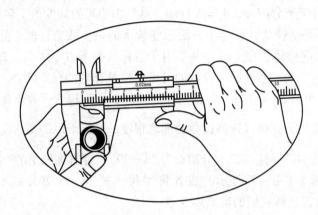

图 2-1-10 游标尺持法

在测量过程中，要特别注意保护量爪不被磨损，使用时轻轻地把物体卡住即可读数，不允许用来测量粗糙的物体，并切忌把被夹紧的物体在量爪内挪动。

使用游标尺测量之前，应先进行零点修正，即把量爪 A、B 合拢，检查游标尺"0"线是否和主尺"0"线重合，如不重合，应记下零点读数，加以修正。即待测长度的结果应写为 $l = l_1 - l_0$，l_1 是未做零点修正前的读数值，l_0 是零点读数，可以是正的，也可以是负的。

2. 螺旋测微计

如图 1-3 所示，使用螺旋测微计测量物体时，反旋棘轮旋柄 6，使螺杆 2 离开测砧 7，再把待测物体放在 2 与 7 之间，然后转动棘轮 6 借助摩擦力带动螺杆将物体夹住。

使用螺旋测微计测量时应注意下面几点。

（1）进行零点修正。先记录下零点读数 l_0，并注意 l_0 的正负。

测量前要检查零点读数。没夹被测物时，测杆和砧台相接时，副尺的零刻线应当与主尺上的横线对齐（即在一条直线上），但是由于调整不充分或使用不当，初始状态往往和上述要求不符，即有一个不等于零的初始值，把它称为零点读数，测量后要从测量值的平均值中加上（或减去）零点读数。其规律是，若副尺零刻线在主尺横线上方则加，在下方则减。如图 2-1-11 所示。

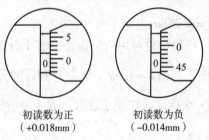

初读数为正　　初读数为负
（+0.018mm）　（−0.014mm）

图 2-1-11 零点修正

（2）记录零点及将待测物夹紧测量时，不能直接拧转螺杆，以免夹得太紧，影响测量结果及损坏仪器，必须转动棘轮旋柄推进螺杆，当转动棘轮时听到"喀、喀"声音时，就不要再推进螺杆，这时即可进行读数了。

（3）要保护仪器的测量面，不要用手去触摸。使用完毕，放入仪器盒时应使两测量面之间留有一小空隙，以免仪器受热膨胀而损坏。

（赵　喆　支壮志）

Experiment 1　Measurement of Length

【Objects】

(1) To become familiar with the two length – measuring instruments commonly used in a physics laboratory – the vernier caliper and micrometer.

(2) Study calculating significant digits and master the error theory and processing the experiment data.

【Apparatus】

vernier caliper, micrometer, copper annulus, metal sphere.

【Theory】

1. Vernier caliper and its reading

The vernier caliper is a kind of more accurate instruments for measuring length. Shown as Fig. 2 – 1 – 1, it consists of a fixed main scale with cm and mm divisions and a movable jaw with the vernier. The vernier is that its 10 divisions are equal to 9 mm. With the jaws closed, the zero error is first read. The object to be measured is then placed between the jaws. The zero line for measurement is the left – hand line on the vernier. The number of whole centimeters and millimeters is measured directly on the stem scale. The number of tenths of millimeters is estimated by counting the number of the line on the vernier (counting the left – hand line as zero) which lines up with one of the millimeter lines on the main scale. This gives the length in centimeters, millimeters, and tenths of millimeters.

The jaws, A and B, are used to close around the outside of an object, and these are the jaws mostly used. Jaws A′ and B′ fit into a hole and measure the inside diameter, and jaw C is used to measure the depth of a hole.

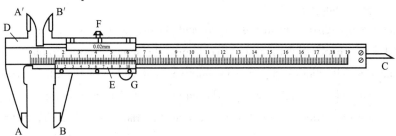

A, B. outside caliper jaws; A′, B′. inside caliper jaws; C. depth probe;
D. main scale; E. vernier; F. coarse locking screw; G. fine adjust thumbscrew

Fig. 2 – 1 – 1　Vernier caliper

The following is a general formula of measurement of l with a vernier caliper, here's an example. Fig. 2-1-2 shows a vernier, or movable scale, it contains ten divisions, each of which is nine tenths as long as the smallest main-scale division.

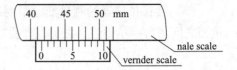

Fig. 2-1-2 Ten-scale cursor

$$l = x + k\delta_x \qquad (2-1-1)$$

Where x represents the reading of integer millimeter on the main scale corresponding to the zero mark on the vernier scale; k represents the kth mark on the vernier coincide with some mark on the main scale, and

$$\delta_x = \frac{y}{n} \qquad (2-1-2)$$

Where n represents total number of divisions on the vernier scale, y is the least graduation on the main scale, δ_x represents accuracy of the device. (For a ten-graduation vernier scale, $n=10$, $y=1$mm). We know that $x = 41$mm, $k = 5$, $\delta_x = 0.1$mm from the Fig. 2-1-2. Therefore the actual length is: $l = 41.5$mm = 4.15cm.

2. A micrometer caliper and its reading

The micrometer caliper (Fig. 2-1-3) is an instrument used for the accurate measurement of short lengths. Essentially, it consists of a carefully machined screw mounted in a strong frame. The object to be measured is placed between the end of the screw and the projecting end of the frame, called the anvil. The screw is then advanced until the object is gripped gently between the two jaws of the instrument. Most micrometers are provided with a ratchet knob arranged to slip on the screw as soon as a light and constant force is exerted on the object. By using the ratchet knob, it is possible to tighten up the screw by the same amount each time and also to avoid using too great a force, If this arrangement is absent, great care should be taken not to force the screw, for the instrument may be easily damaged.

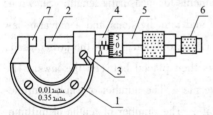

1. frame; 2. spindle; 3. lock nut; 4. barrel; 5. thimble; 6. ratchet knob; 7. anvil

Fig. 2-1-3 Micrometer caliper

The micrometer caliper used in this experiment consists of a screw with a pitch of 0.5mm, a longitudinal scale engraved along a barrel containing the screw, and a circular scale engraved around a thimble which rotates with the screw and moves along the scale on the barrel. The longitudinal scale is divided into millimeters. The circular scale has 50 divisions. Since the pitch of the screw is 0.5mm, which is the distance advanced by the screw in turning through one revolution, it is clear that rotating the thimble through one scale division will cause the screw to move a distance of 1/50 of 0.5mm, or 0.01mm, and this is called measuring accuracy. Hence, readings may be taken directly to one hundredth of a millimeter and by estimating tenths of a thimble-scale division, they may be taken to one thousandth of a millimeter.

The micrometer caliper is read by noting the position of the edge of the thimble on the longitudinal scale and the position of the axial line of the barrel on the circular scale, and adding the two readings. The readings of the main scale gives the measurement to the nearest whole main-scale division; the fractional part of a main-scale division is read on the circular scale. Since two revolutions of the screw are required to make it advance a distance of 1mm, it is necessary to be careful to note whether the reading on the circular scale which refers to the first half or the second half of a millimeter. (In Fig. 2-1-4 (b) the reading is 5.382mm, (c) is 5.882mm).

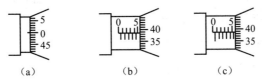

Fig. 2-1-4 Micrometer caliper reading method

【Procedure】

1. Measure volume of the metal annulus with vernier caliper

First, measure the height, inside diameter and outside diameter of the annulus five times at different positions. Then enter the values in Tab. 2-1-1 and calculate the average value, absolute error and relative error. After that, write out the result expressions.

Secondly, calculate the volume, the absolute error and the relative error of the annulus. Finally, write out the final result of the volume, and enter the values in Tab. 2-1-1.

2. Measure the volume of the sphere with micrometer

Measure the diameter of a sphere five times with the micrometer, and record the data in Tab. 2-1-2. Then calculate the average value, absolute error and relative error of the sphere. Then write out the result expression and calculate the volume of the sphere. After that, write out the final result in Tab. 2-1-2.

【Data and Calculations】

(1) Data of measuring the copper annulus with vernier caliper are listed in Tab. 2-1-1.

Tab. 2-1-1 Measurement of copper annulus using a vernier caliper

item / times	outside diameter D (cm)	inside diameter d (cm)	height h (cm)	volume of the annulus V (cm^3)
1				
2				
3				$\bar{V} = \frac{\pi}{4}(\bar{D}^2 - \bar{d}^2) \cdot \bar{h}$
4				
5				
average value				

item \ times	outside diameter D (cm)	inside diameter d (cm)	height h (cm)	volume of the annulus V (cm³)
absolute error				
relative error				$\overline{V} = \dfrac{\pi}{4}(\overline{D}^2 - \overline{d}^2) \cdot \overline{h}$
result expressions				

(2) Data of measuring the metal sphere with the micrometer are listed in Tab. 2-1-2.

Tab. 2-1-2 Data of measuring the metal sphere   unit: _____

item \ times	1	2	3	4	5	average value	absolute error	relative error
diameter d								
volume V	$\overline{V} = \dfrac{\pi}{6}\overline{d}^3 =$							
ΔV	$\Delta V = \dfrac{\pi}{2}\overline{d}^2 \cdot \Delta d =$							
E_V	$E_V = \dfrac{\Delta V}{\overline{V}} \times 100\% =$							
final result	$V = \overline{V} \pm \Delta V =$							

【Note】

(1) Vernier caliper and micrometer are accurate instruments commonly used in a physics laboratory. Great care must be taken to use them, replace them in their special boxes, and not to place them random, which can keep the accuracy and extend their life.

(2) Note the position measured to assure that it has average meaning.

(3) The record and calculation should be done with the rules of significant figures.

【Preview】

(1) What problems should be paid attention to while using a vernier caliper?

(2) What problems should be paid attention to while using a micrometer caliper?

【Questions and Problems】

(1) How much is the accuracy of the vernier of the barometer in the lab? At the same time, read-out the atmospheric pressure accurately this day. Record them in your experiment report.

(2) Try to confirm the accuracy of several kinds of vernier caliper below, and fill in Tab. 2-1-3.

Tab. 2-1-3

number of vernier graduations (the number of graduations)	10	10	20	20	50
number of main graduations corresponding to vernier graduations (mm)	9	19	19	39	49
accuracy (mm)					

[Introduction to Apparatus]

1. Vernier caliper

The structure characteristic of a vernier caliper is, the total number of divisions on the vernier scale n equals $n-1$ divisions on the main scale. Assume y is the length of the least graduation on the main scale, x is the length of the least graduation on the vernier scale, then

$$nx = (n-1)y$$

So the difference between y and x is

$$\delta_x = y - x = \frac{1}{n}y$$

The graduation of a meter ruler is 1mm, namely $y = 1$mm. For a ten-graduation caliper, $n = 10$, and $\delta_x = \frac{1}{10}$mm $= 0.1$mm。

We take a ten-graduation caliper as an example for its reading. When the instrument's jaws are fully closed, the zero line of the main scale coincides with the zero line of the vernier scale, as shown in Fig. 2-1-5. Now the first line of the vernier scale is at 0.1mm left of the first line of the main scale, the second line of the vernier scale is at 0.2mm left of the second line of the main scale⋯, and so on. This is the key point of measurement with a vernier caliper.

Place a piece of paper, whose thickness is 0.1mm, between jaws A and B, now the jaws are opened just 0.1mm, and the vernier marks all move together that far, as shown in Fig. 2-1-6. which brings the first vernier mark into coincidence with the main scale's 1mm mark, and the other lines of the vernier scale can not coincide with any line of the main scale. If the thickness of the paper is 0.2mm, then the second vernier mark coincides with the second line of the main scale⋯; when zero vernier mark coincides with the first main mark, then the length of the object between jaws A and B is 1mm.

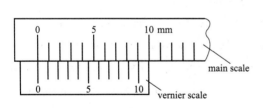
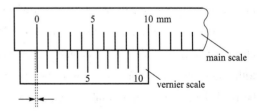

Fig. 2-1-5 Zero alignment   Fig. 2-1-6 Vernier caliper reading

Therefore, we know that the vernier may be used to read the length (less than 1mm) between the zero mark of the vernier scale and nearest mark line of the main scale. As shown in Fig. 2-1-2, the zero line locates between 41mm and 42mm, and the fifth line of the vernier scale lines up best with one of the mark lines on the main scale, so the length of the object between jaws A and B is: $l = 41.5$mm.

Note For a ten-graduation caliper, $\delta_x = 0.1$mm, we get it from the difference between y and x and it is an accurate reading but not an estimated figure, it is the least value read from the vernier, namely the graduation value of the ten-graduation vernier. When we measure the length l of the object between jaws A and B, because of a vernier being used, the digit next to mm is accurate. By the significant digit rules, the last digit of the reading is just where error occurs, so the above reading should be written as: $l = 41.50\text{mm} = 4.150\text{cm}$. The last "0" means the reading error occurs at the last digit. So the reading can be more accurate by using a vernier scale, and the reading error for a vernier calipers is not more than $\frac{1}{2}\delta_x$.

Another common vernier caliper is twenty-graduation vernier ($n = 20$). Its vernier scale is divided into 20 divisions, the total length of the vernier scale equals the length of 19mm of the main scale, as shown in Fig. 1-7, or 20 divisions equals the length of 39mm, as shown in Fig. 2-1-8. Therefore δ_x is

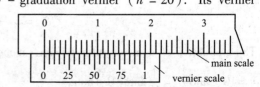

Fig. 2-1-7 Vernier caliper accuracy

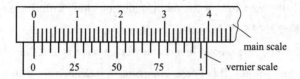

Fig. 2-1-8 Vernier caliper accuracy

$$\delta_x = 1.00 - \frac{19}{20} = 0.05\text{mm} \quad (2-1-3)$$

or

$$\delta_x = 2.00 - \frac{39}{20} = 0.05\text{mm} \quad (2-1-4)$$

In the second case, δ_x is the difference between the length of two divisions (2mm) of the main scale and the length of one division of vernier scale.

It is convenience for reading to mark 0, 25, 50, 75, 1 as special figures on the twenty-graduation vernier scale. For example, when the fifth line of the vernier scale is aligned exactly with one of the marks on the main scale, the mantissa of reading is $5 \times \delta_x = 0.25\text{mm}$, it may be read directly from the number marked at the vernier scale. Its reading error for a twenty-graduation vernier caliper is $\frac{1}{2}\delta_x$, and it occurs at the second number after decimal point. So "0" is not added behind the mantissa of the reading, For example $l = 0.25\text{mm}$, "0" is not added behind 5.

Another commonly used vernier caliper is a fifty-graduation vernier ($n = 50$), which the length of 50 divisions on the vernier scale corresponds to 49mm of the main scale, as shown in Fig 2-1-9. Its graduation is $\delta_x = 1.00 - \frac{49}{50} = 0.02\text{mm}$, and it is convenient for reading to mark figure 0, 1, 2, ⋯, 9 on the vernier scale. The last digit of the reading should be at the second number after decimal point of millimeter.

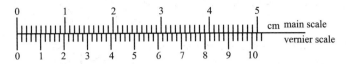

Fig. 2 – 1 – 9　Fifty – scale cursor

Now, we know that the graduation of a vernier caliper is decided by the difference between a division on main scale and the one on vernier caliper, and the reading errors of commonly used vernier caliper are about several hundredths of millimeter. When measuring, it's better for holding object by one hand and holding vernier caliper by another hand, as shown in Fig. 2 – 1 – 10. All the jaws should be protected from being worn – out, the reading is obtained when holding the object gently, and it is not allowed to measure a coarse object or move the object tightened between the jaws.

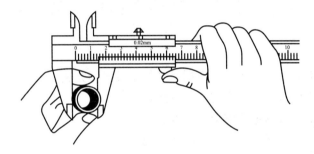

Fig. 2 – 1 – 10　Holding method of vernier caliper

Before making a measurement, the vernier caliper should have been zero corrected, that is, when jaws A and B fold, you should check whether the zero line of vernier scale coincides with the zero line of main scale, if not, the reading should be written down for correcting of all the other measurements. The result can be recorded as $l = l_1 - l_0$, where l_1 is the reading before correcting; l_0 is the reading of zero correcting, and it may be positive or negative either.

2. A micrometer caliper and its reading

As shown in Fig. 1 – 3, when using this instrument to measure the length of an object, close the spindle 2 and the anvil 7 on the object, then tighten the jaws using the ratchet knob 6 only. Then turn the ratchet knob 6 gently to drive the screw by friction force and the object is tighten.

Note

(1) Zero correction. The micrometer should be checked for a zero error, for it may not read zero when the movable spindle and the anvil are completely closed. In such cases a zero correction has to be applied to every reading and may be either positive or negative, then write down the zero reading l_0, and the zero reading may be treated as a correction value. The final reading is obtained by adding (or subtracting) the correcting value to (from) the average

value. As shown in Fig. 2 – 1 – 11, and the final result is $l = l_1 - l_0$, where l_1 is the reading before correcting.

(2) Move the spindle by turning the ratchet knob 6, don't turn the thimble 5 directly, it prevents the spindle being closed so forcefully that the instrument might be damaged or data be influenced. When turning the ratchet knob 6 and clicks are heard, don't turn it any more and the reading can be written down.

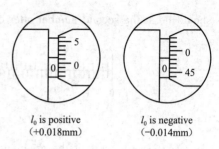

l_0 is positive (+0.018mm) l_0 is negative (−0.014mm)

Fig. 2 – 1 – 11　Zero correction

(3) Don't touch the two measuring jaws' faces. Place it in the apparatus box after measurement, keep a distance between two jaws in case it is damaged for the thermal expansion.

(Zhao Zhe, Wang Xiao – fei)

实验二　物体密度的测定

【实验目的】

(1) 了解天平的结构与原理，掌握天平的正规操作方法。
(2) 掌握规则形状固体密度的测定方法。
(3) 掌握比重瓶法测定液体密度的方法。
(4) 巩固误差的运算和有效数字的运算规则。

【实验器材】

物理天平、比重瓶、待测密度的固体（铝柱或钢球）、电子天平、待测密度的液体（乙醇）、蒸馏水、烧杯、移液管、纱布等。

【实验原理】

1. 规则物体密度的测定

若一物体质量为 m，体积为 V，则其密度 $\rho = \dfrac{m}{V}$。由此可见，只要测出物体的质量，再以测长仪器测出其相关长度，求出其体积，代入上式即可算出该物体的密度。

2. 比重瓶法测定液体（或固体）密度

比重瓶是一个特别的有塞小瓶，玻璃塞中间有毛细管。在比重瓶注满液体后，当用玻璃塞塞住时，则多余的液体就从毛细管溢出，这样瓶内盛有的液体体积就是固定的。

如要测液体的密度时，可先称出比重瓶的质量 m_0，然后再将温度相同的（室温的）待测液体和纯水分别注满比重瓶，称出待测液和比重瓶的总质量 m_2 以及纯水和比重瓶的总质量 m_1。于是，同体积的纯水和待测液体质量分别为 $m_1 - m_0$ 与 $m_2 - m_0$。通过计算可得待测液体的密度为

$$\rho = \dfrac{m_2 - m_0}{\dfrac{m_1 - m_0}{\rho_0}} = \dfrac{m_2 - m_0}{m_1 - m_0} \cdot \rho_0$$

式中，ρ_0 为同温度纯水的密度（见书后附表3）。

另外用比重瓶还可测量不溶于水的小块固体（其大小应保证能放入瓶内）的密度 ρ，可依次称出小块固体的质量 m，盛满水后比重瓶与纯水的总质量 m_1 以及装满纯水的比重瓶内投入小块固体后的总质量 m_2。显然，被小块固体排出瓶外的水的质量是 $m + m_1 - m_2$，排出水的体积是质量为 m 的小块固体的体积。所以，小块固体的密度为

$$\rho = \frac{m}{m + m_1 - m_2}\rho_0$$

【实验步骤】

1. 物理天平使用练习

称量铝柱质量5次，求其质量平均值。

2. 测定规则固体的密度

（1）正确使用物理天平称出固体的质量 m，重复5次取其平均值 $\overline{m}$。

（2）用游标卡尺测出给定规则固体的有关长度，求出其体积 V，代入公式 $\rho = \frac{m}{V}$，求出密度 ρ。将数据填入表 2-2-1 中。

（3）求出密度的相对误差 $\frac{\Delta\rho}{\rho}$ 与绝对误差 $\Delta\rho$，确定物体密度 ρ 的有效数字。

3. 乙醇密度的测定（以下质量用电子天平进行称量）

（1）洗净、烘干比重瓶（注意瓶内外都要干燥），称出其质量 m_0。

（2）称出比重瓶盛满乙醇时的质量 m_2。

（3）回收乙醇，以纯水洗净比重瓶，再装满纯水，称出纯水和比重瓶的总质量 m_1。将数据填入表 2-2-1 中。

（4）由公式 $\rho = \frac{m_2 - m_0}{m_1 - m_0}\rho_0$，计算出乙醇的密度 ρ。

4. 小块固体密度的测定（选做）

（1）取干净的铅粒若干，称出其质量 m。

（2）将称好质量的铅粒投入盛满纯水的比重瓶内，塞上塞子，擦干溢出的水（注意：瓶内不能有残留的气泡），称出盛满水后比重瓶与纯水的总质量 m_1 以及装满纯水的比重瓶内投入小块固体后的总质量 m_2。

（3）由公式 $\rho = \frac{m}{m + m_1 - m_2}\rho_0$，计算出铅粒的密度 ρ。

【数据记录及处理】

数据记录在表 2-2-1 中，并进行处理。

表 2-2-1 测量铝柱及乙醇密度

次数\项目	铝柱			比重瓶		
	高 H (cm)	直径 D (cm)	质量 m (g)	m_0 (g)	m_2 (g)	m_1 (g)
1						
2						

续表

次数\项目	铝柱 高 H（cm）	铝柱 直径 D（cm）	铝柱 质量 m（g）	比重瓶 m_0（g）	比重瓶 m_2（g）	比重瓶 m_1（g）
3						
4						
5						
平均值						
绝对误差						
相对误差						

其中

铝柱密度：

$$\bar{\rho} = \frac{\bar{m}}{\frac{1}{4}\pi\bar{D}^2\bar{H}} = \frac{4\bar{m}}{\pi\bar{D}^2\bar{H}}$$

$$\Delta\rho = \left(\frac{\Delta m}{\bar{m}} + 2\frac{\Delta D}{\bar{D}} + \frac{\Delta H}{\bar{H}}\right)\cdot\bar{\rho}$$

结果表示：

$$\rho = \bar{\rho} \pm \Delta\rho =$$

乙醇密度：

$$\bar{\rho} = \frac{\bar{m}_2 - \bar{m}_0}{\bar{m}_1 - \bar{m}_0}\cdot\rho_0 =$$

【注意事项】

（1）称量时如天平发生故障，请指导教师检修，调整天平时一定要戴好手套。

（2）称量装满液体的比重瓶之前，必须将瓶外液体擦拭干净，称量乙醇的质量时操作一定要快，以免乙醇挥发。

【预习要点】

（1）使用物理天平应注意哪些问题？

（2）怎样消除天平两臂不等而造成的系统误差？

【思考题】

（1）怎样操作天平才能使质量的称量迅速而准确？总结一下调整和使用天平的方法。

（2）用比重瓶法测定物体密度时，可能有哪些原因会引起误差？

（3）请自行设计如何测乙醇和水混合液中乙醇的含量，并写出具体实验步骤。

【仪器介绍】

1. 物理天平

（1）结构和使用：物理天平通常用来称量物体的质量，如图2-2-1所示。它的主要部分是横梁7，梁上有3个用玛瑙或钢制的刀口 B_0、B_1、B_2。中间刀口向下，由中柱11上的刀承支起，两侧刀口上挂吊耳5，吊耳下边悬挂秤盘，3个刀口在同一平面上，且 $B_1B_0 = B_2B_0$，即天平梁是等臂杠杆。在立柱下方有一个止动螺旋13，用以升降横梁。当顺时针转动止动螺旋13时，立柱中升高的刀承将横梁从止动架上托起，天平梁即可灵活地摆动起来，进行称量；不用时，逆时针转动止动螺旋，横梁下降，并由止动架托住，中间刀口和刀承分离，两侧刀口也由于秤盘落在底座上而减去负荷，这就能保护刀口不受损伤。

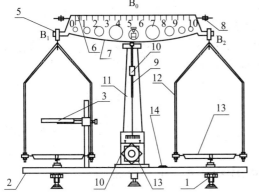

1. 水平调节螺丝；2. 底板；3. 托架；4. 支架；
5. 吊耳；6. 游码；7. 横梁；8. 平衡调节螺丝；
9. 指针；10. 感量调节器；11. 中柱；12. 盘梁；
13. 止动螺旋；14. 水平仪

图 2-2-1 物理天平的结构

天平的性能用最大载量和灵敏度表示。最大载量是天平容许的最大负荷。灵敏度是指天平两侧负载相差一个单位质量时，指针偏转的分度数。有时也用感量来代替灵敏度，感量是指天平指针偏转一个分格时，天平盘上可加的砝码，它正好是灵敏度的倒数。物理天平最大的负载量是 $200 \sim 500g$，灵敏度约为 1 分格/10 毫克，或感量 10 毫克/1 分格。

（2）使用前的调试

①调平：转动底脚螺丝1，使水平仪的气泡位于中央处。

②调零点：天平空载时，调节止动螺旋，升起横梁，观察指针摆动情况，如果两边摆动幅度相同或相差不到一个分格，即为调平。否则可调整平衡调节螺丝使之平衡，调整时必须将横梁落下，以免损伤刀口。

③为了保护天平的灵敏度，操作时必须注意以下几点。

a. 不许在横梁支起的状态下加减砝码或取放物体。

b. 习惯上被测物放在左盘，砝码放在右盘，为的是便于操作。取放砝码时要使用镊子。

c. 称量时，先估计一下被测物的重量，加砝码时要从重到轻，依次更换。

d. 称量后，重新检查横梁是否落下，横梁和吊耳位置是否正常，砝码是否按顺序摆好。

2. 电子天平

（1）工作环境：电子天平为高精度测量仪器，安装平台一定要稳定、平坦，避免震动；避免阳光直射和受热，避免在湿度大的环境工作；避免在空气直接流通的通道上。

（2）使用方法

①调水平：天平开机前，应观察天平后部水平仪内的水泡是否位于圆环的中央，否则通过天平的地脚螺栓调节，左旋升高，右旋下降。

②预热：天平在初次接通电源或长时间断电后开机时，至少需要 30min 的预热时间，因此，实验室电子天平在通常情况下不要经常切断电源。

③称量：按下【ON/OFF】键，接通显示器，等待仪器自检。当显示器显示零时，自检过程结束，天平可进行称量。

④放置称量纸，按显示屏两侧的【Tare】键去皮，待显示器显示零时，在称量纸上加所要称量的试剂，称量。

⑤称量完毕后，按【ON/OFF】键，关断显示器。

（3）注意事项：天平在安装时已经过严格校准，故不可轻易移动天平，否则校准工作需重新进行。严禁不使用称量纸直接称量，每次称量后应清洁天平，避免对天平造成污染而影响称量精度。

（赵　喆　李百芳）

Experiment 2　　Determination of Density

【Objects】

(1) To understand the principle and stucture of the balance, grasp the regular manipulative techniques of the balance.

(2) To master the method for determining the density of regular objects.

(3) To grasp the method for determining the density of liquid by pycnometer.

(4) To master the computing method of the error and the operation rules of significant figures.

【Apparatus】

TW Physics balance, pycnometer (specific gravity bottle), an aluminum cylinder, unknown liquid (alcohol), distilled water, beaker, suction pipet, gauze, etc.

【Theory】

1. Density of a regular object

Density is defined as mass per unit volume. If the mass of an object is m, the volume is V, then its density can be represented by $\rho = \dfrac{m}{V}$. So we can work out the density of the object, as long as mass and volume of the object has been measured.

2. Density determination by pycnometer

The pycnometer is provided with a perforated glass stopper so that when the bottle is filled with a liquid and the stopper inserted, the excess liquid will run out at the top of the stopper. This gives exactly equal volumes of different liquids when put in the bottle.

Determine the mass of an empty clean, dry pycnometer and record as m_0. Then fill it with unknown liquid by pouring very carefully from a small beaker. See that the outside is dry, and determine and record as m_2 (mass of pycnometer + unknown liquid). Record the difference of the two masses as the mass of V cm^3 of water. Now empty the unknown liquid water back into the beaker and rinse the bottle with distilled water, and determine and record as m_1 (mass of pycnometer + water). Record the mass of V cm^3 of the water. From these masses of equal volumes of the two liquids, the density of the unknown liquid can be calculated from

$$\rho = \frac{m_2 - m_0}{\frac{m_1 - m_0}{\rho_0}} = \frac{m_2 - m_0}{m_1 - m_0} \cdot \rho_0$$

Where ρ_0 is the density of water.

3. Determining the density of small irregular solid by pycnometer

Using the pycnometer can also determine the density of small irregular solid that can be put into the bottle. Let the mass of the small irregular solid be m, and let the mass of the pycnometer filled with distilled water be m_1. Then put the small irregular solid into the pycnometer, and excess water will run out at the top of the stopper. Let the mass of the pycnometer + water + small irregular solid be m_2. Then $m_1 + m - m_2$ is equal to the mass of water displaced by the body. Then the density of small irregular solid is then calculated from

$$\rho = \frac{m}{m + m_1 - m_2} \cdot \rho_0$$

Where ρ_0 is the density of water.

【Procedure】

1. Practice using the physical balance

Determine the mass of the aluminum cylinder five times, and calculate its average value.

2. Determine the density of the regular aluminum cylinder

(1) Using the physical balance measure the mass m of the solid correctly for five times, calculate its average value $\bar{m}$.

(2) Measure the length of related quantities of the object by the vernier caliper, calculate its volume V. Then the density of the cylinder is then calculated from $\rho = \frac{m}{V}$, and record the data in Tab. 2 – 1.

(3) Work out the relative error of density $\frac{\Delta\rho}{\rho}$ and the absolute error $\Delta\rho$, pay attention to the number of significance digits of the density ρ.

3. Determine the density of alcohol (The following masses are weighted by electronic balance)

(1) Clean and dry the pycnometer, and then use the balance to determine the mass of the empty pycnometer as m_0.

(2) Fill the pycnometer with alcohol and determine the mass of alcohol + pycnometer, and record it as m_2.

(3) Recover the alcohol, wash the pycnometer with distilled water, then fill it with the distilled water, then measure the mass of the bottle " + " distilled water, and record it as m_1, and enter the data in Tab. 2 – 1.

(4) Calculate the density of the alcohol from $\rho = \frac{m_2 - m_0}{m_1 - m_0} \cdot \rho_0$.

4. Determining the density of small irregular solid by pycnometer.

(1) Take some clean lead in form of tiny granule, and then determine its mass as m.

(2) Put them in the pycnometer that has been filled up with distilled water (the mass recorded as m_1), and insert the stopper. Pay attention to keep the outside dry (note: shake gently to remove air bubbles). Then measure the mass as m_2 (which includes the distilled water, the lead granule and the pycnometer).

(3) Calculate the density of the small solid from $\rho = \dfrac{m}{m + m_1 - m_2} \cdot \rho_0$.

【Data and Calculations】

Measure the density of aluminum cylinder and that of alcohol.

Tab. 2 – 1 Data of the density of aluminum cylinder and that of alcohol

item\times	aluminum cylinder			picknometer		
	high H (cm)	diameter D (cm)	mass m (g)	mass m_0 (g)	mass m_2 (g)	mass m_1 (g)
1						
2						
3						
4						
5						
average						
absolute error						
relative error						

The density of aluminum cylinder is figured out as the following:

$$\bar{\rho} = \dfrac{\overline{m}}{\dfrac{1}{4}\pi \overline{D}^2 \overline{H}} = \dfrac{4\overline{m}}{\pi \overline{D}^2 \overline{H}}$$

$$\Delta\rho = \left(\dfrac{\Delta m}{\overline{m}} + 2\dfrac{\Delta D}{\overline{D}} + \dfrac{\Delta H}{\overline{H}}\right) \cdot \bar{\rho}$$

The result is: $\rho = \bar{\rho} \pm \Delta\rho =$

The density of alcohol is: $\bar{\rho} = \dfrac{\overline{m_2} - \overline{m_0}}{\overline{m_1} - \overline{m_0}} \cdot \rho_0 =$

【Note】

(1) If the balance is out of order, go to the tutor for help. Put on the gloves before adjusting the balance.

(2) Wipe the liquid outside the container before measuring. Act swiftly during the process of measuring alcohol's density so as to minimize alcohol vaporization.

【Preview】

(1) Problems you should pay attention to using a balance correctly.

(2) Ways of eliminating systematic error caused by unequal – arms.

【Questions】

(1) Correct process of operating a balance quickening the measurement and make it accurate? Summarize the method of adjusting and using a balance.

(2) What might be the causes of the errors with the pycnometer method?

(3) Please design by yourself how to determine concentration of alcohol in the mixture of alcohol and water. Please design feasible detailed experiment processes.

【Introduction to Apparatus】

1. Physical balance

(1) Structure and Use

Balances are used to measure the mass or the weight of an object. The most common balance used in the laboratory is the equal-arm beam balance (Fig. 2-1). It consists of a rigid beam 7 with a pointer 9 pivoted in the middle and with a pan hinged at both sides, and the arms are equal in length. The balance therefore compares the weight of an object with a known weight. There are three knife-edges B_0, B_1 and B_2 in the beam which are made of agate or steel, the middle one is downword and supported by the supporter of the stele 11, and the three knife-edges keep that $B_1B_0 = B_2B_0$. The knob 13 below the stele is used to lift or lower the rigid beam 7. When adjust knob 13 clockwise, the beam is elevated by the rising supporter, and the balance can swing freely to be used to measure. When it is not used, turn 13 anti-clockwise, the beam lowers and is supported by the supporter, the middle knife-edge separates from the supporter, the load is subtracted from other two knife-edges, this can protect the knife-edges not to be damaged. The performance of the balance is expressed with the most load and the sensitivity. The most load is the biggest load which the balance allowed. The sensitivity means the number of division which the pointer deflects when loads on two sides of the balance differ a unit quality. Sometimes the feeling-quantity also is used to replace the sensitivity, it refers weights added on the pan when the balance pointer deflects a standard scale, it is right the reciprocal of the sensitivity. The most load of physical balance is 200-500g and its scale sensitivity is about one standard scale per ten milligram or the feeling-quantity is ten milligrams per one minute standard.

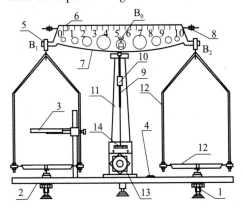

1. base screw; 2. base; 3. tray; 4. spirit level;
5. left (right) edge; 6. rider; 7. rigid beam;
8. leveling screw; 9. pointer; 10. sensitivity adjustment; 11. stele; 12. left (right) pan;
13. locking knob; 14. dial

Fig. 2-1 Structure of physical balance

(2) Pre-use adjusting

①Levelling: Rotate level adjustment screw 1, and cause the air bubble to locate the centre on the level vial.

②Zero correction: While using the balance, first adjust the rider at zero point (the left of the beam). Then touch one of the pans lightly to start the beam swinging. The pans should move far enough so the pointer swings at least three spaces to either side of the center mark. If it swings farther to one side than it does to the other, the leveling screw may be turned until the pointer moves as many spaces to the left as to the right (or at least less than one space). The balance is now in equilibrium and is ready for use. When adjusting, we can lower down the beam for fear of damaging the knife-edge.

③In order to ensure the sensitivity, musts of operations are as follows:

a. Do not add or remove weights or objects while the beam is going up.

b. Practically the objects whose mass is to be determined with the balance is always placed on the left pan of the balance for the convenience of an easier operation. And tweezers are used in adding or removing weights.

c. In the process of weighting, select a weight that you judge to be greater than that of the object being measured and place it on the right pan. If it is greater, remove it and add the next smaller weight. If this weight is less than that of the object, continue to add other weight to the right pan. Try masses successively from the larger to the smaller, until the last weight added is the one gram mass. Then the rider, which is used to indicate mass to the nearest tenth of a gram, should be moved far enough to the right to make the pointer swing as many spaces to the right as to the left, and you do not need to wait for the pointer to come to rest. The mass of the object equals the sum of all the weights on the pan plus the mass represented by the rider. Be careful not to jar the balance when adding or removing the larger masses. Such jarring dulls the knife edges and makes the balance less sensitive. Jarring may be avoided by supporting the pan with one hand while objects are being added to or removed from the pans. The object whose mass is being determined and the large masses should be placed near the center of the balance pan.

d. After the experiment, make sure that all components of the apparatus are in the status of storage, beam being lowered and in its normal place as well as the stirrups, weights being placed back where they are, tec.

2. Electronic Balance

(1) Working environment: the electronic balance is a high-precision measuring instrument, so the installation platform must be stable, flat, avoiding vibration, direct sunlight and heat, and working in the environment with high humidity, and avoiding direct air circulation channels.

(2) Method of use

①Level adjustment: before starting the balance, it is necessary to observe whether the blister in the level at the back of the balance is located in the center of the ring; otherwise it will be adjusted by the anchor bolt of the balance to raise it left-rotation while lower it right-rotation.

②Pre-heating: when the balance is turned on for the first time or after a long time of power failure, the pre-heating time should be at least 30min. Therefore, the power of laboratory electronic balance should not be cut off frequently under normal circumstances.

③Weighing: press ON/OFF button, connect the monitor, and wait for the instrument self-check. When the display shows zero, the self-check process ends and the balance can be weighed.

④Place the weighing paper and peel it according to the Tare key on both sides of the display screen. When the display shows zero, add the reagent to the weighing paper and weigh it.

⑤After weighing, press ON/OFF to turn off the monitor.

(3) Precautions: the balance has been strictly calibrated during installation, so it is not allowed to move the balance easily; otherwise the calibration should be carried out again. Do not use weighing paper to weigh directly. Clean the balance after each weighing to avoid contaminating the balance and affecting the weighing accuracy.

(Zhao Zhe, Zhao Yao)

实验三 液体变温黏滞系数的测定

【实验目的】

（1）学会用落球法测定黏滞系数。

（2）研究液体黏滞系数随温度变化关系。

（3）学会逐差法及图解法处理实验数据的方法。

【实验器材】

多光电门智能计时器、变温黏滞系数测定仪、4位数显恒温加热装置、热水泵、天平、千分尺、温度计、铝球、石头球、甘油等。

【实验原理】

当一个光滑小球在无限深广的液体中下落时，小球在液体中将同时受到三个竖直方向力的作用：小球的重力 mg，竖直向下；液体对小球的浮力 ρgV 及黏滞阻力 f，竖直向上。当该小球的直径 d 及下落速度 u 均小时（即满足小雷诺数条件），则小球在下落过程中所受液体的黏滞阻力可由斯托克斯公式得出

$$f = 3\pi\eta ud \quad (2-3-1)$$

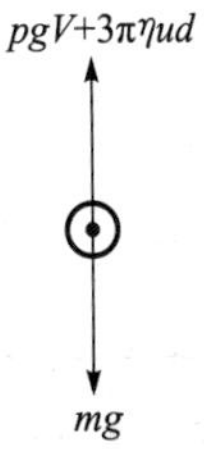

图 2-3-1 小球受力分析

式中，m、d、V、ρ、u 分别是小球的质量、直径、体积、密度和速度；η 是液体的黏滞系数。当小球下落时，由于重力大于竖直向上的浮力与黏滞阻力之和，如图 2-3-1 所示，小球向下做加速度运动。随着小球速度的增加，当上述三个力达到平衡时，即

$$mg = \rho g V + 3\pi\eta u d \qquad (2-3-2)$$

于是小球以速度 v_0 做匀速运动，由上式得：

$$\eta = \frac{(m-\rho V)g}{3\pi u d} \qquad (2-3-3)$$

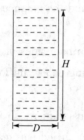

图2-3-2 实验容器简图

实验时由于液体必须盛在容器中，如图2-3-2所示。而小球则沿圆筒中心轴线下降，故不能满足无限深广的条件，因此必须对（2-3-3）式进行如下修正才符合实际情况，即

$$\eta = \frac{(m-\rho V)g}{3\pi u_0 d} \cdot \frac{1}{(1+2.4d/D)(1+3.3d/2H)} \qquad (2-3-4)$$

式中，D 为容器的内直径，H 为液柱的高。

本实验所使用的液体为甘油，当温度 $T=0℃$ 时，$\rho_0 = 1.26 \times 10^3 \text{kg/m}^3$。当温度变化时，密度也随之变化，其变化关系为：$\rho = \rho_0 /(1+\beta T)$，$\beta = 5 \times 10^{-4}/℃$，为甘油体膨胀系数。

【实验步骤】

1. 测定室温下液体的黏滞系数

（1）调节光电变温黏滞系数测定仪上的圆筒使之呈铅直状态，测出盛甘油圆筒的内直径 D 及液体深度 H。

（2）接通计时器并使其进行自检。打开水泵开关，使水箱内水能实现循环，使外筒与内筒之水位与甘油的液面等高。用天平称出 7~8 个小球的质量 M，用 $m=M/n$ 求出每个球的质量，然后用千分尺测小球的直径 3 次，编号，待用。

调节 4 位数显恒温加热装置，使温度指示至 15℃ 左右，如果室温高于 15℃，可在水箱中放入适量冰块，使其降至所需温度。待大约 10min 后，内箱的甘油温度与外筒间的水温基本达到平衡。

将小球先放在甘油内浸一下，然后用镊子夹起放在圆筒顶端的定位盖中心孔上，让其自由下落。当小球下落经过各光电门时，计时器应显示 0、1、2、3、4、…7 各数字，此时说明小球已完成挡光，如图2-3-3所示。

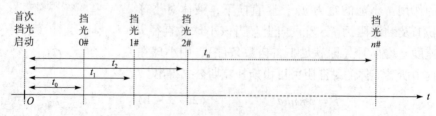

图2-3-3 小球挡光时间参数示意图

按【停止】键后，屏幕循环显示所测时间 t_n。

再按【停止】键，屏幕按"逐差法"循环显示逐次相减的时间数据 $t_{n-(n-1)}$。

又按【停止】键，屏幕按"逐差法"循环显示等间隔相减的时间数据 $t_{n-(n-2)}$；再按【停止】键，就显示 $t_{n-(n-3)}$；依此类推。

【停止】键是屏幕显示切换键，屏幕显示可在 t_n、$t_{n-(n-1)}$、$t_{n-(n-2)}$…值之间切换。

2. 研究液体在不同温度下的黏滞系数与温度的关系

调节 4 位数显恒温加热装置的预定开关，改变测量点的温度，每次上升 3～4℃为 1 个测量点，重复上面各步骤，每个温度点恒温时间应超过 10 分钟，以保证甘油内部温度均匀。共测出 5 个以上不同温度的 η 值。

【数据记录及处理】

1. 测定室温下甘油的黏滞系数

（1）按实验要求独立设计表格。

（2）根据 $v_0 = S/t$ 算出小球匀速下降的速度。

（3）按公式（2-3-4）计算，即可得到甘油的黏滞系数。

表 2-3-1　不同温度下甘油的黏滞系数

温度 T（℃）	20	25	30
黏滞系数 η（Pa·s）	1.5042	0.9488	0.6076

2. 研究液体在不同温度下的黏滞系数与温度的关系

（1）用记录在表 2-3-2 中的实验数据，在坐标纸上作黏滞系数 η 和温度 T（℃）的关系曲线。

（2）从实验曲线上测出 $T = 20℃$ 及 $T = 25℃$ 的实验值，并与表 2-3-1 中所给的公认值比较，算出相对误差。

表 2-3-2　不同温度下的黏滞系数与温度的关系

编码温度 T（℃）									
甘油的有效密度 ρ_L（kg/m³）									
黏滞系数 η（Pa·s）									

以上的温度以实验室要求为准。

【仪器介绍】

（一）VMZ-2 智能变温黏滞系数实验仪

如图 2-3-4 所示是 VMZ-2 智能变温黏滞系数实验仪，该仪器是传统的落球法测液体黏滞系数测试仪的更新换代的新型仪器。其主要性能如下。

1. 主要特点

（1）主机采用光电技术，设置了 8 个光电门使该仪器实现光电计时。

（2）智能数字计时器由于采用了智能化技术，具有同时测得多组时间和存储、调出每组数据功能。可测得任意光电门之间的时间。例如：1、2、3、4、…，1-0、2-1、3-2、4-3…，2-0、3-1、4-2…。

图 2-3-4　实验装置图

（3）该机采用了电子控温，热水循环泵。调温装置控温范围为 5～80℃。

2. 实验内容

（1）测定液体在室温至80℃的黏滞系数。

（2）研究液体黏滞系数随温度的变化关系并可得到液体黏滞系数随温度变化的实验曲线。

（3）小球每下落一次可同时得到多组实验数据：如仪器上共安装8个光电门则可同时测出各光电门之间的挡光时间，如 Δt_{12}、Δt_{34}、Δt_{25}、Δt_{16}、Δt_{57} 等，因此利用此实验可训练学生用逐差法处理实验数据的方法。

（4）可利用此仪器研究在液体内的运动规律（加速 – 匀速 – 减速）。

3. 主要技术参数

（1）四位半 LED 数码管显示；

（2）计时范围：0～99.999ms；

（3）8 路光电门输入；

（4）调温范围：室温至80℃；

（5）调控温精度：0.1℃；

（6）测定精度（误率）≤6%；

（7）温升速度：1℃/min；

（8）整机重量：5kg；

（9）外形尺寸：415mm×300mm×560mm；

（10）电源电压：220V，功率<200W。

（二）多光电门智能计时器

1. 多光电门智能计时器具备 8 个功能

（1）β：角加速度；S_1：挡光间隔；S_2：挡光间隔计时；T：周期；a：加速度；g：重力加速度；col：碰撞；η：黏滞系数。

（2）智能计时器采用流行开关面板，微处理器及智能化技术，操作十分简便，只需按 4 只操作开关就可完成全部功能。

（3）智能计时器共有 8 个光电门输入，在实验中可任意选择光电门数，仪器可自动判定无需设置。

（4）具有自检功能及存储功能，仪器可靠性高，使用方便。

（5）当待测设备安装 2～8 路光电门时，可同时测出每个光电门的挡光时间及任意两只光电门挡光时间间隔（例 1 – 2、1 – 3、1 – 6、2 – 4…）。利用该功能演示平均速度与瞬时速度关系会得到非常好的效果。

（6）使用两路、两只光电门做气垫导轨实验时，可测 4 个时间值，并可得到 4 个滑行器速度值。

（7）利用该仪器还可完成（变温黏滞系数、智能刚体转动惯量、气垫导轨）多项实验内容，详见《多光电门智能计时器使用说明书》。

2. 主要技术参数 ① 5 位 LED 数码管显示；② 8 路光电门输入；③ 计时范围：0～99.999ms；④外形尺寸：275mm×225mm×80mm。

（孙宝良　马　骄）

Experiment 3 Determination of the Coefficient of Viscosity for Liquid of Varied Temperature

【Objects】

(1) Learn to use falling – ball method to measure coefficient of viscosity.

(2) To study the relations of liquid coefficient of viscosity with temperature changes.

(3) To deal with experiment data by the method of successive minus and graphic.

【Apparatus】

Photoelectric door more smart timer, variable temperature coefficient of viscosity meter, four digital constant temperature heating device, heat – exchanger pump, balance, microcalliper, thermometer, aluminum ball, stone ball and so on.

【Theory】

When a smooth ball falls in the infinite vast liquid, the ball in the liquid is stressed by three vertical force at the same time, the gravity of the ball is mg, fluid buoyancy of the ball is $\rho g V$ and viscous resistance is f. When the ball's diameter d and falling speed u are average hour meeting the conditions of small Reynolds number, the ball in the process of falling by the viscous coefficient of liquid resistance can be got by Stokes formula

$$f = 3\pi\eta u d \qquad (2-3-1)$$

Among them, m, d, V, ρ, u are the quality, diameter, volume, density and speed of the ball, η is the coefficient of viscosity.

When the ball falling due to gravity is greater than the vertical downward buoyancy and the sum of the viscous drag, as show in the Fig. 2 – 3 – 1, the ball runs down with acceleration motion conditions. With the increase of small ball speed, the above three forces turn to equilibrium, so

$$mg = \rho g V + 3\pi\eta u d \qquad (2-3-2)$$

the ball moves with uniform motion speed v_0.

$$\eta = \frac{(m - \rho V)g}{3\pi u d} \qquad (2-3-3)$$

Experiments with liquid must be filled in the container, as shown in Fig. 2 – 3 – 2. But the falling

Fig. 2 – 3 – 1 Force analysis of the ball   Fig. 2 – 3 – 2 Diagram of the experimental container

ball is along the center of a cylinder, it can't satisfy the infinite vast conditions. So you must correct (2-3-3) to conform to the actual situation.

$$\eta = \frac{(m-\rho V)g}{3\pi u_0 d} \cdot \frac{1}{(1+2.4d/D)(1+3.3d/2H)} \quad (2-3-4)$$

D is the diameter of container, H is the height of liquid column. In this study, we use glycerin as liquid, when the temperature is $T = 0°C$, $\rho_0 = 1.26 \times 10^3 \text{kg/m}^3$. When the temperature changes, the density changes. The following relation: $\rho = \rho_0/(1+\beta T)$. $\beta = 5 \times 10^{-4}/°C$ as glycerin body expansion coefficient.

【Procedure】

1. Determination of the coefficient of viscosity of liquid at room temperature

(1) Adjust the photoelectric variable temperature coefficient of viscosity meter on the state of cylinder to make it into a straight. Within the measure of glycerin cylinder diameter D and liquid depth.

(2) Switch on the timer and make it to self-check. Switch on the pump. Make the water in the tank cycle. Within the outer cylinder and the cylinder of water and glycerin liquid surface contour.

In the balance of seven or eight balls' quality M, calculate the quality of each ball with $m = M/n$. Then using micrometer to measure the diameter of the ball three times, the number is to use. Adjust the four digital display constant temperature heating device, make temperature indicating to around 15°C, If the room temperature is higher than 15°C, put the right amount of ice in the tank to make it to the required temperature. After waiting for about ten minutes, inside the box of glycerin basic balance between temperature and external temperature。

Put the ball in glycerin first, then use tweezers to pick up on the positioning cover central hole at the top of the cylinder, let it free fall. When the ball fall through the photoelectric door, timer displays 0, 1, 2, 3, 4, ⋯ 7 all numbers, at this point the ball's light blocking has been completed, as shown in Fig. 2-3-3.

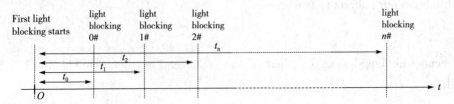

Fig. 2-3-3 Light blocking time parameters of small ball

After press "stop" button, the screen cycle displays measured time t_n.

Then press "stop" button, the screen displays successive subtraction by "gradual deduction method" cycle time data of $t_{n-(n-1)}$.

Press "stop" button again, the screen displays successive subtraction by "gradual deduction method" cycle time data of $t_{n-(n-2)}$. The next time press "stop" button, the screen displays $t_{n-(n-3)}$, and so on.

"Stop" button is the screen display switch button, the screen can switch between t_n, $t_{n-(n-1)}$ and $t_{n-(n-2)}$ and so on.

2. Study the relationship between viscous coefficient of liquid at different temperatures and temperature

Adjust the four digital display reservation at constant temperature heating device switch, changing the temperature measurement point. Every time 3 to 4℃ is a measurement point, and repeat the above steps. Each temperature point should keep constant temperature more than 10 minutes to make sure the evenness of internal temperature of the glycerol, then five η values at different temperatures are measured.

[Data and Calculations]

1. Determination of the coefficient of viscosity of liquid at room temperature

(1) The special table is designed according to the experiment for independence.

(2) According to $v_0 = S/t$, calculate the rate of decline at a constant speed under different temperature respectively.

(3) According to the formula (2-3-4), measure the viscous coefficient of glycerol at different temperatures.

Tab. 2-3-1 Viscosity coefficients of glycerol at different temperatures

temperature T (℃)	20	25	30
coefficient of viscosity (Pa·s)	1.5042	0.9488	0.6076

2. Study the relationship between viscous coefficient of liquid at different temperatures and temperature

(1) According to the recorded data, please draw the $\eta - T$ relation curve.

(2) Measured from the experiment curve $T = 20$℃ and $T = 25$℃ of the experimental data, and compared with recognized value given in Tab. 2-3-2, calculate the relative error.

Tab. 2-3-2 The relationship between viscosity coefficient and temperature at different temperatures

coding temperature T (℃)											
the effective density of castor oil ρ_L (kg/m^3)											
coefficient of viscosity (η)											

Above the temperature of the laboratory requirements shall prevail.

[Introduction to Apparatus]

1. VMZ -2 experimental apparatus of intelligent variable temperature viscosity coefficient

As shown in Fig. 2 -3 -4, VMZ -2 experimental apparatus of intelligent variable temperature viscosity coefficient is an update of the traditional falling - ball method experiment instrument. The main performance is as follows:

(1) Main characteristics

① Host using the photoelectric technology, which set up eight photoelectric doors, makes the photoelectric timing realized.

② The intelligent digital timer can measure more groups of time at the same time and storage, data function in each group. It can be measured at any time between the photoelectric doors. Such as 1、2、3、4、……, 1 -0、2 -1、3 -2、4 -3……, 2 -0、3 -1、4 -2……

Fig. 2 -3 -4 Experimental device

③ The machine uses the electronic temperature control, hot water circulating pump. The control range of thermal control device is 5℃ ~80℃.

(2) Experimental contents

① Measure the viscous coefficient of liquid at room temperature to 80℃.

② Study the changing relation between liquid viscosity coefficient and temperature. Get the liquid coefficient of viscosity with temperature change curve of experiment.

③ Ball every fall more groups of experimental data can be obtained at the same time. Such as the light blocking time the eight photoelectric door Δt_{12}, Δt_{34}, Δt_{25}, Δt_{16}, Δt_{57}, and so on. So using this experiment, we can train students to use method of data processing through "gradual deduction method".

④ It can be used to study the law of motion.

(3) The main technical parameters

① Four and a half LED digital tube display.   ② Measurement range: 0 ~99.999ms.

③ 8 - input photoelectric door.   ④ Temp. range: room temperature ~80℃.

⑤ Accuracy of temperature control.   ⑥ Measurement accuracy(Error rate) ≤6%.

⑦ Speed of temperature rise: 1℃/min.   ⑧ Total weight: 5kg.

⑨External dimension: 415mm ×300mm × 560mm.   ⑩ Supply voltage:220V, Power <200W.

2. Multi - photoelectric - door intelligent timer

(1) Function

① β: angular acceleration; S_1: light blocking interval; S_2: light blocking time interval; T: cycle; a: accelerated velocity; g: gravitational acceleration; col: collision; η: coefficient of viscosity.

② The operation of this intelligent timer is very convenient. With only four switch operation can be completed all functions.

③ Smart meters with a total of eight photoelectric door type, is the present domestic alone. In the experiment photoelectric gate number can be chosen optionally. Instrument can automatically judge does not need to set.

④ Instrument has self-checking function and storage function, which make you feel the reliability of the instrument height. At the same time it can bring great convenience to you to use.

⑤ When the installation of equipment under test 2-8 photoelectric door, each photoelectric door light blocking time and any two of the photoelectric door light blocking time interval can be measured simultaneously. Use of the function relationship between average velocity and the instantaneous velocity demonstration will get very good effect.

⑥ Use two way, two photoelectric door air track experiment, can measure four time value, and get four glide speed value.

⑦ With the instrument we can complete a number of experiment content (temperature coefficient of viscosity, smart moment of inertia of rigid body, air track). See details in "Much Intelligent Photoelectric Timer Operation Instruction Handbook".

(2) The main technical parameters

① Five LED digital tube display.　② 8 road input photoelectric door.

③ Measurement range：0~99.999ms.　④ External dimension：275mm×225mm×80mm.

(Ma Jiao)

实验四　液体表面张力系数的测量

测量液体的表面张力系数有多种方法，如最大泡压法、拉普拉斯法、毛细管上升法、焦利秤法、扭力天平法等。本实验主要介绍焦利秤法和毛细管上升法。

一、焦利秤法

【实验目的】

（1）了解焦利秤独特的设计原理。

（2）学会用焦利秤测定液体的表面张力系数。

（3）研究溶质对液体表面张力系数的影响。

【实验器材】

焦利秤、Π形金属（铂）框、烧杯、酒精灯、游标卡尺、温度计、镊子、蒸馏水、砝码。

【实验原理】

液体中的分子在各方向上以大小相等的力相互吸引。然而，处在液体表面的分子没有受到向外的吸引力，因此液面的分子会受到指向液体内部的力。由于该原因，液体表面有趋于尽量最小的趋势。这种存在于液体表面使液体具有收缩倾向的张力被称为液体的表面张力。如果在液面上设想有一条分界线 MN，表面张力的方向与液面相切且与所选取的分界线垂直，其大小与该分界线 MN 的长度 l 成正比，即

$$F = \sigma l \tag{2-4-1}$$

式中，σ 称为该液体的表面张力系数，它表示单位长度液面分界线上的表面张力，在国际单位制中单位为 $N \cdot m^{-1}$。表面张力系数的大小与液体的性质有关，密度小而易挥发的液体 σ 小，反之 σ 较大；表面张力系数还与杂质和温度有关，液体中掺入某些杂质可以增加 σ，而掺入另一些杂质可能会减少 σ；温度升高，表面张力系数 σ 将降低。

将 Π 形金属框浸入液体中，通过弹簧将其缓慢拉起，在框内将形成一层液膜，如图 2-4-1 所示。此时金属框在竖直方向上受到三个力的作用：弹簧的拉力 F'，框的重力 W，液膜表面张力 $2F$（有两个液面），如图 2-4-2 所示。设两侧液面与竖直方向成 θ 角，则表面张力在竖直方向上的分力为 $2F\cos\theta$。若不计框所受浮力及水膜的重力，金属框在竖直方向的平衡条件为

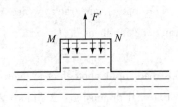 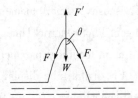

图 2-4-1　Π 形金属框拉膜情况　　　图 2-4-2　Π 形金属框受力示意图

$$F' = 2F\cos\theta + W \tag{2-4-2}$$

当缓慢地拉起金属框时，随着 Π 形框的上升，θ 角将逐渐减小，而弹簧的拉力 F' 将不断增大。在水膜破裂的瞬间，$\theta = 0$，F' 达到最大值 F_m。因此由式（2-4-1）和式（2-4-2）可得

$$\sigma = \frac{F_m - W}{2l} \tag{2-4-3}$$

式中，l 为 Π 形框宽度。利用焦利秤可测得式（2-4-3）中 $F_m - W$ 的大小。若弹簧在 Π 形框重力 W 作用下伸长为 x_0，则 $W = kx_0$，被拉脱时伸长为 x，则 $F_m = kx$，因此

$$(F_m - W) = k(x - x_0) = k\Delta x$$

代入式（2-4-3）可得

$$\sigma = \frac{k\Delta x}{2l} \tag{2-4-4}$$

由式（2-4-4）可知，只要测出 k，Δx 及 l 各量，即可求出液体的表面张力系数 σ。

【实验步骤】

1. 测定弹簧的劲度系数

（1）按图 2-4-3 挂好弹簧、指标镜、指标管和砝码盘，并调节三脚底脚螺丝 7，使指标镜 10 处于指标管中间，保持竖直方向，且与指标管不接触。调节螺旋 6 使指标

管上的刻线及其在指标镜中的像与指标镜中间刻线达到"三线"重合。这时刻度尺的读数即为初始读数 x_0，将其记录在表 2-4-1 中。

（2）逐次添加相同的砝码 m（$m=0.5g$），即分别将 $0.5g$、$1.0g$、$1.5g$、…$5.0g$ 的砝码加在砝码盘上，每添加一次砝码，调节螺旋 6 使"三线重合"，并将相应的读数 x_1，x_2，…，x_{10} 记录在表 2-4-1内；再逐次减少相同的砝码 m（$m=0.5g$），每减少一次砝码，调节螺旋 6 使"三线重合"，分别将相应的读数 x_{10}，x_9，…，x_1 记录在表 2-4-1内。

（3）先用逐差法处理所测数据，求出弹簧的劲度系数。再用作图法求出弹簧的劲度系数：将测量数据以弹簧伸长量为纵坐标，所加砝码质量为横坐标，在坐标纸上作图，由图即可求出弹簧的劲度系数。最后，将两种方法所得劲度系数的平均值作为最后结果。

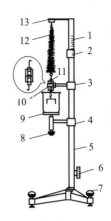

1. 金属管；2. 读数游标；3. 紧固夹；
4. 紧固夹；5. 套筒；6. 高度调节螺旋；
7. 底脚螺丝；8. 平台升降调节螺旋；
9. 平台；10. 指标镜；11. 指标管；
12. 精细弹簧；13. 横梁

图 2-4-3 焦利秤

2. 测定水的表面张力系数

（1）用游标尺测量 Π 形金属框的宽度 l 五次，取其平均值。用镊子夹住 Π 形金属框在乙醇里擦洗干净，然后挂于指标镜下。

（2）用蒸馏水冲洗玻璃烧杯，然后倒入待测蒸馏水并置于平台 9 上，用温度计测出水温 T。

（3）调节平台升降旋钮 8，使 Π 形金属框浸入水中后，再使其上端与水平面齐平。调节升降旋钮 6，使焦利秤达到"三线对齐"，记下游标所示的位置坐标 x_0。

（4）调节升降旋钮 6，使 Π 形金属框缓慢上升，同时调节旋钮 8，使水面缓慢下降，并保持"三线对齐"。当水膜刚被拉破时，记下游标所示的位置坐标 x。

（5）重复步骤（3）和（4）四次，把相应的游标尺读数 x 记入表 2-4-2 内。

（6）用上述方法测量添加活化剂后水的表面张力系数。

【数据记录及处理】

1. 测量弹簧的劲度系数 k

（1）由作图法测量弹簧的劲度系数　将表 2-4-1 中的测量数据，以弹簧伸长量为纵坐标，所加砝码质量为横坐标，在坐标纸上作图。得弹簧的劲度系数 $\bar{k}_1 =$

（2）由逐差法测量弹簧的劲度系数

表 2-4-1　测量弹簧劲度系数 k 的实验数据　　　单位：_____

次数 i	砝码质量 m（g）	s_i（增重时读数）	s_i'（减重时读数）	平均值 $\bar{s}_i$	弹簧伸长量 $\Delta s_i = s_{i+5} - s_i$
1	0.5				$\Delta s_1 = s_6 - s_1$
2	1.0				

续表

次数 i	砝码质量 m (g)	s_i (增重时读数)	s_i' (减重时读数)	平均值 $\bar{s}_i$	弹簧伸长量 $\Delta s_i = s_{i+5} - s_i$
3	1.5				$\Delta s_2 = s_7 - s_2$
4	2.0				
5	2.5				$\Delta s_3 = s_8 - s_3$
6	3.0				
7	3.5				$\Delta s_4 = s_9 - s_4$
8	4.0				
9	4.5				$\Delta s_5 = s_{10} - s_5$
10	5.0				

数据处理：$\Delta m = 2.5$ (g)，弹簧的伸长量 $\overline{\Delta s} = \dfrac{\Delta s_1 + \Delta s_2 + \Delta s_3 + \Delta s_4 + \Delta s_5}{5}$

得 $\bar{k}_2 = \dfrac{\Delta mg}{\overline{\Delta s}} =$

则弹簧的劲度系数：$k = \dfrac{\bar{k}_1 + \bar{k}_2}{2} =$

2. 测量水的表面张力系数 σ

水温 $T =$ _____ ℃

表 2-4-2 测量水的表面张力系数 σ 的实验数据　　　　单位 _____

次数 i	初始位置 x_0	破裂位置 x	弹簧伸长量 $\Delta x_i = x - x_0$	平均值 $\overline{\Delta x}$	框宽 l	平均值 $\bar{l}$
1						
2						
3						
4						
5						

数据处理

$$\bar{\sigma} = \dfrac{k \cdot \overline{\Delta x}}{2\bar{l}} =$$

$$\Delta \sigma = E \cdot \bar{\sigma} =$$

实验结果

$$\sigma = \bar{\sigma} \pm \Delta \sigma =$$

【思考题】

(1) 影响实验结果的因素有哪些，为什么？

(2) 如何测量某种浓度 NaCl 溶液的表面张力系数，请设计具体实验步骤，并说明溶液浓度对表面张力系数有无影响。

【注意事项】

(1) 焦利秤的弹簧十分精密，实验时切勿使其超负荷，以免损坏。

(2) 实验所用烧杯、镊子尖端及Π形框的清洁与否直接影响实验结果，请切勿用手触摸。

(3) 拉膜过程中动作要缓慢，观察时眼睛应与刻线处于同一水平面，以减少误差。

【仪器介绍】

焦利秤实际上是一个精细弹簧秤，常用来测微小的力，其结构如图2-4-3所示。其中5为固定金属支架的套筒，它的上端装有一0.1mm刻度的游标2。5内套装一毫米刻度的金属管1，1与游标2组成游标尺。调节螺旋6可使1杆上下移动，1杆上主尺伸出5管的长度可从游标尺上读出。精细弹簧12悬挂于1杆上端的横梁13上，可随1杆上下移动。在弹性限度内，12的伸长服从胡克定律。弹簧下端挂一平面反射镜10，叫作指标镜，镜面上有三条刻线。指标镜处于玻璃管11内，11叫作指标管，其上有一条水平刻线。指标镜下端挂钩可挂砝码盘、Π形金属框等。9为平台，其高度可由螺旋8及紧固夹4调节，紧固夹3可调节指标管的高度，调节7可使秤体竖直。指标管11上的刻线实际上就是测量弹簧长度的参照线，实验时要使11上的刻线和其在指标镜10中的像与指标镜10的刻线达到"三线重合"，这样弹簧下端的位置才保持不变，此时方可从游标尺上读数。

二、毛细管法

【实验目的】

(1) 掌握用毛细管测量液体表面张力系数的方法。
(2) 了解读数显微镜的结构与原理。
(3) 学习用读数显微镜测量微小长度。

【实验器材】

读数显微镜、玻璃毛细管、烧杯、支架、小砂轮、温度计、蒸馏水、乙醇。

【实验原理】

任何液体的表面都存在表面张力。若液面是水平的，则表面张力也沿着水平方向；若液面是弯曲的，则表面张力和液面相切，结果使弯曲的液面对液体内部施以附加压强。对于凸面，附加压强为正，对凹面，附加压强为负。附加压强可解释毛细现象。

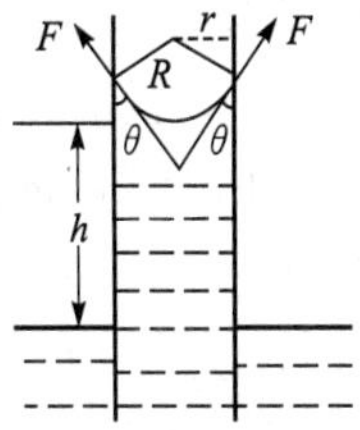

把一根玻璃毛细管插入液体中，若液体能润湿管壁，则管内液面呈凹球形，附加压强为负，管内液面将高于管外液面。若液体不能润湿管壁，则管内液面呈凸球形，附加压强为正，管内液面将低于管外液面。图2-4-4所示是半径为r的毛细管插入水（或乙醇）中的毛细现象。F为表面张力，其方向与凹球面相切，大小为

图2-4-4 润湿液体的现象

$$F = \sigma 2\pi r \qquad (2-4-5)$$

式中，σ为表面张力系数，数值上等于作用在周界单位长度上的力；$2\pi r$为凹球面周界长度；θ为接触角，即液固交界处液体表面的切线与固体表面间的夹角。液体不润湿固

体时，$\theta > \pi/2$；完全不润湿时，$\theta = \pi$；液体润湿固体时，$\theta \leq \pi/2$；完全润湿时，$\theta = 0°$。例如，水、乙醇与玻璃的接触角 $\theta = 0°$，而汞与玻璃的接触角 $\theta = 140°$。

若设凹球面的半径为 R，h 为平衡时管内凹球面下端至管外液面的高度。由图 2-4-4 可得 $\cos\theta = \dfrac{r}{R}$，则由表面张力 F 产生的垂直向上提高液面的力为 $F\cos\theta = 2\pi r^2 \sigma / R$，其与高度为 h 的液柱重力平衡，因而有

$$\sigma = \frac{r\rho g h}{2\cos\theta} \tag{2-4-6}$$

如玻璃管壁和水都非常清洁，则 $\theta = 0°$，$R = r$ 则式（2-4-6）变为

$$\sigma = \frac{r\rho g h}{2} \tag{2-4-7}$$

在推导式（2-4-7）时，忽略了凹球面下端以上液体的质量，而这部分体积约等于半径为 r，高为 h 的圆柱和半径为 R 的半球体体积之差，即为 $\pi r^3 - 2/3\pi r^3 = 1/3\pi r^3$，故忽略的液体重力为 $1/3\pi r^3 \rho g$，考虑这一修正项后，可得到更为精确的计算公式，即

$$\sigma = \frac{1}{2}r\rho g\left(h + \frac{1}{3}r\right) = \frac{1}{4}d\rho g\left(h + \frac{d}{6}\right) \tag{2-4-8}$$

式（2-4-8）为本实验的理论依据，只要精确测定毛细管内径 d 和液柱高度 h，就能计算出表面张力系数 σ。式（2-4-8）中各量的单位：d、h 的单位为 m，密度 ρ 为 $kg \cdot m^{-3}$，g 为 $9.80 N \cdot kg^{-1}$，σ 为 $N \cdot m^{-1}$。

【实验步骤】

1. 测毛细管内径

（1）把读数显微镜（图 2-4-6）以 E 面为底面安放在桌面上，把毛细管用夹子固定在水平位置上，并处于物镜的正前方。

（2）缓慢转动目镜，使十字叉丝成像清晰。

（3）调节旋钮 A，使毛细管端面在目镜中成像清晰，旋转目镜镜筒使十字叉丝横丝和主尺平行，调节旋钮 C 使竖丝沿着毛细管直径移动，横丝与孔的圆周相切。在两个相切点上的读数之差，就是毛细管的内径。转动毛细管，在不同方位测量三次，以其平均值作为毛细管的内径 d。

2. 测量水和乙醇的表面张力系数

（1）将玻璃容器清洗干净后，装满蒸馏水，以便读数显微镜读取数据。

（2）实验装置如图 2-4-5 所示。将洗净的干燥玻璃毛细管用夹头固定在支架的金属杆上，并插入玻璃容器中，以铅锤 Q 来调节毛细管，使其处于竖直方向。在容器中插入一温度计测量水温。

（3）调节目镜，使十字叉丝成像清晰。调节旋钮 C 使十字叉丝的横丝与容器内水面重合，读取一个数据；再使横丝与毛细管内凹球面底部相切，读取另一个数据。两读数之差，即为液柱的高度 h。

（4）在凹球面底部作一标记，取下毛细管，用读数显微镜按步骤（1）的方法测量标记处毛细管内径 d。

（5）另取两根毛细管，分别按步骤（2）~（4）重复测量一次，所测数据均记入表

2-4-3内。

(6) 按上述方法，另取三根毛细管。将其插入乙醇中，测量液柱上升的高度 h 毛细管内径 d。

【注意事项】

(1) 所用的毛细管、玻璃容器和液体必须保持清洁，否则将影响测量结果。

(2) 若玻璃管是非均匀的，测定毛细管某一截面的内径时，要先用小砂轮在该处割一道痕，再轻轻一折即可断开。断开的截面是否平整将直接影响测量结果，因此实验前最好先试折几次，以掌握其要领。

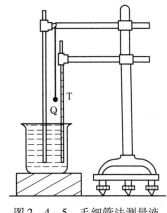

图 2-4-5 毛细管法测量液体表面张力系数的装置

【数据记录及处理】

表 2-4-3 水和乙醇的表面张力系数的测量数据

被测液体	毛细管序号	毛细管内径 d (m)				上升高度 h (m)	温度 T (℃)
		I	II	III	平均值		
水	1						
	2						
	3						
乙醇	1						
	2						
	3						

水温 $T=$ _____ ℃，查表 $\rho=$ _____ kg·m^{-3}，$\sigma=$ _____ N·m^{-1}。

得 $\bar{\sigma}=$ _____ N·m^{-1}，$E=\dfrac{|\sigma-\bar{\sigma}|}{\sigma}\times 100\%$

实验结果：$\sigma=\bar{\sigma}(1+E_r)=$

乙醇温度 $T_0=$ _____ ℃，查表：$\rho_0=$ _____ kg·m^{-3}，$\sigma_0=$ _____ N·m^{-1}。

得 $\overline{\sigma_0}=$ _____ N·m^{-1}，$E=\dfrac{|\sigma_0-\overline{\sigma_0}|}{\sigma_0}\times 100\%$

实验结果：$\sigma_0=\bar{\sigma}_0(1\pm E_r)=$

【仪器介绍】

读数显微镜是光学测量仪器之一，主要用来精确测定微小的或不能用夹持量具测量的物体的线度，如测量毛细管内径、刻线宽度、狭缝宽度等。图 2-4-6 所示为一读数显微镜，其主要部分为放大待测物体的显微镜 M 和读数用的主尺及附尺。附尺有两种形式，一种是游标尺的形式，另一种是螺旋测微器的形式，其读数的原理与游标尺或螺旋测微器相同。显微镜由目镜、物镜和十字叉丝组成，显微镜能够通过拖板固定在螺母套管上，转动旋钮 C 即转动丝杆，将使套在丝杆上的螺母套管和固定在套管上的显微镜

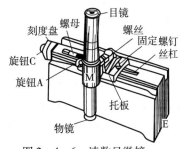

图 2-4-6 读数显微镜

左右移动。若以端面 E 作为底面，转动旋钮 C，显微镜将上下移动，此时可测量高度。

（邓岩浩　赵　喆）

Experiment 4　Determination of Surface Tension Coefficient of Liquid

There are many measuring techniques for determining surface tension coefficient of liquids. Jolly Balance method and capillary method are the most commonly used in this experiment for being selected.

（Ⅰ）　**Jolly Balance Method**

【Objects】

(1) To master the structure and the application of Jolly Balance.

(2) Learn to determine surface tension coefficient of liquids with Jolly Balance method.

(3) To research on the effect of the solute to the surface tension.

【Apparatus】

Jolly Balance, Π metal (platinum) frame, beaker, spirit lamp, vernier caliper, thermometer, tweezers, distilled water, weights.

【Theory】

Within liquids, molecules attract each other equally in all directions. At the surface, however, there is no force attracting the liquid. For this reason, liquid surfaces tend to become as small as possible. The tension occurs in liquid surface and causes the liquid surface to shrink, and it is called surface tension. Image there is a line MN on the liquid level, then the direction of the surface tension is tangential with the liquid surface and vertical to the boundary selected, and its magnitude is directly proportional to the frame's width l, which is the length of MN, namely

$$F = \sigma l \tag{2-4-1}$$

Where σ is called surface tension coefficient of liquid, it expresses surface tension per unit length of the liquid, $N \cdot m^{-1}$ is its SI unit. Its magnitude is related with the nature of the liquid, it is less for liquid which the density is small and it is easy to volatilize, and larger inversely. Surface tension coefficient also has relation to impurity and temperature, it may be increased when some impurity is added into liquid, and it also may be decreased when other impurity is added into liquid. Moreover with temperature rising, the surface tension coefficient will decrease.

A Π metal frame is immersed into the liquid, and then pulled up slowly through a spring, a liquid film will be formed in the frame, as shown in Fig. 2-4-1. This time three forces act on the frame in vertical direction: pulling force F' from the spring, gravity W of the frame, the liquid film's surface tension $2F$ (it has two liquid surface), as shown in Fig. 2-4-2. Assume the angle between each side liquid surface to the vertical direction is θ, then component of force in the vertical direction of the surface tension is $2F\cos\theta$. If buoyancy of the frame and gravity of the water film are neglected, then the condition for equilibrium in the vertical direction is:

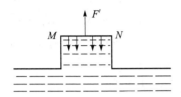 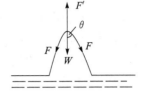

Fig. 2-4-1 Forces acting on the frame   Fig. 2-4-2 Forces acting on the frame

$$F' = 2F\cos\theta + W \qquad (2-4-2)$$

When the metal frame is being pulled up slowly, with the Π frame rising, the angle θ reduces gradually, but the pulling force F' of the spring increases unceasingly. At the moment that the water film ruptures, $\theta = 0$, F' has the maximum value of F_m. Therefore from formula (2-4-1) and (2-4-2), we may get σ as

$$\sigma = \frac{F_m - W}{2l} \qquad (2-4-3)$$

Where l is the width of the Π frame. The magnitude of $F_m - W$ in formula (2-4-3) can be measured using Jolly Balance. If the spring's elongation is x_0 under action of gravity of the Π-shaped frame, then $W = kx_0$, and the spring's elongation is x when it is pulled to escape, then $F_m = kx$, therefore

$$(F_m - W) = k(x - x_0) = k\Delta x$$

Substitute them to formula (2-4-3), we may get σ as

$$\sigma = \frac{k\Delta x}{2l} \qquad (2-4-4)$$

Known from formula (2-4-4), so long as k, Δx and l are determined, the surface tension coefficient σ can be calculated.

【Procedure】

1. Determine the spring's coefficient stubborn

(1) Gently hang the spring, the index mirror, the index tube and the weights pan refer to Fig. 2-4-3, and adjust the leveling screws 7, keep the mirror 10 vertical. Turn screw 6 to make the engraved line of the index tube, its image reflected from the mirror and the engraved line in the mirror coincide, and the reading is taken as x_0, the initial reading, and write it down in Tab. 2-4-1.

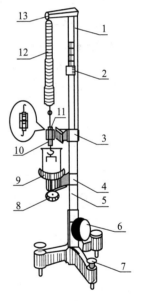

1. metal rod; 2. vernier;
3. clamp; 4. clamp; 5. sleeve;
6. adjusting screw; 7. screw;
8. adjusting screw; 9. platform;
10. index mirror; 11. index tube;
12. fine spring; 13. beam

Fig. 2-4-3 Jolly balance

(2) Successively increasing weights 0.5g, 1g, 1.5g, ⋯ 5.0g, which is added respectively on the weights pan, turn screw 6 in accordance with "three lines coincide" for each time, and record the readings x_1, x_2, ⋯x_{10} in Tab. 2-4-1. Then successively reduced by the same weight of m ($m = 0.5g$), by

adjusting screw 6 to make the "three lines coincide" for each time, and record the corresponding readings in Tab. 2 - 4 - 1.

(3) First processing of the measured data by the difference method to find the spring's coefficient stubborn. Next by mapping method: from the data obtained to plot a curve in coordinate paper, taking elongation as Y - axis, the added weights as X - axis, and then the spring's coefficient of elasticity is obtained. Finally, the average value of the spring's coefficient stubborn obtained by the two methods are taken as the final result.

2. Determination of the surface tension coefficient of water

(1) Measure the width of the Π metal frame with a vernier scale for five times, take theirs mean value. Grip the Π metal frame with a tweezers and clean it in alcohol, then hang it under the indicating mirror.

(2) Clean the glass beaker with distilled water, then fill it with distilled water and placed it on the platform 9, measured the water temperature with a thermometer as T.

(3) Turn the platform adjustment screw 8, make the Π - shaped frame to immerge into the water, and then make its upper edge is flush with the surface of the water. Adjust the knob 6 to achieve "three lines coincide", and the reading on the scale as x_0 is then taken and set down in Tab. 2 - 4 - 2.

(4) Turn the screw 6 slowly to make the Π - shaped frame to rise slowly while adjusting knob 8, to make the surface of the water descending and to achieve "three lines coincide". When the water film has just been pulled broken, the reading on the scale as x is then taken and set down in Tab. 2 - 4 - 2.

(5) Repeat steps (3) and (4) four times, and write down the corresponding readings as x in Tab. 2 - 4 - 2.

(6) Determine water's surface tension coefficient after adding activating agent into it.

[Data and Calculations]

1. Determine the spring's coefficient stubborn

(1) Measure the spring's coefficient stubborn by the mapping method

From Tab. 2 - 4 - 1, plot a curve in coordinate paper, taking elongation as Y - axis, the added weights as X - axis, and then the spring's coefficient of elasticity is obtained as following

$\bar{k}_1 =$

(2)

Tab. 2 - 4 - 1 Data of the spring's coefficient stubborn

unit: _____

times i	mass m (g)	s_i (increasing)	s_i' (decreasing)	average value $\bar{S}_i$	spring's elongation $\Delta s_i = s_{i+5} - s_i$
1	0.5				$\Delta s_1 = s_6 - s_1$
2	1.0				

续表

times i	mass m (g)	s_i (increasing)	s_i' (decreasing)	average value $\overline{S}_i$	spring's elongation $\Delta s_i = s_{i+5} - s_i$
3	1.5				$\Delta s_2 = s_7 - s_2$
4	2.0				
5	2.5				$\Delta s_3 = s_8 - s_3$
6	3.0				
7	3.5				$\Delta s_4 = s_9 - s_4$
8	4.0				
9	4.5				$\Delta s_5 = s_{10} - s_5$
10	5.0				

Data processing: $\Delta m = 2.5$ (g), the spring's elongation is

$$\overline{\Delta s} = \frac{\Delta s_1 + \Delta s_2 + \Delta s_3 + \Delta s_4 + \Delta s_5}{5}$$

Then, the spring's coefficient stubborn is

$$\bar{k}_2 = \frac{\Delta mg}{\overline{\Delta s}} =$$

So the average value is

$$k = \frac{\bar{k}_1 + \bar{k}_2}{2} =$$

2. Determination of the surface tension coefficient of water

Water temperature: $T = $ _____ ℃

Tab. 2-4-2 Data of the Surface Tension Coefficient of Water σ unit: _____

times i	initial position x_0	final position x	elongation $\Delta x_i = x_i - x_0$	average value $\overline{\Delta x}$	width of the frame l	average value $\bar{l}$
1						
2						
3						
4						
5						

Data processing:

$$\bar{\sigma} = \frac{k \cdot \overline{\Delta x}}{2\bar{l}} =$$

$$\Delta \sigma = E \cdot \bar{\sigma} =$$

The result is

$$\sigma = \bar{\sigma} \pm \Delta \sigma =$$

【Questions】

(1) What factors may influence the result of the experiment? Why?

(2) How will you determine the surface tension coefficient with a certain concentration NaCl solution, please design concrete steps, and explain whether solution concentration affect the surface tension coefficient.

【Note】

(1) The spring of Jolly Balance is extremely precise, in the experiment be sure not to exceed its elastic limit for fear damaging it.

(2) The cleanness of the beaker, the tweezers and the Π-shaped metal frame may influence experiment result directly, please not touch them with your hands.

(3) In the process of pulling the film, the movement should be slow, when observing your eyes should be parallel to the graduation to minimize the error.

【Introduction to Jolly Balance】

Jolly Balance is actually a fine spring scale, which is often used to measure the tiny forces, its structure is shown in Fig. 4 – 3. Where 5 is the sleeve for fixing the metal pillar, its upper is attached a 0.1 scale's vernier 2. Within 5 is a one millimeter scale's metal pipe 1, and a vernier is consisted of 1 and 2. Adjusting screw 6 can move rod 1 up and down, and the extending length of rod 1 from 5 can be read from the vernier. The fine spring 12 is hung on the beam 13 of the upper end of rod 1, which can move up and down with rod 1. Within the elastic limit, the elongation of 12 obeys Hooke's law. A mirror 10, called index mirror, is hanging at the end of the spring, and there are three score lines in the mirror. The index mirror is located within a glass tube 11, 11 is called the index tube, on which there is a horizontal score line. The hook under the index mirror can be used to hang a weights pan, a Π-shaped frame and so on. 9 is a platform, the height of which can be adjusted by screw 8 and tightening clamp 4, the height of the index tube can be adjusted by tightening clamp 3, adjust screw 7 can make the body of the balance vertical. The engraved line in 11 is actually the reference line for the length of the measured spring, in the experiments, you should ensure exactly that the engraved line in 11 coincides with its image in mirror 10 and the engraved line of the mirror, so the position of the lower end of the spring remains unchanged, and the reading from the vernier can be set down at the moment.

(Ⅱ) **Capillary Method**

【Objects】

(1) Grasp the method of determining the liquid surface tension coefficient with a capillary.

(2) To understand the structure and principle of a reading microscope.

(3) Study to measure minuteness length with the reading microscope.

【Apparatus】

Reading microscope, a glass capillary, beaker, bracket, grinding wheel, thermometer, distilled water, alcohol.

【Theory】

Each liquid surface has surface tension. If liquid level is horizontal, then surface tension is also along horizontal direction; If the liquid surface is curving, then surface tension is tangential with the liquid surface, and it cause the curving liquid surface exert attachment intensity of pressure to the liquid interior. The attachment pressure is positive for convexity, but negative for concavity. The attachment intensity of pressure may be used to explain capillarity.

Insert a glass capillary into liquid, if the tube wall can be wetted by the liquid, then the liquid in the tube assumes on shape of concave sphere, the attachment intensity of pressure is negative, and the liquid level will be higher than outside liquid level. If the liquid cannot be wetted by the liquid, then the liquid in the tube assumes on shape of convex sphere, the attachment intensity of pressure is positive, and the liquid level will be lower than outside liquid level. Consider the situation depicted in Fig. 2-4-4, in which the end of a capillary tube of radius, r, is immersed in water (or alcohol), F is the surface tension, its direction is tangential to the concave sphere, it is

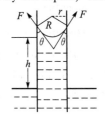

Fig. 2-4-4 Phenomenon of wetting liquid

$$F = \sigma 2\pi r \qquad (2-4-5)$$

Where σ is surface tension coefficient, its value equals strength function per unit length of the boundary, $2\pi r$ is border length of the concave spherical surface, θ is the angle between tangent of liquid surface at liquid solid intersection and solid surface, When solid is not wetted by liquid, $\theta > \pi/2$, When completely non-wetted, $\theta = \pi$; When solid is wetted by liquid, $\theta \leq \pi/2$, completely wetted, $\theta = 0°$。For example, water, for alcohol and glass, the angle $\theta = 0°$, but for mercury and glass, the angle $\theta = 140°$.

Suppose R is the radius of the concave sphere, h is the capillary rise. As shown in Fig. 4-4, we may get $\cos\theta = r/R$, then a substantial rise of h can be observed in the capillary, because of the force F exerted on the liquid due to surface tension, it is equal to $F\cos\theta = 2\pi r^2 \sigma/R$. Equilibrium occurs when the force of gravity on the volume of liquid balances the force due to surface tension. The balance point can be used to measure the surface tension:

$$\sigma = \frac{r\rho g h}{2\cos\theta} \qquad (2-4-6)$$

If the glass tube and the water are extremely clean, then $\theta = 0°$, $R = r$, and formula (2-4-6) becomes

$$\sigma = \frac{r\rho g h}{2} \qquad (2-4-7)$$

When deducing formula (2-4-7), we have neglected the liquid quality above the lower concave spherical surface, and its volume is approximately equal to the volume difference between the column of radius r, height h and the half spheroid of radius for R, namely $\pi r^3 - 2/3\pi r^3 = 1/3\pi r^3$, Therefore the neglected liquid gravity is $1/3\pi r^3 \rho g$, after considering this correction term, the more precise formula may be obtained, namely

$$\sigma = \frac{1}{2}r\rho g(h + \frac{1}{3}r) = \frac{1}{4}d\rho g(h + \frac{d}{6}) \qquad (2-4-8)$$

The formula (2-4-8) is theory basis in this experiment, so long as the inner diameter d of the capillary and the height h of the liquid column are determined accurately, we can calculate the surface tension coefficient. In the formula (2-4-8), each quantity's unit is: d and h are meter, ρ is $kg \cdot m^{-3}$, g is $9.80 N \cdot kg^{-1}$, σ is $N \cdot m^{-1}$。

【Procedure】

1. Measure the inner diameter of the capillary

(1) Place the reading microscope on the tabletop taking E as the bottom surface, fix the capillary with clip on horizontal position, and locate it in front of objective lens right.

(2) Rotate the eyepiece slowly, and make the cross hairs' image clear.

(3) Adjust knob A, cause the capillary's end surface image to be clear in the eyepiece, revolve the eyepiece to cause the cross hair's horizontal hair parallel to the main ruler, adjust knob C to cause the vertical hair moving along the capillary's diameter, the horizontal hair is tangential with circumference of the hole. The difference of the two readings of tangent points is the capillary's inner diameter. Rotate the capillary, make three measurements at different direction, and then take its mean value as the inner diameter of the capillary.

2. Determine the surface tension coefficient of water and alcohol

(1) After cleaning the glass container, fill it with the distilled water in order to take the reading using the reading microscope.

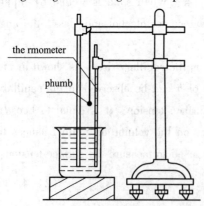

Fig. 2-4-5 Device for measuring liquid surface tension coefficient by capillary method

(2) Fix the clean glass capillary with the clamp on the metal rod, insert it in the glass container, and adjust the capillary with a plumb to make it in the vertical direction. Insert a thermometer to survey water temperature in the container. The test installation is shown as Fig. 2-4-5.

(3) Adjust eyepiece to make cross hairs image clearly. Adjust knob C, cause its horizontal hair coincide with the water surface of the container, reading the datum; Again make the horizontal hair tangential with the concave spherical surface base, read another datum. The difference of the two readings is the liquid column's height h.

(4) Make a mark at the bottom of the concave spherical surface, remove the capillary, measure the capillary's inner diameter at the mark position using the reading microscope according to step (1).

(5) Take another two capillaries, repeat step (2) to (4) respectively, the measured

data are recorded in Tab. 2 – 4 – 3.

(6) By the above method, take other three capillaries. Insert them in alcohol, then determine the rising height h of the liquid column and capillary's inner diameter d.

【Note】

(1) The capillary, the glass container and the liquid must be clean, otherwise the measurement result will be affected.

(2) If glass tube is heterogeneous, when determine capillary's inner diameter at the certain section, cut a trace using a small grinding wheel at this place first, then separate it gently. Whether the separation section is smooth will affect the measurement result directly, therefore you should try several times to grasp essentials.

【Data and Calculations】

Tab. 2 – 4 – 3 Data of surface tension coefficient of water and alcohol

	capillary serial number	capillary inner diameter d (m)				rising height h (m)	temperature T (℃)
		I	II	III	average value		
the measured liquid	1						
	2						
	3						
alcohol	1						
	2						
	3						

The result of the water is

T_{water} = _____ ℃, From appendix: ρ = _____ kg·m^{-3}; σ = _____ N·m^{-1}

$\bar{\sigma}$ = _____ N·m^{-1}, $E = \dfrac{|\sigma - \bar{\sigma}|}{\sigma} \times 100\%$

$\sigma = \bar{\sigma}(1 \pm E_r)$ =

The result of the water is

$T_{alcohol}$ = _____ ℃, From appendix: ρ_0 = _____ kg·m^{-3}; σ_0 = _____ N·m^{-1}

$\overline{\sigma_0}$ = _____ N·m^{-1}, $E = \dfrac{|\sigma_0 - \overline{\sigma_0}|}{\sigma_0} \times 100\%$

$\sigma_0 = \bar{\sigma}_0(1 \pm E_r)$ =

【Introduction to Reading Microscope】

A reading microscope is one kind of optical measuring instruments, which is used to determine accurately the object whose degree is tiny or cannot be griped to measure, such as the inner diameter of a capillary, the width of an engraved line, the width of a slit and so on. Fig. 2 – 4 – 6 is a reading microscope, consisting of a microscope M to magnify the measured object and a main ruler and a attached feet to take readings. There are two forms for the attached feet, one is vernier, and the other is in the form of a micrometer, which principle for reading is the same as vernier or micrometer. The microscope consists of eyepiece, objective and crosshairs, which can be fixed on the screw nut casing, turn the casing by turning the knob C can make the microscope attached to the sleeve to move around. Take E as underside, turn the knob C to make the microscope moving up and down for measuring heights.

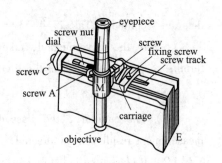

Fig. 2-4-6 Reading microscope

(Ma Jiao, Zhi Zhuang-zhi)

实验五　万用电表的使用

【实验目的】

(1) 了解万用电表的结构原理。
(2) 学会正确使用万用电表的方法。
(3) 学习用万用电表检查电路故障。

【实验器材】

万用电表、数字万用表、电流计、二极管、三极管、滑线变阻器和交直流电源等。

【实验原理】

万用电表是常用的电工测量仪表。它的主要部分是一个电流计，还装配着一套大小不同的电阻和整流器等，利用一个多档选择开关的控制，可以测量交直流电压、电流和电阻，还可以对半导体器件进行简单的测试。图2-5-1为万用电表的原理示意图，图中G为直流微安计，ε为直流电源，D为半导体二极管，$R_1 \sim R_3$为阻值较小的电阻，$R_4 \sim R_{12}$为阻值较大的电阻，W为电位器，K为选择开关，a、b分别表示正负两插孔，"+"孔与

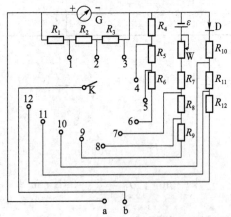

图2-5-1　万用电表原理示意图

微安计的正极相连，插红色表棒。"-"孔与微安计的负极相连，插黑色表棒。下面就万用电表的几种应用说明其原理。

1. 直流电流和电压的测量

一般磁电式电表表头的灵敏度较高，即满度电流值I_g很小，能承受的电压值（$U_g = I_g R_g$，式中，R_g为电流计的内阻）也很小。若要测量较大的电流（或电压），只需给表头并联一合适的分流电阻R_f，或串联一合适的分压电阻R_z，其电路分别如图2-5-2和2-5-3所示。分流电阻R_f和分压电阻R_z的阻值分别由下式给出，即

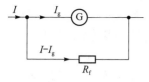
图 2-5-2 电路（1）

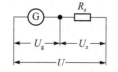
图 2-5-3 电路（2）

$$R_\mathrm{f} = \frac{R_\mathrm{g}}{n-1} \qquad (2-5-1)$$

式中，n 为扩大量程的倍数

$$R_\mathrm{z} = \frac{U}{I_\mathrm{g}} - R_\mathrm{g} \qquad (2-5-2)$$

式中，U 为扩大电压量程。

从图 2-5-1 可知，如果把选择开关 K 分别转向 3、2 或 1，则万用电表就成为如图 2-5-2 所示不同量程的直流安培计；如果 K 分别转向 4、5 或 6，则成为图 2-5-3 所示不同量程的直流伏特计。

2. 电阻的测量——欧姆计

测量电阻用的欧姆计的电路如图 2-5-4 所示，图中虚线框内为欧姆计，a、b 为两个表棒插孔。使用时将待测电阻 R_x 接在 a、b 处。若用 R' 表示欧姆计内的限流电阻（R 与 W 之和），则回路中的电流强度为：

$$I = \frac{\varepsilon}{R_\mathrm{g} + R' + R_X} \qquad (2-5-3)$$

可见，对给定的欧姆计（即 ε，R_g，R' 一定）而言，I_x 与 R_x 有一一对应的关系。这样，在表头刻度上标出相应的 R_x 值即成为一欧姆计，而欧姆计的刻度是非线性的（为什么?）。

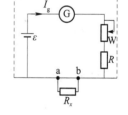
图 2-5-4 欧姆计电路图

从图 2-5-1 可知，如果把选择开关转向 7、8 或 9，则万用表就成为如图 2-5-4 所示不同量程的欧姆计。

3. 交流电压的测量

如果要测量交流电压，就把选择开关分别转向图 2-5-1 中 10、11 或 12，这时电表电路中串有一个二极管 D，由于二极管的整流作用，就将正弦交流电变成脉动直流电，电流的方向由二极管 D 决定，图 2-5-1 中所示流经二极管电流的方向是与电表 G 所允许的方向一致的，所以电表就可用来测量高低不同的交流电压。

4. 晶体管极性的判别

当万用电表转换开关拨至电阻档时，若黑、红表笔依次插入标有"-""+"插孔，则黑笔接电表内部电池正极，红笔接电池负极。由 PN 结单向导电特性和晶体三极管的放大原理，用电表的电阻档（×100 或 ×1k）可辨别晶体二极管和三极管的管脚极性。

（1）判定二极管极性：把电表转换开关拨至电阻档，选取 ×1k 倍率，把两表笔分别接二极管的二个电极，记下读数；把两表笔对调再测一次，记下读数。比较两次测量数值，读数小的一次，黑表笔接的是二极管阳极，红表笔接的是二极管的阴极。

若两次读数均为零，则二极管已短路；若两次读数均为无穷大，则二极管已断路；若两次读数接近，则二极管性能不好。

（2）判定三极管极性及管型：把电表转换开关拨至电阻档，选用×1k的倍率。先假定某一管脚为基极b，并与黑表笔相接，红表笔先后接其余两管脚，若两次读数均小，则假定正确，所测三极管为NPN型，如图2-5-5（a）所示。若两次读数均大，而改用红表笔接假定的基极，黑表笔接其余两极后，读数均小，则假定正确，该管为PNP型，如图2-5-5（b）所示。若两次读数差别较大，则假定错误，需另假定进行测定，直至符合上述判定要求。

确定管型和基极b后，假定余下的某一管脚为集电极c，并与红笔相连，另一极接黑笔，同时用手捏住b、c两极（两极不接触，相当于b、c间接入一个电阻R），记下读数；将两表笔对调，再测一次，记下读数。对PNP型三极管，读数小的一次红笔所接是c，黑笔接的是e，如图2-5-6（b）所示。对NPN型三极管则相反，如图2-5-6（a）所示。

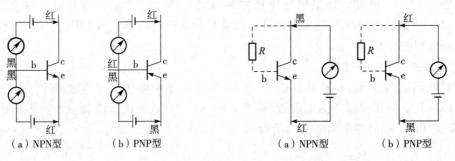

（a）NPN型　　（b）PNP型　　　　　（a）NPN型　　（b）PNP型

图2-5-5　判别管型和基极　　　　图2-5-6　判别三极管的集电极和发射极

【实验步骤】

（一）万用表的使用

1. 电阻的测量

（1）用万用电表检查实验台上所有的导线，将内部断线的导线交实验室。检查单刀双掷开关的接触是否良好。

（2）用电阻箱校准万用电表的中值电阻（等于电流计内阻与限流电阻之和），要求每档校5个点。

（3）测量实验台上的待测电阻R_1和R_2。

（4）用万用电表确定二极管的极性，并测量其反向电阻。

2. 直流电压的测量

（1）测量实验台上所放干电池的端电压。

（2）测量直流稳压电源的输出电压，并将测量值与表头读数进行比较。

（3）将滑线变阻器按图2-5-7接成分压电路，用万用电表测量分压电压值。调节滑线变阻器滑动端的位置，观察分压的变化情况。

3. 直流电流的测量

将滑线变阻器按图2-5-8接成限流电路（此电路与分压电路的区别是什么？），用万用电表测量回路中的电流。调节滑线变阻器滑动端的位置，观察小灯泡亮度的变化和电表读数的变化情况。

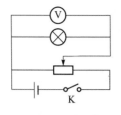 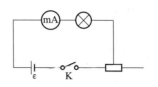

图 2-5-7　分压电路　　　　　图 2-5-8　限流电路

4. 交流电压的测量

（1）用万用电表的交流电压档测量实验室所用的电源电压。

（2）将一交流变压器与电源接通，用万用电表测量变压器的初级及各次级的极间电压。

万用电表使用时往往不是固定连接在待测电路上，而是在测量时连上，读数后即撤离，所以经常要考虑接入误差的问题。接入误差是指未接入电表时的测量值与接入电表后读数值之差。例如在图 2-5-9 所示的电路中，需测量的电压是 U_{BC}，接入伏特计后，由于伏特计有一定的内阻，则回路电压分配会产生变化，使读数为 U'_{BC}。利用电路方程可以证明，U_{BC} 与 U'_{BC} 的差值（即接入误差）ΔU 与 U'_{BC} 的关系是

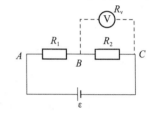

图 2-5-9　交流电压的测量

$$\frac{\Delta U}{U'_{BC}} = \frac{R_1 R_2}{R_V (R_1 + R_2)}$$

式中，R_V 是伏特计的内阻，$\frac{R_1 R_2}{R_1 + R_2}$ 正是以伏特计接入点 BC 为参考点的等效电阻 $R_{等效}$，（此时，把电源看作短路，故 R_1 与 R_2 并联），故得

$$\frac{\Delta U}{U'_{BC}} = \frac{R_{等效}}{R_V}$$

同样，在测量电流时也有接入误差，若万用电表内阻为 R_A，以电表接入点为考察点，电路的电阻为 $R_{等效}$，则接入误差为：

$$\frac{\Delta I}{I'} = \frac{R_A}{R_{等效}}$$

式中，I' 是电表读出的电流值。

5. 晶体管极性的判别（选做）

辨别晶体二极管和三极管的管脚极性。

（1）判定二极管极性。

（2）判定三极管极性及管型。

（二）用万用表检查电路故障

按图 2-5-7 连接电路，相邻两组同学互相设置故障，使用伏特计法及欧姆计法检查电路故障，并排除故障，使电路正常工作。

【注意事项】

（1）使用万用电表进行测量前，首先要检查万用电表平放时指针是否停在表面刻度线左端"0"位处，否则需用小螺刀旋转"指针零位调节器"，使指针指在"0"位处。

（2）测量电阻时，被测电路不能带电；测电流时不能把电表直接连在电源两端，以防短路烧坏电表。

（3）测量高压时手指切勿触及表笔金属部分，以免触电。

（4）用万用电表测电流时，须串联接入电路；测电压时，须并联在待测电路的两端。

（5）当被测电压和电流的数值无法估计时，应把电表转换开关拨至最大量程处。测量时用瞬时点接法试测，依据指针偏转大小选择适当量程。

（6）使用转换开关选择项目和转换量程时，表笔要离开被测电路。在每次测量前必须认真检查转换开关位置是否合适，不得搞错。牢记：一档二程三正负，正确接入再读数；调换量程断开笔，切断电源测电阻。

（7）测量结束后，应将万用电表转换开关拨至最大交流电压量程处，以保证电表安全。

【思考题】

（1）万用电表主要由哪几个部分组成？

（2）为什么不能用万用电表的欧姆档测量带电的电阻？误测将会有什么后果？

（3）为什么使用万用电表可以判别晶体三极管的管型和极性？在判别三极管集电极和发射极时为什么基极与集电极不能直接接触？

【仪器介绍】

（一）万用电表

万用电表的种类很多，使用方法大致相同，现以 MF-30 型万用表为例加以说明，图 2-5-10 为 MF-30 型万用电表的外观图。

图 2-5-10　MF-30 型万用电表外观图

1. 表盘上符号的意义

（1）-2.5~4.0Ω：2.5 表示测量直流量时准确度等级为 2.5 级（即 $a = \pm 2.5\%$）；测量交流量时准确度等级为 4.0 级；电阻档准确度等级为 2.5 级（"V"表示计算准确度是以标度尺总长为分母）。

（2）0dB=1mW600Ω：表示 0 分贝的标准是在 600Ω 负载上产生 1mW 的功率。

（3）45~1000Hz：表示电表中交流档对被测电源的要求是正弦式交流电，其频率在 45~1000Hz 之间。

（4）20000Ω/V~；5000ΩV~：分别表示直流电压档和交流电压档的电压灵敏度。电压灵敏度是指表头灵敏度的倒数，即 $\frac{1}{I_g}$。灵敏度越高的电压表对被测电压的反应越灵敏。用电压灵敏度乘以该档量程即可求得相应的电压表的内阻，如该表直流 25V 档的内阻应为 20000Ω/V × 25V = 500kΩ。可见，对电压表来说，电压灵敏度越高，内阻越大，则测量时接入被测电路后对电路的影响就越小。

（5）测量机构保护和线路熔丝保护是指该表对表头和线路两部分都有保护措施。

2. 表盘读数线

由表针顶端开始,第一条线,右端标有Ω,测各档电阻时读数用。第二条线,左端标有≃,测各档交直流电压(交流10V除外)和直流电流时读数用。第三条线测交流10V以下电压时专用,由于表内整流二极管在低压工作时估计为非线性,因而刻度是不均匀的,为能与高压档共用。第四条线用于测量增益。

3. 操作规程

(1) 准备:认清所用万用电表的面板和刻度,根据待测量的种类(交流或直流;电压、电流或电阻)和大小,将面板上选择开关拨至合适的位置(如待测量的大小未知,一般应选择最大量程先行试测),接好表棒。

(2) 测量:用作伏特计或安培计时应注意以下事项。

① 安培计是测量电流的,它必须串联在电路中。伏特计是测量电压的,它应该与待测对象并联。

② 表棒的正负极不能接反。

③ 执表棒时,手不能接触表棒上任何金属部分。

④ 测试时应采用跳接法,即在用表棒接触测量点的同时,注意电表偏转情况,并随时准备在出现不正常现象时,使表棒离开测量点。

(3) 使用完毕:务必将万用电表选择开关拨离欧姆档,应拨到空档或最大电压档,以保安全。

(4) 用万用表检查电路:实验中经常有这种情况,电路连接无误,但合上开关后不能正常工作,这说明电路有故障。它可能是由导线内部断线、开关或接线柱接触不良、元件内部损坏等因素造成的。我们可以用万用电表来检查线路的故障。其方法有二。①欧姆计法:将电路逐段拆开,特别注意要将电源和电路断开,而且应使用待测部分无其他分支电路。然后用万用电表检查无源部分的电阻分布,特别要检查导线和触点是否接通。②伏特计法:在电源接通的情况下,从电源两端开始,按对接点依电流顺序用万用电表的伏特计检查电压分布,出现电压分布反常之点就是产生故障处。

(二) 数字万用电表

数字万用电表是由集成电路组装而成以液晶显示数字的多功能电表。用此表可以测量交直流电压,电流和电阻,并可检验晶体管等。与普通万用电表相比,它的优点是精度高、读数容易,而且在强磁场下也能正常工作。

DT-830液晶显示数字万用表的面板功能如图2-5-11所示。使用前应仔细阅读说明书,使用时先将电源开关置于"ON"位置接通电源,再按测量信号将选择开关置于合适的位置,插好表棒即可进行测量,所测得数字显示在荧光屏上,改变量程时,表棒应

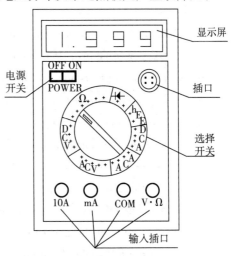

图2-5-11 数字万用电表面板

离开测试电路，测量完毕时将电源开关置于"OFF"位置，断开电源。图中 h$_{FE}$ 插口为晶体管插座。

试用数字万用电表测量电阻和交、直流电压。

（孙宝良　李百芳）

Experiment 5　Manipulation of Multimeter

【Objects】

(1) Know the mechanism of the multimeter.
(2) Learn the correct technique of using the multimeter.
(3) Master the technique of diagnosing electrical problems using the multimeter.

【Apparatus】

Multimeter, digital multimeter, amperemeter, diode, transistor, slide – wire rheostat, DC and AC power supply, etc.

【Theory】

Multimeter is a kind of commonly used electric meter. Its main unit is an amperemeter, and fitted a set of different resistors and commutators together. Controlled by a multi – position optional switch, it can be used not only to measure AC and DC voltage, AC and DC electric current and resistance, but also to make a test for semiconductor devices. Fig. 2 – 5 – 1 shows the mechanism of the multimeter, G is a DC microammeter, ε is a DC power supply, D is a semiconductor diode, $R_1 \sim R_3$ are resistances with less resist-

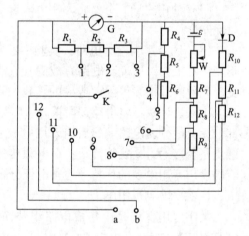

Fig. 2 – 5 – 1

ance values, $R_4 \sim R_{12}$ are resistances with larger resistance values, W is a potential device, K is the optional switch, a、b are positive and nagative jacks, " + " jack connects with the positive terminal of the microammeter, it is plugged by the red test lead. The " – " jack connects with the negative terminal of the microammeter, and it is plugged by the black test lead. We can illustrate the mechanism by some kinds of applications of the multimeter below.

1. Measuring DC current and voltage with a multimeter

As the sensitivity of the gauger of the ordinary magnetic electricity ammeter is high relatively, namely the full scale of the electric current I_g is very small, the voltage ($U_g = I_g R_g$, R_g is the internal resistance of the amperemeter) that it can carry is small also. If we want to measure a larger electric current (or voltage), we only need to connect one suitable shunt resistance R_f in parallel to the gauger, or connect one suitable divided-voltage resistance R_z, as shown in Fig. 2-5-2 and Fig. 2-5-3. The values of shunt resistance R_f and divided-voltage resistance R_z are expressed in the equations below, namely

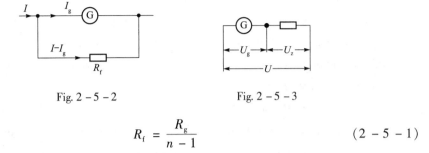

Fig. 2-5-2   Fig. 2-5-3

$$R_f = \frac{R_g}{n-1} \quad (2-5-1)$$

Where n is the multiple of the expand range.

$$R_z = \frac{U}{I_g} - R_g \quad (2-5-2)$$

Where U is the range of the expand voltage.

We can conclude that from Fig. 2-5-1, if we turn respectively the optional switch K to 3, 2 or 1, the multimeter will become a DC ammeter with difference ranges, as shown in Fig. 2-5-2; if we turn respectively the optional switch K to 4, 5 or 6, the multimeter will become a DC voltmeter with difference ranges, as shown in Fig. 2-5-3.

2. Measuring the resistance——Ohmmeter

The circuit of the ohmmeter which is used to measuring the resistance is showed as the Fig. 2-5-4, the ohmmeter is in the dotted line frame of Fig. 2-5-4, a and b are jacks for the two test leads. Connecting the resistance to be measured R_x to a and b when using it. If we use R' to express the limiting-current resistance (it equals the summation of R and W), the current in the circuit is

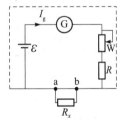

Fig. 2-5-4

$$I = \frac{\varepsilon}{R_g + R' + R_x} \quad (2-5-3)$$

It is obvious that, to the given ohmmeter (ε, R_g, R' are fixed), I_x has a corresponding relation to R_x. So, the corresponding value of R_x are marked on the scale of the gauger, and then it becomes an ohmmeter, but the scale of the ohmmeter is nonlinear (why?).

We can conclude from the Fig. 2-5-1 that, if we turn the optional switch to 7、8 or 9, the multimeter will become ohmmeter with difference ranges as shown in Fig. 2-5-4.

3. Measuring AC voltage

If we want to measure AC voltage, just to turn the optional switch to 10, 11 or 12 in Fig. 2-5-1. There is a diode D in series connection to the circuit of the ammeter. The sine alternating current becomes the pulsating direct current because the rectification of the diode. The direction of the electric current is decided by the diode D. The direction of the electric current that flows diode coinciding with the direction that the permitted by the ammeter G in the Fig. 2 - 5 - 1, so the ammeter can be used to measure different voltage of alternating current.

4. Testing a transistor with a multimeter

When the rotary switch is turned to resistance range, if the black and red test leads are plugged the jacks " - " and " + ", the black test lead connects the positive terminal of the battery in ammeter, the red test lead connects the negative terminal of the battery. Because of unilateral conduction characteristic of PN junction and the magnify principle of the semiconductor, we can distinguish the pin polarity of crystal diode and dynatron by use of the resistance range of the meter (×100 or ×1k).

(1) Testing polarity of a diode: Turn the rotary switch of the meter to resistance range, choose the multiplying factor such as ×1k, connect the two test leads to the two poles of the diode, write down the reading; measure again by exchanging the two test leads, write down the reading. Comparing the two values, which, the reading is small, means the black test lead is connected to the anode of the diode, and the red test lead is connected to cathode.

If the two readings are both zero, it expresses that the diode is in short circuit; if the two readings are both infinite, it expresses that the diode is open circuit; if the two readings is near, the function of the diode is bad.

(2) Testing polarity of a transistor dynatron and its type: Turn the rotary switch of the meter to the resistance range, choose the multiplying factor such as ×1k. Assume one of the three pobes is the b, connect it to black test lead, connect the red test lead to the other pobe, if the twice readings are both small, then the assumption is right, and the type of the transistor measured is an NPN type, showed as in Fig. 2 - 5 - 5 (a). If the twice readings are both big, and use the red test lead to connect the base that be assumed, the black test lead is connected to other two poles, and the readings are both small, then the assumption is right, and it is a PNP transistor, as shown in Fig. 2 - 5 - 5 (b). If the difference between the two readings is relatively big, then the assumption is wrong, and we need to make a new assumption to test it, till the result coincides with the demand above.

After the type and the base b have been confirmed, we assume the pole left is the collector c, and it is connected to the red test lead, the other pole is connected the black test lead, at the same time we nip poles b and c (the two poles don't contact, just like we link a resistance R between b and c), write down the reading; reverse the two test leads, measure it again, write down the reading. To a PNP type transistor, the second reading is small while the red test lead connected to c and the black test lead connected to e, as shown in Fig. 2 - 5 - 6 (b). The condition is reversed to a transistor of NPN type, as shown in Fig. 2 - 5 - 6 (a).

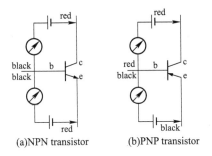 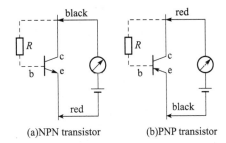

Fig. 2-5-5 Distinguish type and base

Fig. 2-5-6 Distinguish the collector and emitter of the dynatron

【Procedure】

1. Using a multimeter

(1) Measuring resistance with a multimeter

① Checking all wires on the experimental desk with a multimeter and give the broken wires to your teacher. Checking the contraction of the single knife double-throw switch.

② Calibrating the median resistance (it equals to the summation of the internal resistance of the amperemeter and the current limiting resistance) of the multimeter with resistance box, it is required that 5 points are calibrated for each range.

③ Measuring resistances R_1 and R_2 on the experimental desk.

④ Testing polarity of a diode with the multimeter, and measuring its reverse resistance.

(2) Measuring DC voltage

① Measuring the terminal voltage of the battery on the experimental desk

② Measuring the output voltage of the direct current stabilized voltage supply, and comparing the measured value with the reading on its gauger.

③ Connecting a slide-wire rheostat to the circuit as a divided-voltage circuit as shown in Fig. 2-5-7, to measure the divided-voltage with the multimeter. Observe the varied cases of divided-voltage by adjusting the position of the slider of the side-wire rheostat.

(3) Measuring DC current: Connect the slide-wire rheostat to the circuit as a limiting-current circuit, as shown in Fig. 2-5-8 (What is the difference between the kind of circuit and the divided-voltage circuit?). Measure the electric current of the circuit with the multimeter. Adjust the position of the slider of the slide-wire rheostat. Observe the change of the brightness of the small bulb and the reading of the ammeter.

(4) Measuring AC voltage

① Measure voltage of the power supply used in the laboratory with the AC voltage range of the multimeter.

② Connect the power source to a AC transformer, and measure the voltages between the primary and other subordinates of the transformer with the multimeter.

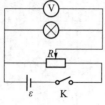
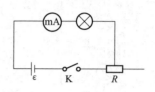

Fig. 2-5-7   Fig. 2-5-8

The multimeter is not often fixed connection to the circuit to be measured, it is only connected while our measuring, it will be disconnect after you having written down the reading, so we often need to consider the cut-over error. The cut-over error is the difference between the measured value when the ammeter is not connected and the reading when the ammeter is connected. For example, in a circuit shown as Fig. 2-5-9, the voltage to be measured is U_{BC}, after the voltameter V has been connected, for there is some internal resistance in V, and the distribution of the circuit voltage will change, this make the reading become U'_{BC}. With the help of the circuit equation, we can prove that the relation between ΔU and U'_{BC} is, (The difference between U_{BC} and U'_{BC} is the cut-over error)

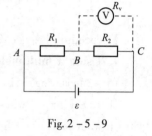

Fig. 2-5-9

$$\frac{\Delta U}{U'_{BC}} = \frac{R_1 R_2}{R_V(R_1 + R_2)}$$

Where R_V is the internal resistance of the voltameter, $\frac{R_1 R_2}{R_1 + R_2}$ is just the equivalent resistance $R_{equivalent}$ that it takes BC where the voltameter connects as the reference points, (here, we consider the power source in short circuit, so R_1 and R_2 are parallel connection), we can get

$$\frac{\Delta U}{U'_{BC}} = \frac{R_{equivalent}}{R_V}$$

Sameness, there is the cut-over error when measuring the electric current. If the internal resistance of the multimeter is R_A, taking the cut-over point of the ammeter as the reference point, the resistance of the circuit is $R_{equivalent}$, thus the cut-over error is

$$\frac{\Delta I}{I'} = \frac{R_A}{R_{equivalent}}$$

Where I' is the value of the electric current that is read by the ammeter.

(5) Testing polarity of a transistor (choose to do): Testing polarity of a diode and a transistor.

① Testing polarity of the diode.

② Testing polarity and the type of a transistor.

2. Diagnosing electrical problems using a multimeter

Connect circuit as Fig. 2-5-7, two groups students set up the malfunction each other, check the circuit malfunction using the methods of voltameter or the ohmmeter, and exclude the malfunction, then make the circuit regular work.

【Note】

(1) Before measuring with a multimeter, touch the meter probes together and check that the meter reads zero first, if it doesn't read zero, turn the switch to "Set Zero" and adjust it with a small screw – driver.

(2) When measuring resistance, there is no electric current flowing in the circuit being measured; when measuring the electric current we can't connect the ammeter with the amphi – power source directly for fear burning out the ammeter for short circuit.

(3) When measuring the high voltage, the fingers don't touch the metal part of the test leads for fear getting an electric shock.

(4) When measuring the electric current with the multimeter, we must put – in circuit in series, when measuring the voltage we must connect it with the amphi – circuit in parallel.

(5) When the values of the voltage and the electric current measured can not be can't be estimated, we should turn the rotary switch of the ammeter to the position where the range is max. Using the method of point contacting to measure, and select the suitable range by deflexion of the needle.

(6) When selecting function and changing range with the rotary switch, the test leads must leave the circuit being measured. We must check the position of the rotary switch to be suitable or not before measuring every time.

(7) After measuring, we should dial the rotary switch of the multimeter to the position where the range is max so that we can be sure the security of the ammeter.

【Questions】

(1) What do the multimeter consist of?

(2) Why cannot resistance with electricity on it be measured with the ohm range of the multimeter? What the effect will happen by mistaking measurement?

(3) Why can test polarity of a diode and a transistor with the multimeter? Why the base and the collector can not be connected directly when testing the collector and the emitter of the transistor?

【Introduction to Instruments】

1. Multimeter

There are many kinds of multimeters, but the methods of using are same roughly, We will illustrate it by taking type of MF – 30 multimeter as an example, as shown in Fig. 5 – 10.

(1) The meaning of the symbol on the dial

① $-2.5 \sim 4.0\Omega$: 2.5 means the grade of accuracy is 2.5 (namely $a = \pm 2.5\%$) when measuring DC quantities, the grade of accuracy is 4.0 when measuring AC quantity, the grade of accuracy of the resistance range is 2.5 ("V" expresses that the grade of the computational accuracy taking total length of the scale ruler as the denominator).

② 0dB = 1mW600Ω: The standard of 0 DB is the 1mW power produced on a load of 600Ω.

③ 45 ~ 1000Hz: AC range of the ammeter demands the power supply to be measured is sine alternating current, and its frequency is between 45 hertz and 1000 hertz.

④ 20000Ω/V ~ ; 5000Ω/V ~ : Respectively voltage – sensitivity of the DC voltage range and AC voltage range. The voltage – sensitivity is the reciprocal of sensitivity of the gauger, namely $\frac{1}{I_g}$. Higher the sensitivity of the ammeter, more sensitive the meter is.

Fig. 2 – 5 – 10

We can get the internal resistance by multiplying voltage – sensitivity by the grade range. For example, the internal resistance of DC 25V range should be 20000Ω/V × 25V = 500kΩ. So, for a voltmeter with higher voltage – sensitivity and larger internal resistance, there will be a less influence to the circuit while connecting it in the circuit.

⑤ The protection to structure of measurement and the fuse of the circuit indicates that the meter can protect both the gauger and the circuit.

(2) The reading line of the dial: From the top of the dial, the first line, which a symbol Ω is marked at the right, can be used to get the reading when measuring different resistances. The second line, there marked ≃ at the left, which can be used to get the reading when measuring different DC and AC voltages range (except AC 10V) and DC current. The third line is special purpose to measure AC voltage less than 10V. For joint use with high voltage, the scale is not uniform because the internal commutation diode is nonlinear under low voltage. The fourth line is used to measure the gain.

(3) Service regulations

1) Preparation: Familiar with the facial plate and the scale of the avometer, adjust the optional switch to the proper position (if the quantity to be measured is not known, we usually choose the largest range to measure) and connect the test leads according to the sorts (DC or AC, voltage, electric current or resistance) and the size of the quantity to be measured.

2) Measurement: Pay attention as voltameter or ammeter:

①When ammeter is used to measure electric current, it must be connected to the circuit in series. And measuring the voltage with voltmeter, it must be connected in parallel.

②The positive and negative test leads should be linked right.

③Don't touch any metal part of the test leads during holding them.

④Jumper connection method should be adopted when measuring, it means, while contacting the test leads to the measuring points, pay attention to the deflexion of the ammeter, and make a preparation for the test leads to leave the measuring points at any moment when abnormal phenomenon occurs.

3) After measuring: We must turn the optional switch of the multimeter away ohm range,

and to the empty or the largest voltage range to make it safety.

4) Checking the circuit with the multimeter: Some circumstances often appear in the experiment that the connection of the circuit is normal, but when you close the switch it does not work normally, it shows that there is malfunction in the circuit. It is perhaps caused by broken wires, bad contractions of switch or wiring terminal and internal damage of component etc. We can check the malfunction of the circuit with a multimeter. The two methods are:

①Ohmmeter method: Taking apart the circuit every segment, in particular open the power source and the circuit, and make parts to measured have not other subcircuit. Then check the resistance distribution of the passive part with the multimeter, especially checking if the wires and the contacts are switch on.

②Voltmeter method: On the condition that the power is facilitory, from the amphi - power supply checking the voltage distribution with the multimeter according to the sequence of the electric current, the position where appear abnormal point in the voltage distribution is the malfunction position.

2. Digital multimeter

Digital multimeter is a versatile meter which is come up of the integrated circuit and has a liquid - crystal display the number. It can be used to measure AC and DC voltage, AC and DC current, resistance, and it can also be used to make diode and continuity tests, etc. Its precision is higher than normal multimeter and it is easy to read the reading, and it can work normally in a strong magnetic field.

For a digital multimeter of a DT - 830 type, its facial plate is showed as Fig. 2 - 5 - 11. Please read the instruction book carefully before

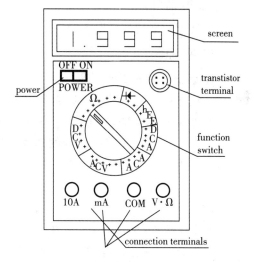

Fig. 2 - 5 - 11

using it. Put the power to the position "ON" while power is applied first, then turn the rotary switch in the correct position according to the measured signal. And you can begin to measure after you have plugged the test leads, then the number will display on the fluorescent screen, when you change the range, the test leads should get away the circuit being measured. Turn the mains switch to the position "OFF" after your measurement..

Try to measure the resistance and the voltage of alternating current and direct voltage with the digital multimeter.

(Sun Bao - Liang, Wang Xiao - fei)

实验六　示波器的使用

阴极射线示波器（或电子射线示波器），简称示波器，由示波管和电子线路组成。它能直接观察电信号随时间的变化情况，并能测量信号的电压和频率。凡是能转换为电压的电学量和非电学量都可用示波器来观测，示波器是一种用途极为广泛的现代测量工具，在医学诊断、监护及科研方面是不可缺少的仪器。

本实验以通用示波器为例，介绍其基本结构和主要功能。

【实验目的】

（1）了解示波器的基本结构及基本工作原理，学会正确使用示波器。

（2）掌握用示波器观察波形，测量电压及时间。

（3）学习用示波器观察李萨如图形，加深对相互垂直振动合成的理解，测定正弦交变电压的频率。

【实验器材】

示波器、低频信号发生器等。

【实验原理】

1. 示波器基本工作原理

示波器通常是由示波管、扫描及整步装置、XY 轴放大器及电源四部分组成。电路方框如图 6-1 所示。下面我们只叙述前两部分的作用，以及如何使用示波器得到稳定的图形，至于具体的线路在此不做介绍。

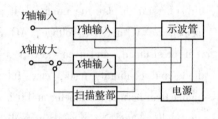

图 2-6-1　示波器组成

（1）示波管：如图 2-6-2 所示，左端为一电子枪，电子枪加热后发出一束电子，电子经电场加速后以高速打在右端的荧光屏上，在屏上形成一个亮点。在电子枪和荧光屏间装有两对相互垂直的平行板，称为偏转板。如果偏转板上加有电压，则电子束通过偏转板时受正电极吸引，受负电极排斥，从而使电子束在荧光屏上的亮点位置随之改变，即偏转板是用来控制亮点位置的。两对偏转板的符号如图 2-6-2 所示，其中横方向的一对称为 X 轴偏转板（简称横偏），纵方向的一对称为 Y 轴偏转板（简称纵偏）。在一定范围内，亮点的位移与偏转板上所加电压成正比。

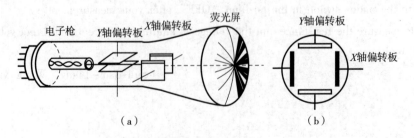

图 2-6-2　示波管

（2）扫描、整步装置：若水平偏转板上加一波形为锯齿形的电压［图 2 - 6 - 3 (a)］，锯齿电压的特点是电压从负开始（$t = t_0$）随时间成正比地增加到正（$t_0 < t < t_1$），然后又突然返回到负（$t = t_1$），再从此开始与时间成正比地增加（$t_1 < t < t_2$）……重复前述过程，这时电子束在荧光屏上的亮点就会做相应的运动：亮点由左（$t = t_0$）匀速地向右运动（$t_0 < t < t_1$），到右端后马上回到左端（$t = t_1$）；然后再从左端匀速地向右运动（$t_1 < t < t_2$）……不断重复前述过程，亮点只在横方向运动，当锯齿波频率较高时，我们在荧光屏上看到的便是一条水平线，如图 2 - 6 - 3 (b)。

如果在纵偏转板上加正弦电压［波形如图 2 - 6 - 4 (a)］，而横偏不加任何电压，则电子束的亮点在纵向随时间做正弦振动（横向不动），我们看到的是一条垂直的亮线，如图 2 - 6 - 4 (b)。

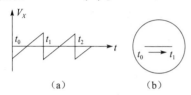

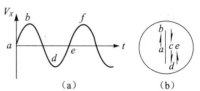

图 2 - 6 - 3　锯齿形电压波形图　　　　　图 2 - 6 - 4　正弦电压波形图

如果在纵偏上加正弦电压，同时在横偏上加锯齿形电压，则荧光屏上的亮点将同时进行方向互相垂直的两种位移，我们看到的将是亮点的合成位移，即正弦图形。其合成原理如图 2 - 6 - 5 所示，对于正弦电压的 a 点，锯齿形电压是负值 a′，亮点在荧光屏上 a″ 处，对应于 b 是 b′，亮点在 b″ 处……故亮点由 a″ 经 b″、c″、d″ 到 e″，描出了正弦图形。如果正弦波与锯齿波的周期相同（即频率相同），则正弦电压到 e 时，锯齿波电压也刚好到 e′，从而亮点描完整个正弦曲线，由于锯齿形电压这时马上变负，故亮点回到左边，重复前过程，亮点第二次在同一位置描出同一根曲线……这时我们将看到这根曲线稳定地停在荧光屏上。但如果正弦波与锯齿波的周期稍有不同，则第二次所描出的曲线将和第一次的曲线位置稍微错开，在荧光屏上将看见不稳定的图形，或不断移动的图形，甚至很复杂的图形。

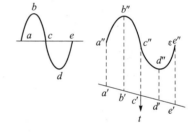

图 2 - 6 - 5　波形图

由此可见：

① 要想看见纵偏电压的波形，必须加上横偏电压，把纵偏电压产生的垂直亮线"展开"来，这个展开过程称为"扫描"。如果扫描电压与时间成正比变化（锯齿形波扫描），则称为线性扫描。线性扫描能把纵偏电压波形如实地描绘出来；如果横偏加非锯齿形波，则为非线性扫描，描出来的波形将不是原来的波形。

② 只有纵偏电压与横偏电压振动周期严格地相同，或后者是前者的整数倍，图形才会简单而稳定，换言之，获得简单而稳定的示波图形的条件是：纵偏电压频率与横偏电压频率的比值是整数，也可表示为公式：

$$\frac{f_y}{f_x} = n \qquad (n = 1,2,3\cdots) \qquad (2 - 6 - 1)$$

实际上，由于产生纵偏电压和产生横偏电压的振荡器是互相独立的振荡器，它们之间的频率比不会自然满足整数比，所以示波器中的锯齿扫描电压的频率必须是可调的。细心调节它的频率，就可以大体上满足（2-6-1）式，但要准确地满足（2-6-1）式，光靠人工调节是不够的，特别是待测电压的频率越高，问题就越突出。为了解决这一问题，在示波器内部加装了自动频率跟踪的装置，称为"整步"，在人工调节到接近满足（2-6-1）式的条件下，再加入"整步"的作用，扫描电压的周期就能准确地等于待测电压周期的整数倍，从而获得稳定的波形。

2. 利用李萨如图形测未知频率

如果纵偏加正弦电压，横偏也加正弦电压，那么得出的图形就是李萨如图形，如图2-6-6所示，李萨如图形可用于测量未知频率。令f_y、f_x分别代表纵偏和横偏电压的频率，n_x代表x方向的切线和图形相切的切点数，n_y代表y方向的切线和图形相切的切点数，则有：

$$\frac{f_y}{f_x} = \frac{n_x}{n_y} \tag{2-6-2}$$

如果已知f_x，就可由李萨如图形和关系式（2-6-2）求出未知频率f_y。

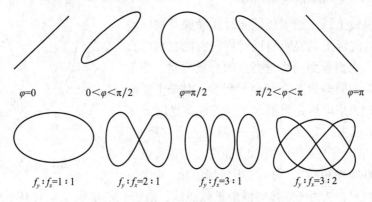

图 2-6-6　李萨如图形

【实验步骤】

（一）校准示波器

将"V/div"旋钮旋至"⊓⊔"位置，"t/div"旋钮旋至2ms/div，它们的"微调"均旋到校准位置，并在测量过程中保持不动。荧光屏上应显示一个周期的方波，如不稳定，逆时针方向缓缓旋转"触发电平"旋钮至方波稳定。方波峰-峰间Y坐标格数应为5div，一个周期方波的X坐标格数为10div。若有误差，可用螺丝刀分别旋转"Y轴灵敏度校准"和"扫描灵敏度校准"电位器进行校准。

（二）测量电压

1. 测量直流电压

（1）将垂直信号输入耦合开关置于"⊥"，触发电平旋至"自动"，使荧光屏上出现一条扫描基线，并按被测信号的大小将"V/div"和"t/div"旋至适当档级。然后调节"⇕"，使扫描基线位于某一选定的基准位置（0V）。

（2）将输入耦合开关改置于"DC"位置，并将被测信号直接或经10∶1衰减探极（信号被衰减到1/10，称探极的衰减倍率为10）接到 Y 轴输入端，然后调节"触发电平"使信号波形稳定。

（3）观察并记录扫描基线与基准线相比上升的格数 b（div），读出 Y 轴输入的灵敏度选择 a（V/div），代入下式计算待测直流电压值，即 $U = a \times b \times c$（c 为探极的衰减倍率）。反复测量多次，取平均值 $\bar{U}$ 为待测电压值。

2. 测量交流电压

（1）将垂直信号的输入耦合开关置于"AC"，根据被测交流信号电压的大小适当选择"V/div"和"t/div"旋钮的档级，从低频信号发生器选择适当频率和电压值的交流信号，然后直接或通过10∶1探极输入到 Y 轴，调节"触发电平"使波形稳定。

（2）读出荧光屏上信号波形的峰-峰值 b(div)，如 Y 轴灵敏度选择 a（V/div），且 Y 轴输入使用了10∶1探极，即 $c = 10$，则被测信号的峰-峰值为

$$U_{\text{p-p}} = a \times b \times c$$

反复测量多次平均值 $\bar{U}_{\text{p-p}}$，计算出待测交流电压的有效值 $U = \dfrac{\sqrt{2}}{4}\bar{U}_{\text{p-p}}$。

（3）用万用电表测量待测电压值，并与用示波器所测得的有效值比较。

（三）测量时间

时间测量，即对被测信号波形上任意两点的时间参数进行定量测量，步骤如下。

（1）将被测信号输入至已被校准过的示波器 Y 轴，根据被测信号的大小将"V/div"和"t/div"旋至适当档级，调节"触发电平"，使波形稳定、幅度适中。为提高测量精确度，调节"t/div"旋钮的档级，使电压波形上被测两点间的距离在荧光屏有效面积内为最大距离。

（2）记下被测两点间所占的水平格数 M 及"t/div"旋钮所用档级的标称值 A，由 $t = A \times M$ 算出该两点间的时间 t。

（四）测定频率

（1）根据待测信号周期的大小选择适当的"t/div"扫描档级。

（2）读出荧光屏上待测信号的一个波长所占的水平格数 B（div），如选择 X 轴灵敏度选择 A（t/div），则待测信号的周期为

$$T = A \times B$$

（3）根据公式 $f = \dfrac{1}{T}$，计算出待测频率为

$$f = \dfrac{1}{AB}$$

【思考题】

（1）如果有一台示波器，由于各旋钮的位置未调适当，荧光屏上看不到亮点，请分析出现这种情况是哪几个旋钮位置不合适。应如何调节才能看到亮点？

（2）若示波器上观察到的正弦波图形不断向右移动，锯齿波频率偏高还是偏低？试加以说明。

(3) 在利用李萨如图形测频率的实验中,已知 X 轴与 Y 轴输入的正弦电压频率相同,而荧光屏上的图形还时刻转动,请解释其原因。

【仪器介绍】

1. 脉冲示波器

脉冲示波器又称为同步示波器,它不但具有普通示波器的功能,而且特别适宜于观察脉冲信号电压。如常用的 ST-16 型(其面板如图 2-6-7 所示)等,这类示波器上设有"Y 轴灵敏度"(V/div)选择档级和"扫描灵敏度"(t/div)选择档级。因此,用它来测量电压和频率非常简便。其各旋钮的功能如下:

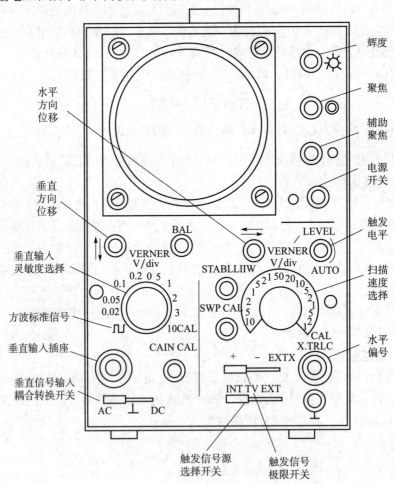

图 2-6-7 脉冲示波器面板图

(1) ☼辉度:用来调节荧光屏上亮点和波形线的亮暗。

(2) ◎聚焦:用来调节荧光屏上亮点和波形线的粗细。

(3) ○辅助聚焦:调节该旋钮可使荧光屏上亮点和波形线更为清晰。

(4) Y 轴位移:用来调节波形在荧光屏上做上下移动。

(5) X 轴位移:用来调节波形在荧光屏上做左右移动。

(6) V/div Y 轴灵敏度选择:这是一个信号电压的分压衰减装置,自 0.02~10V/div

共分九档,可根据被测信号的电压幅度选择适当的档级,以利于观察。第一档级"⊓⊔"为100mV、50Hz的方波校准信号。当与它同轴的小旋钮——"微调"旋钮顺时针旋足听到"嗒"的一声时,旋钮指示即达"校准"位置,此时"V/div"档级的标称值为示波器的Y轴灵敏度。

(7) AC⊥DC Y 轴输入耦合开关:"AC"测量交流信号,"DC"测量直流或缓慢变化的信号,"⊥"输入端处于接地状态。

(8) t/div 扫描速度选择:调节扫描速度,由 10ms/div ~ 0.1 μs/div,共分 16 个档级。可根据被测信号频率的高低,选取适当的档级。当与它同轴的小旋钮—"微调"旋钮顺时针旋足至"校准"位置时,"t/div"档级的标称值即为示波器的时基扫描速度。

(9) LEVEL 触发电平:调节该旋钮可出现信号波形或使信号波形稳定。若将它顺时针旋足,即处于自动(AUTO)位置,此时扫描电路在没有触发信号输入的情况下,能自动扫描。

(10) + − EXTX 触发信号极性开关:用以选择触发信号的上升或下降部分来触发扫描。当开关置于"外接X"时,则使"水平信号"插孔成为 X 轴信号输入端。

(11) INT TV EXT(内、电视场、外)触发信号源选择开关:开关置于"外"时,触发信号来自"水平信号"插孔。

2. FJ – XD7AS 型低频信号发生器

(1) 使用前的准备:开机前应把面板上各输出旋钮逆时针旋至最小。若想得到足够的频率稳定度,需预热 30min 后使用。

(2) 使用

① 频率调节:面板左下方五档琴键是频段选择,频段上方具有频段调整和细调调整,频率显示由上方 LED 显示,此时请注意发光管所指示的单位。

② 功能选择:功能选择为右下方四档互锁琴键,请注意正弦波的电压输出和功率输出不能混淆。

③ 脉宽调节:正弦波与脉冲幅度是分别调节的。正弦波衰减分 20、40、60、80dB 四档,正弦波衰减在功率输出时不保证指标。当使用 TTL 电平时请把脉冲幅度调到最大。

④ 频率计使用:频率计可进行内测和外测。内测功能键按下时为外测,抬起时为内测。频率单位由面板上的发光二极管显示。

⑤ 输出电压测量:抬起内、外测频率键,按下"V/F"键,显示器即显示正弦波衰减前的有效电压值。

(赵 喆 李百芳)

Experiment 6　Usage of Oscilloscope

The cathode ray oscilloscope (or electron – ray oscilloscope), is called oscilloscope, consists of oscillograph tube and electronic circuit. It can be used to observe how electrical signals change over time, and it can determine the voltage and frequency of a signal. All signals, including electricity and non-electric quantity, if only can be transformed into voltage, the oscilloscope may be used to observe or measure, so it is widely used in medicine diagnosis, guardianship and scientific research, etc.

This experiment takes the general oscilloscope as an example, its basic structure and the main function are introduced.

【Objects】

(1) Understand the basic structure and principle of oscilloscope, learn to use the oscilloscope correctly.

(2) Observe waveform with the oscilloscope and determine the voltage and time.

(3) Learn to observe Lissajous' figures using the oscilloscope. To better understand the mutual vertical vibration synthesis, and determine the frequency of sine alternating voltage.

【Apparats】

Oscilloscope, low-frequency signal generator and so on.

【Theory】

1. Basic operational principle of oscilloscope

Oscilloscope is usually composed by the oscillograph tube, the scanning and the synchronization installation, the XY axis amplifier and the power source. The simple block diagram is shown in Fig. 2-6-1. We only introduce two former parts of functions as following, and how to obtain the stable graph from the oscilloscope.

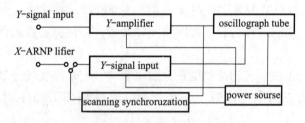

Fig. 2-6-1

(1) Oscillograph tube: As shown in Fig. 2-6-2 (a), the left end is a heater, after it is heaten, a beam of electrons are emitted, the electrons are accelerated by the electric field and strike on the fluorescent screen at a high speed, and then on the screen fluorescent substance forms a glowing spot. Two pairs of parallel plates, which are vertical to each other, are placed between the heater and the screen, they are called deflection plates. If applying voltage to these deflection plates, when passing through the deflection plates, the electron beam will be attracted by anelectrode and repelled by negative electrode, it causes the position of the spot on screen to move, namely these deflection plates are used to control the position of the spot. These deflection plates are marked as shown in Fig. 2-6-2, the horizontal deflec-

tion plates are called X – plates, the vertical deflection are called Y – plates. In the certain scope, the displacement of the spot is directly proportional to the voltage applied to the deflection plates.

(2) Scanning and synchronization: With applying a sawtooth waveform voltage to horizontal deflection plates [as shown in Fig. 2 – 6 – 3 (a)], that is: the voltage starts from negative ($t = t_0$) and then rises directly proportional to the time to a particular maximum (positive) over the course of almost the whole cycle ($t_0 < t < t_1$), and then suddenly drops to a particular minimum (negative) ($t = t_1$), then it starts from the time and rises directly proportional to the time ($t_1 < t < t_2$) $\cdots$, repeat the process, and the spot of electron beam will make the corresponding movement on the screen: the spot moves from left ($t = t_0$) to right at uniform speed ($t_0 < t < t_1$), after arriving at the right end it returns left end immediately ($t = t_1$). Then moves from left end to right end at uniform speed again ($t_1 < t < t_2$) $\cdots$, repeat the former process unceasingly, and the spot only moves in horizontal direction, when the frequency of the sawtooth waveform voltage is large enough, we can see a level line on the screen, as shown in Fig. 2 – 6 – 3 (b).

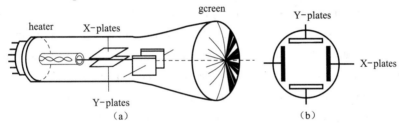

Fig. 2 – 6 – 2

If we apply a sinusoidal waveform voltage to vertical deflection plates [the waveform is shown in Fig. 2 – 6 – 4 (a)], but no voltage is applied to horizontal deflection plates, then the spot of the electron beam (horizontal direction is motionless) make a sinusoidal vibration in longitudinal direction, we will see a vertical bright line, as shown in Fig. 2 – 6 – 4 (b).

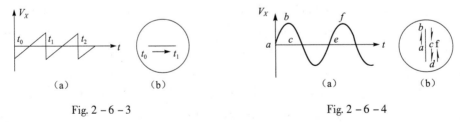

Fig. 2 – 6 – 3   Fig. 2 – 6 – 4

If we apply a sinusoidal waveform voltage in vertical direction, and at the same time apply a sawtooth waveform voltage to horizontal deflection plates, then the spot will participate in the displacements in two directions that are vertical to each other, what we see is a synthesis displacement of the spot, namely a sine graph. Its synthesis principle is shown in Fig. 2 – 6 – 5, for point a of sinusoidal voltage, the sawtooth voltage is negative a', the spot is a'' on the screen, corresponding b is b', the spot is b''. Therefore the spot travels from a'' to b'' to c'' to

d'' until to e'', traces the sine graph. If the cycle of the sinusoidal waveform equals one of the sawtooth waveform (namely their frequencies are the same), when the sinusoidal waveform voltage arrives to e, and the sawtooth waveform voltage also arrives to e', thus the glowing spot traces the entire sine curve, because of the sawtooth voltage immediately drops negative by now, therefore the spot returned left end, and continues to repeat the former process, the spot traces the identical graph in the identical position for second time…by now we will see this graph stopping steadily on the screen. But if the cycle of sine waveform is a little difference with that of the sawtooth waveform, then the graph traced for the second time will stagger slightly with the first graph, on the screen will display an unstable graph, or a moving unceasingly graph, even a very complex graph.

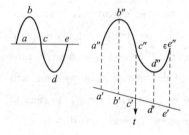

Fig. 2-6-5

Thus we can conclude:

① To display the waveform of vertical voltage, a horizontal voltage must be applied, deploy the vertical bright line produced by the vertical voltage, this process is called horizontal scanning. If the scanning voltage is directly proportional to the time (sawtooth waveform scanning), then it is called the linear scanning. The linear scanning can shows truthfully the vertical voltage wave, if a non-sawtooth wave is applied to the deflection plates, then it is the non-linear scanning, the waveform traced will not be the original one.

② Only when vibration period of the vertical voltage is strictly equal to that of horizontal voltage, or the latter is integral multiple times than the former, then the figure will be simple and stable, in other words, the condition of obtaining a simple and stable figure is – the ratio of frequency of vertical voltage and the frequency of horizontal voltage is an integer, it may be expressed as the following formula:

$$\frac{f_y}{f_x} = n \qquad (n = 1, 2, 3\cdots) \qquad (2-6-1)$$

In fact, because of the oscillators, which produce vertical voltage and horizontal voltage, are independent mutually, so their frequencies cannot satisfy the integer ratio by nature, therefore the frequency of scanning voltage of the oscilloscope must be allowed to adjust. To adjust its frequency carefully, formula (2-6-1) may be satisfied on the whole, but to satisfy formula (2-6-1) accurately, it is insufficient only to depend on the manual regulation, specially for a higher frequency voltage measured, the question will be more prominent. In order to solve the problem, a automatic frequency tracer installment is installed in the oscilloscope, it is called "the synchronization", when the manual regulation approaches to satisfy equation (2-6-1), by the function of "the synchronization", the scanning voltage cycle can accurately be equal to the integral times of the cycle of voltage measured, thus a stable waveform is obtained.

2. Measure frequency by Lissajous' figures

Apply sinusoidal voltage to both the vertical deflection plates and the horizontal deflection plates, and then we get Lissajous' figures. As shown in Fig. 2 – 6 – 6, it may be used to determine the unknown frequency. Suppose f_x and f_y represent separately vertical and horizontal voltage frequency, n_x and n_y represent the number of the tangent points which are in direction of x – axis and y – axis, which a tangent contacting the figure, then we have:

$$\frac{f_y}{f_x} = \frac{n_x}{n_y} \quad (2-6-2)$$

If f_x is known, we may calculate the unknown frequency by Lissajous' figures or the relational expression (2 – 6 – 2).

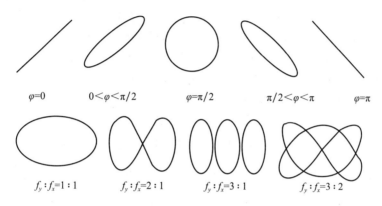

Fig. 2 – 6 – 6

【Procedure】

1. Calibrate the oscilloscope

Adjust the "V/div" to the position "⊓⊔", the "t/div" to 2 ms/div, turn their "fine tuning" to the calibration position, and fix their position in the measurement. A cyclical square – wave will display on the screen, if it is unstable, adjust the "trigger level" slowly along counter – clockwise direction to keep the square – wave stable. Its grid number of peak – peak value of Y – axis should be 5 div, and the corresponding grid number of X – axis is 10div. If the error exists, we may calibrate it using a screwdriver to screw separately "Y – axis sensitivity calibration" and "scanning sensitivity calibration".

2. Determine the voltage

(1) Determine the D. C. voltage

① Position the coupling switch of vertical signal input to "⊥", turn the trigger level to "automatic", make a scanning base line appear on the screen, and turn "V/div" and "t/div" to suitable level according to the measured signal. Then adjust the knob "↕", cause the scanning base line to locate at certain selected position (0 V).

② Turn the input coupling switch to position "DC", and connect the measured signal directly or pass through 10:1 decay (signal is decayed to its 1/10, then the decay multiplying factor is 10) to the Y – axis input, then adjust the "trigger level" to keep the signal waveform stable.

③ Observe and record the grid number between the rising scanning base line and the reference axis, and record it as b (div), and read the Y - axis input sensitivity selection for a (V/div), then calculate the D. C voltage value measured by substituting it in the following formula, it is $U = a \times b \times c$ (where c is the decay multiplying factor). Repeat the measurement many times, take the average value as the measured voltage value.

(2) Determine the A. C. Voltage

① Position the coupling switch of vertical signal input to "AC", and turn "V/div" and "t/div" to suitable level according to voltage value of the measured A. C. signal. Choose a signal that has a suitable frequency and the voltage value from the low - frequency signal generator, and connect the measured signal directly or pass through 10∶1 decay to the Y - axis input, then adjust the "trigger level" to keep the signal waveform steady.

② Read out the waveform's peak - peak value b (div) on the screen, the Y - axis sensitivity is a (V/div), and the Y - axis input has a decay of 10∶1, so $c = 10$, then the measured signal's peak - peak value is

$$U_{p-p} = a \times b \times c$$

Repeat the measurement many times, take the average value $\overline{U}_{p-p}$ as the measured value, and calculate the effective value of the measured A. C. voltage by $U = \frac{\sqrt{2}}{4}\overline{U}_{p-p}$.

③ Determine the measured voltage value with the multimeter, and compare it with the effective value which obtains from the oscilloscope.

3. Determine the time

Time determination, it means to measure the time between any two points of the waveform, the steps are as follows:

(1) Input the measured signal to Y - axis of the oscilloscope which has already calibrated, turn "V/div" and "t/div" to suitable level according to the measured signal, adjust the "trigger level" to keep the waveform stable and the amplitude suitable. In order to enhance the measuring precision, adjust the "t/div" level and cause the distance between two points in the waveform to be the maximum distance in the effective area on screen.

(2) Record the horizontal grid number M occupied by the two measured points and the nominal value A used level of "t/div", and then calculate the time t by formula $t = A \times M$.

4. Determinate the frequency

(1) Choose suitable scanning level "t/div" according to the cycle of the measured signal.

(2) Read out the measured signal's horizontal grid number B (div) occupied by the length of a waveform on the screen, if select X - axis sensitivity as A (t/div), then the cycle of the measured signal may be computed

$$T = A \times B$$

(3) By formula $f = \frac{1}{T}$, the measured frequency is

$$f = \frac{1}{AB}$$

【Questions】

(1) If there is an oscilloscope, with all sections not been adjusted suitably, and the glowing spot does not appear on the screen, please analyze what sections' positions are not suitable? How should you adjust them to see the spot?

(2) If the sine waveform on the screen of the oscilloscope moves to right unceasingly, then is the frequency of the sawtooth wave higher or lower? Try to explain it.

(3) In the experiment using Lissajous' figures to determine the frequency, with input frequencies of sinusoidal voltage of X – axis and Y – axis' being the same, but the graph on the screen rotates all the time, please explain it?

【Introduction to Apparatus】

1. Pulse oscilloscope

The pulse oscilloscope is called the sycnhronoscope, it not only has the functions of ordinary oscilloscope, but also is specially suitable in observing pulse signal voltage. Such as common – used ST – 16 type (its front panel is shown in Fig. 2 – 6 – 7) and so on, this kind of oscilloscope is equipped with the "Y sensitivity" (V/div) level and the "scanning sensitivity" (t/div) level. Therefore, it is extremely simple to determine the voltage and the frequency with it. Its various section' functions are as follows:

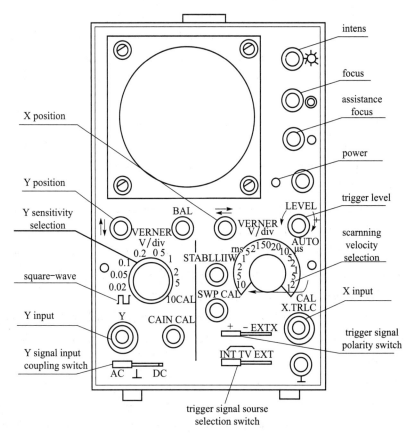

Fig. 2 – 6 – 7

(1) Intensity: It is used to adjust the intensity of the glowing spot and waveform on the screen.

(2) Focus: It is used to adjust the thickness of the spot and waveform.

(3) Assistance focus: Adjust it to make the spot and the waveform on the fluorescent screen clear.

(4) $Y-\text{pos}$: To cause the waveform to move up and down on the screen.

(5) $X-\text{pos}$: To cause the waveform to move left and right on the screen.

(6) V/div (Y axis sensitivity choice): This is a voltage decay device, it is divided into nine grades from 0.02V/Div to 10V/Div altogether, and a suitable level may be selected according to the amplitude of the voltage measured signal for your observation. The first level "⊓⊔" is a square – wave calibration signal whose voltage is 100mV and frequency is 50Hz. When turning the small knob clockwise, which is coaxial with it, full to the position with a sound "clatter", the knob reaches the position "calibration", this time nominal value of the "V/div" is Y – axis sensitivity level of the oscilloscope.

(7) AC⊥DC: The "AC" is used to measure alternating signal, the "DC" is used to measure direct current or slow – changed signal, the "⊥" input end is at the earth condition.

(8) t/div Adjust the speed, it is divided into 16 levels from 10ms/div to 0.1ms/div altogether. You may select the suitable level according to the measured signal's frequency. When turning the small knob "fine tuning" clockwise, which is coaxial with it, until the position "the calibration", the nominal value of the "t/div" level is the time base speed of the oscilloscope.

(9) LEVEL (trigger level): Adjust it to make the signal waveform display and keep it stable. If we turn it clockwise as much as possible, namely it is at the automatism position (AUTO), this time the scanning circuit can scanning in the state of no trigger signal input.

(10) + – EXTX (trigger signal polarity switch): It is used to choose the rising or the dropping part of the trigger signal to trigger scanning. When adjust the switch to "outside X", then cause the "horizontal signal" jack to act as X signal input.

(11) INT TV EXT (trigger signal source selection switch): When put the switch to "EXT", trigger signal is produced from "the horizontal signal" jack.

2. FJ – XD7AS low – frequency signal generator

(1) Preparation for use: Turn anti – clockwise each output knob on the front panel to its minimum before turn on the power. Time of reaching enough frequency stability should be more than 30min.

(2) Use

① Frequency modulation: The five buttons locating left underneath on the front panel are the frequency band option, the frequency band adjustment and the fine adjustment are above the frequency band, and the frequency is indicated by the upper LED, please note the unit which the illuminated tube indicates.

② Function option: The four interlock buttons located right underneath are for function option, please note the voltage output of the sine wave and the power output cannot be confused.

③ Pulse width modulation: The sine wave and the pulse amplitude are adjusted separately. The sine wave attenuator is divided into 20, 40, 60 and 80 dB four steps, the sine wave attenuator does not guarantee the target when the power makes output. While using TTL, please adjust the pulse amplitude modulation to its maximum.

④ Usage of the frequency meter: Frequency meter may be used for interior measurement and exterior measurement. To press the inner functional key is for exterior measurement, and uplift is for interior measurement. Frequency unit is displayed by the light-emitting diode on the front panel.

⑤ Measurement of output voltage: Lift the inner and outer frequency key, then press the button V/F, the indicator will display the active voltage of the sine wave before it is attenuated.

(Zhao Zhe, Deng Yan-hao)

实验七 惠斯通电桥测电阻

【实验目的】

（1）掌握惠斯通电桥的原理和特点。
（2）学会使用惠斯通电桥测量电阻的方法。
（3）了解非平衡电桥的作用及测量热敏电阻阻值和温度的关系。

【实验器材】

滑线板、箱式惠斯通电桥、直流稳压电源、检流计、电阻箱、滑线变阻器、待测电阻、万用表、热敏电阻、b形管、温度计、酒精灯、铁架台等。

【实验原理】

1. 惠斯通电桥原理

惠斯通电桥的电路如图 2-7-1 所示，四个电阻 R_1、R_2、R_0、R_x 联成一个四边形，每一条边称为电桥的一个臂。对角 A 和 B 加电源 E，对角 C 和 D 之间连接检流计 G。

所谓桥就是指 CD 对角线而言，它的作用是将桥两端端点的电位直接进行比较。当 CD 两点电位相等时，检流计中无电流流过，则电桥就达到了平衡。这时电桥电路就完全等效为一个简单的串并联电路，不难证明：

$$\frac{R_1}{R_2} = \frac{R_x}{R_0} \quad 即 \quad R_X = \frac{R_1}{R_2}R_0 \quad (2-7-1)$$

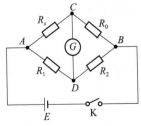

图 2-7-1 电桥电路

若 R_1、R_2、R_0 均为已知，或 $\frac{R_1}{R_2}$ 和 R_0 为已知，则 R_x 由上式可以求出。

(1) 滑线式惠斯通电桥

①滑线式惠斯通电桥原理：滑线式惠斯通电桥原理如图 2-7-2 所示。它与图 2-7-1 的区别仅在于 R_1 和 R_2 是由一根粗细均匀的金属电阻丝组成。D 点是一个可以滑动的电键。当 D 滑动时就可以改变 R_1 和 R_2 的大小，以达到电桥平衡。

在电桥平衡时，测得 AD 和 DB 的长度分别为 l_1 和 l_2，根据电阻定律可知：$\frac{R_1}{R_2} = \frac{l_1}{l_2}$

把上式代入（2-7-1）式，则有

$$R_x = \frac{l_1}{l_2} \cdot R_0 \qquad (2-7-2)$$

滑线式惠斯通电桥实验装配电路图如图 2-7-3 所示。滑线板是在一块木板上装有一固定米尺和三个铜片，铜片上配有若干接线柱，以便分别联接待测电阻、标准电阻箱、检流计和电源等。一根均匀电阻丝 AB 固定在米尺上，滑动电键 D 可沿米尺滑动。滑线变阻器 R' 用来控制电路中电流，以防止电流过大而损坏检流计和电桥。

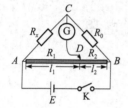

图 2-7-2 滑线式电桥

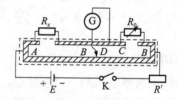

图 2-7-3 滑线式电桥实验线路图

②影响滑线式电桥测量精度的因素有很多，我们这里仅讨论一点：采取等臂电桥减小测量的相对误差。

所谓等臂电桥，就是 $R_1 = R_2$（或 $l_1 = l_2$），$R_x = R_0$。现在证明，当 $l_1 = l_2 = l/2$（$l = l_1 + l_2$）时误差最小。

已知

$$R_x = \frac{l_1}{l_2} R_0 = \frac{l_1}{l - l_1} R_0$$

可以推导出它的相对误差为

$$\frac{\Delta R_x}{R_x} = \frac{l}{l_1(l - l_1)} \Delta l_1$$

分母 $f(l_1) = l_1(l - l_1)$ 为极大值的条件是

$$\frac{df(l_1)}{dl_1} = l - 2l_1 = 0$$

则 $l_1 = l/2$ 时，分母有极大值，相对误差最小。

(2) 箱式惠斯通电桥：箱式惠斯通电桥的基本电路与滑线式电桥相同，只是为了便于携带，将整个仪器都装在箱内。QJ23 型箱式惠斯通电桥内部电路和面板图分别如

图 2-7-4，图 2-7-5 所示。

为了便于测量，在箱式电桥中 R_1 和 R_2 的比值用一个可调的十进制旋钮来指示，我们把它叫作比例臂旋钮。其指示的比值为 0.001、0.01、0.1、1、10、100 和 1000 七个档位。标准电阻 R_0 用一个电阻箱。测量时，应根据待测电阻阻值的范围选择适当的比例臂旋钮的位置，调节电阻箱的四个可调旋钮。当电桥达到平衡时，待测电阻阻值 R_x 可由下式求得：

$$R_x = （比例臂指示值）\times（电阻箱四个旋钮指示值）$$

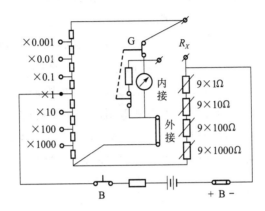

图 2-7-4　箱式电桥内部电路图

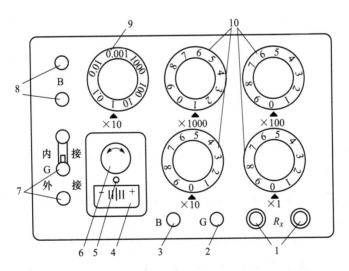

1. 待测电阻 R_x 接线柱；2. 检流计开关按钮 G；3. 电源开关按钮 B；
4. 检流计；5. 检流计调零旋钮；6. 检流计扣锁；7. 外接检流计接线柱；
8. 外接电源接线柱；9. 比例臂旋钮；10. 标准电阻（本处为四个转盘）

图 2-7-5　箱式电桥面板图

2. 非平衡电桥的作用

惠斯通电桥测量电阻是应用电桥达到平衡时的特点，桥路中检流计指示为零，得

到 R_x 的阻值。如果待测电阻 R_x 的阻值具有随某种因素（如温度、压力等）而变化的特性，则电桥的平衡状态将随着该种物理因素的变化而遭到破坏，使电桥的桥路上有电流流过，检流计指针指示电流的方向和大小，这就是非平衡电桥。利用电桥的非平衡状态，根据桥路检流计指示电流的方向和大小就可以研究和分析待测电阻随该因素的变化规律，这就是非平衡电桥的作用。

热敏电阻是一种对温度变化非常敏感的半导体元件，当其周围温度发生微小变化时，它的电阻阻值就发生相应变化。半导体热敏电阻阻值和温度的关系可表示为：

$$R_T = R_0 e^{B/T}$$

式中，R_T 为热敏电阻的阻值；R_0、B 为常数，由热敏电阻的形状和材料决定。

将热敏电阻 R_T 作为待测电阻接入惠斯通电桥的一个臂上，在某一温度（0℃或室温）调节电桥达到平衡，然后改变热敏电阻的温度，利用非平衡电桥来研究热敏电阻的温度与桥路电流的关系，定性地分析热敏电阻阻值和温度的变化规律。根据电桥在非平衡状态下热敏电阻的温度与电流的关系可以制成半导体热敏电阻温度计，利用温度和电流关系绘制校准曲线，然后用测电流的方法在标准曲线上就可迅速得到待测的温度。

【实验步骤】

1. 利用滑线式惠斯通电桥测量电阻

（1）按图 2-7-3 连接线路，电源（直流稳压电源）输出电压调为 4.5V。

（2）调节 R'，使 U_{AB} 在 1.5~2V 之间（用万用表测量）。

（3）对应一个待测电阻 R_x，选择 R_0（为了减小测量中的相对误差，取 R_0 分别等于 R_x、$\frac{1}{2}R_x$、$2R_x$），移动电键 D，使电桥平衡，记录 l_1、l_2、R_0。

（4）在待测电阻板上任选三个不同的 R_{x1}、R_{x2}、R_{x3}，（分别为几十欧姆，几百欧姆和几千欧姆，可利用万用表粗测），然后按照步骤3，每个各测5次，由公式（7-2）计算待测量电阻 R_x。

注意：在利用滑线式惠斯通电桥测电阻时，应区分由于电路中有断路而导致的检流计指针不动及真正的电桥平衡。另外寻找平衡点时，可先估计大致位置，移动电键 D 采用跃接法。跃接法的优点是避免电流长时间流过电桥，发现异常情况能迅速切断电源，同时根据跃接时检流计指针的左右摆动情况，能较快地找到电桥平衡的位置。

2. 利用箱式惠斯通电桥测电阻

（1）熟悉面板上各旋钮，接线柱及按键的作用及使用方法。

（2）电源仍用直流稳压电源，输出 4.5V，检流计用箱式电桥自带的内部检流计。接好各旋钮，选择适当的比例臂值，对前面测过的 R_{x1}、R_{x2}、R_{x3}，再各测 1 次。

（3）记录箱式电桥的测量准确度（在箱盖的内侧或箱底处），按下式计算误差 ΔR_x

$$\Delta R_x = \overline{R_x} \cdot 准确度$$

3. 使用箱式惠斯通电桥研究热敏电阻的温度和桥路电流的关系（选做）

（1）将热敏电阻两端联上导线，然后将其固定在温度计底部的乙醇液泡或汞液泡附近与温度计一起插入 b 形管水中，使热敏电阻位于 b 形管中心处，如图 2 - 7 - 6 所示。

（2）将热敏电阻用导线接入电桥面板接线柱，在室温温度调节电桥达到平衡，记录平衡时热敏电阻阻值和温度。

（3）用酒精灯在 b 形管弯曲部分的末端加热（图 2 - 7 - 6），改变热敏电阻温度，每升高 5℃，记录该温度与该温度时电桥桥路电流。直到温度上升到 55℃ 为止，注意控制温度变化的快慢。

（4）当温度升高到控制温度以上（一般在 60℃ 左右）停止加热，使温度自然下降，每降到上面记录的温度时，迅速调节电桥平衡，测量该温度时热敏电阻的阻值。

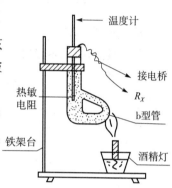

图 2 - 7 - 6

【数据记录及处理】

1. 滑线式惠斯通电桥测电阻（表 2 - 7 - 1）

表 2 - 7 - 1　滑线式惠斯通电桥测电阻

R_x 标准值（Ω）	l_1（cm）	l_2（cm）	R_0（Ω）	R_x（Ω）	$\overline{R_x}$（Ω）	ΔR_x（Ω）
1						
2						
3						

测量结果：$R_x = \overline{R_x} \pm \Delta R$

2. 箱式惠斯通电桥测电阻（表 2-7-2）

表 2-7-2　箱式惠斯通电桥测电阻

准确度：_____

R_x 标准值（Ω）	比例臂值	R_0（Ω）	R_x（Ω）	ΔR_x（Ω）

测量结果：$R_x = \bar{R}_x \pm \Delta R$

3. 测量热敏电阻温度-电流、阻值（表 2-7-3）

表 2-7-3　热敏电阻温度-电流、阻值

温度（℃）	室温	20	25	30	35	40	45	50	55
电流（μA）									
平衡电阻（Ω）									

根据测量数据在坐标纸上画出温度与电流校准曲线和温度与电阻值的校准曲线。

【思考题】

（1）当电桥达到平衡后，若互换电源和检流计的位置，电桥是否仍然保持平衡？试证明之。

（2）使用惠斯通电桥测量电阻时，若几次接入同一个被测电阻和比较臂电阻 R_0 的接线松紧不同，对测量结果有什么影响？

（3）在惠斯通电桥电路中，当滑动端在 AB 电阻丝上滑动时，无论位于何位置，检流计指针都不偏转，而将一伏特计并联于 AB 间，检流计则有指示，试分析哪条导线断了？

【仪器介绍】

1. QJ23 型箱式惠斯通电桥

箱式电桥内部电路及面板图参见图 2-7-4 和图 2-7-5。

检流计左侧的 3 个接线柱是检流计连接端，当连接片接通"外接"时，检流计被接入桥路，当连接"内接"时，检流计被短路，此时既"锁住"了检流计指针，又可以从"外接"处接入灵敏度更高的检流计。

测量前，通常需将检流计连接片从"内接"换到"外接"，调整检流计的指针到零点，接入 R_x，根据 R_x 的粗略值，选择适当的比例臂值，并使 R_0 也先调至相应的读数，然后操作按钮 B 和 G，再细调 R_0 至电桥平衡。

使用时需注意以下几点。

① 为了充分利用电桥的精度，比例臂值的选取一定要使待测电阻保持四位读数。

② 操作按钮 B 和 G 时，应先接通 B，后接通 G；先断开 G，后断开 B，以防止过大

电流通过检流计。按钮 B 和 G 只要按下，再旋一角度，即被锁住，使之长时间接通。

③ 调节电桥平衡时，必须对按钮 G 采用"跃接法"，若检流计指针偏向"+"，表示 R_0 需加大，反之需减小。

④ 测量完毕，应放开 B、G 按钮，并将连接片重新接到"内接"上以锁住检流计指针。

2. JWD-2 型直流稳压电源

仪器的外形请参见实物。

本仪器有 A、B 两档，分别独立，电压为 0~15V 可调，电流最大为 1.5A 的直流输出，能符合多方面的要求。仪器采用分档粗调，另配有电位器微调，并装有过载短路保护自动复原装置，使用方便可靠。

如需用电压为 0~15V，电流为 1.5A 以下电源时，可用 A、B 两档之任何一档。根据需要调节粗调开关及微调电位器。表头指示开关拨向 A，表头指示为 A 档输出；拨向 B，表头指示为 B 档输出。

如需用电压为 15~30V，电流为 1.5A 以下电源时，可将中间两个接线柱的叉片连接起来串联使用。电表开关拨到 A+B，表头指示为两档串联电压。

如需用 0~15V，输出电流为 1.5~3A 电源时，先应分别调节两档的输出电压为需要值，然后再并联使用。

接上负载后，如发现输出电压偏低，或者为零，此时证明有超载或短路现象。应切断电源，排除负载中的故障后再使用。

3. AC5 型直流指针式检流计

本实验中检流计采用 AC5 型直流指针式检流计（其外形参见实物）。

该检流计属于便携型磁电式结构，其可动部分固定在张丝上面，当线圈内通入微小的电流时，带有电流的线圈与永久磁铁的磁场相互作用产生了转动力矩，使指针偏转，检流计的反作用力矩由起导电作用的张丝产生。检流计的可动部分用短路阻尼的方法制动，这样可防止可动部分张丝等因机械振动而引起的变形，当面板上的小旋转钮移向红色圆点位置时，线圈即被短路。

使用检流计时，先将其"+""-"接线柱端钮接入电路。

将小旋钮移向白色圆点位置，并用"零位调节器"将指针调到零位。

按下"电位"按钮，检流计即被接入电路，如需将检流计长时间接入电路时，可将"电计"按钮按下并转一角度，则"电计"按钮即被锁住。

若使用中检流计指针不停地摆动时，将"短路"按钮按一下，指针便立即停止运动。

注意：当检流计应用完毕后必须将小旋钮移向红色圆点位置，并将"电计"及"短路"按钮放松。

（李玉娟　孙宝良）

Experiment 7 Measurement of Resistance with Wheatstone Bridge

【Objects】

(1) To grasp the principles and characteristics of Wheatstone bridge.

(2) To learn to measure the resistance with Wheatstone bridge.

(3) To understand the function between unbalance bridge and the relationship between thermal resistor and temperature.

【Apparatus】

Slide wire board, box type Wheatstone bridge, direct current stabilized voltage supply, galvanometer, resistance box, slide-wire rheostat, the resistance to measure, multimeter, thermal resistor, b-type tube, thermometer, alcohol burner, hob stage, etc.

【Theory】

1. The principle of the Wheatstone bridge

The circuit of the Wheatstone bridge is shown in Fig. 2-7-1, the resistances R_1、R_2、R_x、R_0 are connected to a quadrangle, the edge is called an arm of the bridge. The power source E is connected with A and B, and the galvanometer G is connected with C and D.

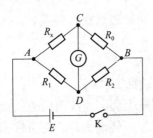

Fig. 2-7-1 Circuit diagram of Wheatstone bridge

The so-called bridge is to point the diagonal line of CD, and its function is comparing the potential of the ends of the bridge. When the potentials of them are equalized, there are no current through the galvanometer, then the electric bridge reaches its balance. The bridge becomes a simple serial-parallel circuit, and it is easy to prove:

$$\frac{R_1}{R_2} = \frac{R_x}{R_0} \quad \text{namely} \quad R_X = \frac{R_1}{R_2} R_0 \qquad (2-7-1)$$

If we have known the value of R_1、R_2、R_0 or the value of $\frac{R_1}{R_2}$ and R_0, the value of R_x can be worked out from the equation above.

(1) The slide wire Wheatstone bridge

①The principle of slide wire Wheatstone bridge: The principle of slide wire Wheatstone bridge can be shown in Fig. 2-7-2. The difference with Fig. 2-7-1 only lies in R_1 和 R_2 which is composed of a metal wire with a uniform thickness. D is a sliding key. When D is sliding, the size of R_1 and R_2 can be changed in order to achieve the balance of bridge.

As the bridge is balance, the measured AD and DB in length were l_1 and l_2, according to the resistance law: $\frac{R_1}{R_2} = \frac{l_1}{l_2}$

The type substitution (2-7-1), is

$$R_x = \frac{l_1}{l_2} \cdot R_0 \qquad (2-7-2)$$

The experimental assembly circuit of slide wire Wheatstone bridge can be shown in fig. 2-7-3. The sliding plate is a fixed ruler and three copper sheets arranged on a board, copper with a plurality of wiring terminals, so that the combined reception measured resistance, standard resistance box, galvanometer and power etc. A uniform resistance wire AB is fixed on the meter, the sliding key D sliding along the ruler. The slide wire rheostat is to control the current in the circuit in order to prevent the current which is too large to damage the galvanometer and bridge.

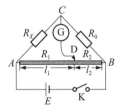
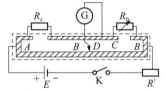

Fig. 2-7-2 Sliding-line Bridge   Fig. 2-7-3 Experimental Circuit Diagram of Sliding Bridge

②There are many factors affecting the measurement precision of slide wire bridge. Here we only discuss the point: using the arm bridge to reduce the measuring error.

The so-called equal arm bridge, is $R_1 = R_2$ (or $l_1 = l_2$), $R_x = R_0$. Now that, when $l_1 = l_2 = l/2$ ($l = l_1 + l_2$), the error is the smallest.

Known
$$R_x = \frac{l_1}{l_2} \cdot R_0 = \frac{l_1}{l - l_1} \cdot R_0$$

It can derive the relative error for it

$$\frac{\Delta R_x}{R_x} = \frac{l}{l_2 (l - l_1)} \Delta l_1$$

The denominator $f(l_1) = l_1(l - l_1)$ for the maximum condition

$$\frac{df(l_1)}{dl_1} = l - 2l_1 = 0$$

When L1 = l/2, the denominator is maximum, while the minimum relative error can be obtained.

(2) The box Wheatstone bridge: The basic circuit for box Wheatstone bridge and slide wire bridge is the same, just in order to facilitate the carrying, the instruments are installed in the box. QJ23 type Wheatstone bridge internal circuit and panel respectively can be shown in Fig. 2-7-4 and Fig. 2-7-5.

In order to facilitate the measurement, the ratio of R_1 and R_2 in the box bridge with an adjustable decimal knob to instruct, we call it the arm knob. The ratio is . A resistance box with a standard resistor R_0. When measuring, we should select the proportion arm knob proper position according to the scope of resistance to be measured and adjust four knobs in the resistance box. When the bridge is balanced, the measured resistance value can be obtained by the formula:

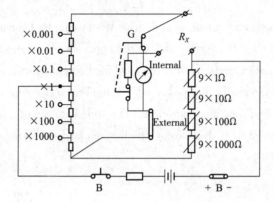

Fig. 2 – 7 – 4   Internal Circuit Diagram of Box Bridge

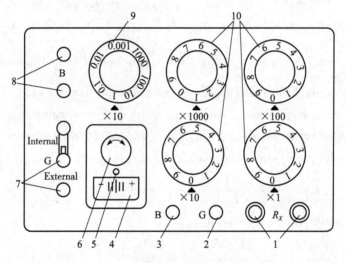

1. Resistance R_x junction column to be tested; 2. Flowmeter switch button G; 3. Power switch button B; 4. Flow meter; 5. Flip meter adjustment zero knob; 6. Flowmeter lock; 7. External flowmeter junction post; 8. External power wiring column; 9. Proportional arm knob; 10. Standard resistance (There are four turntables in this office)

Fig. 2 – 7 – 5   Box Bridge Panel Diagram

RX = (proportional arm indicator values)(box four knob indicators)

2. The function of non – balance electric bridge

The Wheatstone bridge measuring resistance is the application characteristics of bridge balance, indicating the galvanometer bridge to zero, get the value of RX. If the resistance to be measured with the RX resistance of some factors (such as temperature, pressure, etc.) and changes, the dynamic equilibrium of bridge will change along with the physical factors and the destruction. When the current flows along the way, the galvanometer pointer indicates the direction and size of current, which is non – balance electric bridge. Using non – balance bridge, according to the current direction and size of bridge galvanometers indicator, we can study and analysis of resistance to be measured with variations of factors, which is the function of non-balance electric bridge.

Thermal resistor is a temperature change of sensitive semiconductor element, when the ambient temperature changes, its resistances change. The relationship between semiconductor thermistor resistance and temperature can be expressed as:

$$R_T = R_0 e^{B/T}$$

Where R_T is the resistance of thermistor, R_0、B is the constant, by the shape and material of thermistor.

The thermal resistance R_T as tested resistance connected with a Wheatstone bridge arm, in a certain temperature (0℃ and room temperature) regulating bridge is balanced, then change the thermistor temperature. We research the relationship with the temperature of thermistor resistance and bridge current with unbalanced bridge, a qualitative analysis of variation of thermistor and temperature. According to the relationship between thermistor temperature and current bridge in the non-balance state, semiconductor thermistor thermometer can be made. The calibration curve was plotted using temperature and current relationship, and then the method for measurement of current in the standard curve can be obtained rapidly the measured temperature.

【Procedure】

1. Measure the resistance with the slide-wire Wheatstone bridge

(1) Interconnect the circuit according to Fig. 2-7-3, and adjust the output voltage of the current source to 4.5V.

(2) Adjust R' to make U_{AB} at the range of 1.5V to 2.0V (measure with multimeter).

(3) Select a value for R_0 corresponding R_x (In order to diminish the relative error, the value of R_0 is respectively equal to R_x, $1/2R_x$, $2R_x$), and move D to make the bridge balanced, then record the value of I_1、I_2、R_0 in Tab. 1.

(4) Select the resistances R_{x_1}、R_{x_2}、R_{x_3} at discretion (The order of magnitude of them is respectively equal to 10, 10^2, 10^3. Loose measure them with multimeter.), and measure each of them five times according to the third step. Then calculate the value of R_X.

Note: When measuring resistance with slide-wire Wheatstone bridge, if the indicator on galvanometer stops moving, one should determine if it is caused by open circuit or a real bridge balance. When searching for balance point, one can estimate the position first. When moving the electricity key D, a method called leap-and-connect should be used. The benefit of this method is to avoid long-time current flow through electric bridge, and to be able to shut off power supply promptly in case of unusual situation. When using this method, bridge balance position can be quickly found based on the swinging director of galvanometer indicator.

2. Measure the resistance with the box type Wheatstone bridge.

(1) Be familiar with the usage method and action of knobs, wiring terminals and keys.

(2) Select a direct-current-stabilized voltage supply to output a voltage of 4.5V. We can measure current with the galvanometer of the box type electric bridge. Connect every

knob, and select a proper ratio arm, then measure R_{x_1}、R_{x_2}、R_{x_3} in Tab. 2.

(3) Record the accuracy of the bridge (on the bottom of the box or the inside of lid), then calculate the error ΔR_x with the equation below:

$$\Delta R_x = R_x \cdot \text{accuracy}$$

3. Study the relationship between the current of the bridge and the thermal resistor with box type Wheatstone bridge (optional).

【Data and Calculations】

1. Measure the resistance with the slide-wire Wheatstone bridge (Tab. 2-7-1)

Tab. 2-7-1 Data of measurement with the slide-wire Wheatstone bridge

standard value of R_x (Ω)	l_1 (cm)	l_2 (cm)	R_0 (Ω)	R_x (Ω)	$\bar{R}_x$ (Ω)	ΔR_x (Ω)
1						
2						
3						

The result: $R_x = \bar{R}_x \pm \Delta R$

2. Measure the resistance with box type Wheatstone bridge (Tab. 2-7-2).

Tab. 2-7-2 Data of measurement with box type Wheatstone bridge

standard value of R_x (Ω)	value of ratio arm	R_0 (Ω)	R_x (Ω)	ΔR_x (Ω)

The result: $R_X = \bar{R}_X \pm \Delta R$

【Questions】

(1) When the positions of the power source and galvanometer are changed with each other, can the electric bridge still keep its balance or not? Try to prove it.

(2) If connect the same resistance to measure and the same R_0 with different degree of tightness, what will happen to the result?

(3) In Wheatstone bridge circuit, when gliding end slides on resistance AB, galvanometer indicator does not move at all. If parallel connects a voltmeter between AB, galvanometer begin to indicate changes. Try to analyze which lead is broken.

(Li Yu-juan, Zhao Yao)

实验八 RLC 交流电路

【实验目的】

（1）测量电容器的电容和线圈的电感。

（2）测量日光灯电路的功率因数，了解提高功率因数的方法。

【实验器材】

日光灯、可变标准电感箱、电阻箱、电容器、交流安培计、交流伏特计、瓦特计、自耦变压器等。

【实验原理】

1. RLC 串联电路

图 2-8-1 为 RLC 串联电路。当电路接通 50Hz 的交流电源后，设电路的总电流和总电压的有效值各为 I 和 U，则电路的总阻抗

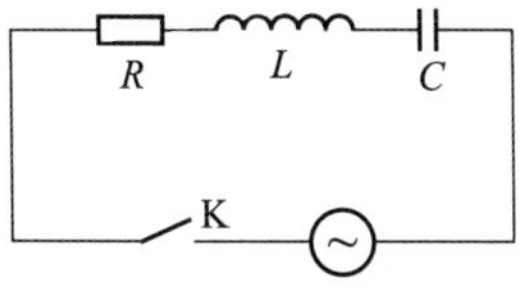

$$Z = \frac{U}{I} = \sqrt{R^2 + \left(\omega L - \frac{1}{\omega C}\right)^2} \quad (2-8-1)$$

图 2-8-1 RLC 串联电路

其中 $\omega = 2\pi f = 2 \times 50\pi = 314$（$s^{-1}$）。总电压与各元件上电压有下列关系：

$$U = IZ = \sqrt{(IR)^2 + \left(I\omega L - \frac{I}{\omega C}\right)^2} = \sqrt{U_R^2 + (U_L - U_C)^2} \quad (2-8-2)$$

当电路仅有电阻、电感串联时，则总电压

$$U_{RL} = \sqrt{(IR)^2 + (I\omega L)^2} = \sqrt{U_R^2 + U_L^2}$$

所以

$$L = \frac{1}{\omega}\sqrt{\left(\frac{U_{RL}}{I}\right)^2 - R^2} \quad (2-8-3)$$

实验线路中若不另外串联电阻，则上式中的 U_{RL} 为 L 两端的电压，R 为 L 线圈本身的直流电阻。当电路仅有电阻、电容串联时，则总电压

$$U_{RC} = \sqrt{(IR)^2 + \left(\frac{I}{\omega C}\right)^2} = \sqrt{U_R^2 + U_C^2}$$

所以

$$C = \frac{I}{\omega\sqrt{U_{RC}^2 - U_R^2}} \quad (2-8-4)$$

式中，U_{RC} 为 R、C 串联后两端的电压。

2. 日光灯电路、功率和功率因数

图 2-8-2 为常用的日光灯电路，图中 L 为日光灯管，D 为镇流器，S 为启辉器（又叫起动器），C 为电容器。日光灯管等值阻抗一般可以认为是纯电阻性，设其阻值为 R。而镇流器是有铁芯的电感线圈，它在电路中同时有电阻 r 和电感 L 两个控制电流的因素，因此其等效电路相当于电阻和电感的串联。当开关 K_C 未按下时，接通交流电

源的两端,就组成一个 RLr 串联电路。设总电压为 U,灯管的电压为 U_R,镇流器的电压为 U_{rL},则日光灯电路的电压由电压三角形矢量图 2-8-3 得

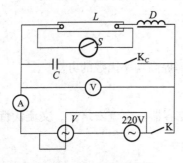
图 2-8-2 日光灯电路

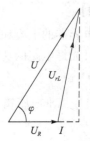
图 2-8-3 电压三角形矢量图

$$U_{rL}^2 = U^2 + U_R^2 - 2UU_R\cos\varphi \qquad (2-8-5)$$

式中,$\cos\varphi$ 为日光灯电路的功率因数;φ 为流过负载的电流 I(即线路上的电流)与总电压之间的位相差。这时电路的有功功率为

$$P = IU\cos\varphi \qquad (2-8-6)$$

无功功率为

$$Q = IU\sin\varphi = I^2 X_L \qquad (2-8-7)$$

式中,X_L 为镇流器的感抗。因 $X_L = \omega L$,所以由上式得电感

$$L = \frac{U\sin\varphi}{I\omega} = \frac{U\sqrt{1-\cos^2\varphi}}{I\omega} \qquad (2-8-8)$$

当按下开关 K_C 时,即将电容 C 并联到电路中,这时由于增加了一相位超前于电压 U 为 $90°$ 的电流 I_C,所以线路上的总电流不再是 I,而是 I 与流过电容器的电流 I_C 的矢量和 I',如图 2-8-4 所示,即 $I' = I + I_C$,φ' 为 I' 滞后电压 U 的位相差,故有功功率为

$$P' = I'U\cos\varphi' \qquad (2-8-9)$$

又如图 2-8-4 可见,$I' < I$,$\varphi' < \varphi$,因而功率因数 $\cos\varphi' > \cos\varphi$。从供电角度来看,要求负载的功率因数愈大愈好。由于线路上电流的减小,在传输导线上损耗的电能也就减少了。

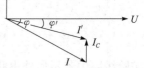
图 2-8-4 电流与电压的相位关系

在实验时,若测得电源电压 U,灯管电压 U_R,镇流器电压 U_{rL} 及每一次的总电流 I 和 I',由式(2-8-5)、(2-8-6),即可算出上述两种情况下的功率因数 $\cos\varphi$ 和 $\cos\varphi'$,继而了解提高功率因数的意义和方法。

【实验步骤】

1. 测量电感和电容

(1)将各元件按图 2-8-1 联接,开关 K_C 暂不按下,检查线路是否正确地接好。

(2)将电路中电容器的两端 C、D 短路,接通电源。调节变阻器,使电压渐增,保持电流不超过负载的额定值。测出电流 I 和 L 两端电压 U_{rL} 的数值,连同 L 本身的已知电阻 R 代入式(2-8-3)求出电感。

（3）将电路中 L 的两端 B、C 短路，并接入已知电阻 R，再接通电源。调节变阻器，使电压渐增，保持电压不超过负载电容器的额定值。测出电流 I 和串联电路的端电压 U_{RC} 的数值，连同串联电阻 R 代入式（2-8-4）中求出电容 C。

2. 测日光灯的功率和功率因数

（1）按图 2-8-2 日光灯电路接好线路，检查无误后接通电源，按下开关 K，使日光灯发光。

（2）首先断开开关 K_C，不接入电容器 C，测出总电压 U、灯管电压 U_R、镇流器电压 U_L、及电流 I 值。代入式（2-8-5）、（2-8-6）、（2-8-8）算出电路的功率因数 $\cos\varphi$、有功功率 P 和镇流器的电感 L。再用瓦特计测电路的功率，比较两种方法的结果，分析其原因。

（3）将 K_C 按下，接入电容器 C，测出电压 U 及电流 I' 值，代入式（2-8-9）算出具有并联电容器时的功率因数 $\cos\varphi'$。

【思考题】

（1）为提高日光灯电路的功率因数，在线路上并联电容器，其数值是否愈大愈好？

（2）若已知电路的功率为 P，欲使功率因数由 $\cos\varphi$ 提高到 $\cos\varphi'$，应该并联多大的电容 C（已知电源频率为 f，电压为 U）？

（3）结合实验六，如何用示波器测量一般交流电路的电压、频率、电感和电容？

【仪器介绍】

瓦特计的结构原理和使用方法：电动式瓦特计内部装有两个线圈（图 2-8-5），其中线圈 A 是固定线圈，测量时与待测负载串联正如一安培计，故叫作电流线圈；线圈 B 是可转动的，它和指针相连，本身的电阻也较大，而且往往串联上扩大量程用的高电阻，测量时它与待测负载并联，正如一伏特计，故叫作电压线圈。使用时电路联接如图 2-8-6 所示。

设负载两端电压 $U = U_0\sin\omega t$，通过负载的电流 $I = I_0\sin(\omega t + \varphi)$，线圈 A 与负载串联，通过它的电流也为 I，而线圈 B 与负载并联，通过它的电流为 $\dfrac{U}{r}$，式中 r 为电压线圈的电阻（或者还加上扩大量程的电阻）。活动线圈所受力矩与两个线圈的电流的乘积成正比，故某一时刻所受的力矩

$$M = KI\frac{U}{r} = KI_0U_0\frac{1}{r}\sin\omega t\sin(\omega t + \varphi)$$

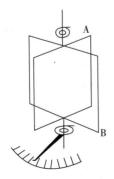

图 2-8-5　瓦特计线圈

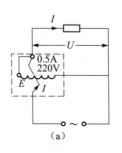 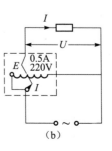

图 2-8-6　瓦特计电路联接图

平均力矩

$$M = K\frac{IU}{r}\cos\varphi = K'P$$

式中，$K' = \dfrac{K}{r}$是一常数，故所受平均力矩和负载的平均功率成正比，从瓦特计的刻度，可以读出负载的电功率数值。如果同时用瓦特计、交流安培计、交流伏特计测出平均功率 P、电流 I 和电压 U，由式（2-8-6）可算出功率因数。

图 2-8-7 是一种瓦特计的外观图。瓦特计的面板上都标准注有额定电压和额定电流。不管待测的功率大小如何，加在电压线圈上的电压和通过电流线圈的电流不得超过各自的额定值，否则会烧毁线圈。

图 2-8-7 瓦特计接线端钮图

（孙宝良　李百芳）

Experiment 8　*RLC* Alternating Current Circuit

【Objects】

(1) To measure capacitor's capacitance and coil's inductance.

(2) To determine power factor of daylight lamp and find out methods of enhancing power factor.

【Apparatus】

Daylight lamp, various standard inductance boxes, resistance box, capacitor, an ac ammeter, an ac voltmeter, wattmeter, autotransformer, etc.

【Theory】

1. Series *RLC* circuit

A series $R-L-C$ circuit is shown in Fig. 2-8-1. After the switch is closed with the ac source (50 Hz), suppose the ac current and the ac voltage are I and U (meter reading), so the impedance of the circuit is

$$Z = \frac{U}{I} = \sqrt{R^2 + \left(\omega L - \frac{1}{\omega C}\right)^2} \quad (2-8-1)$$

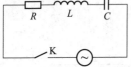

Fig. 2-8-1

where $\omega = 2\pi f = 2 \times 50\pi = 314$ (s^{-1}). The relation between total voltage and voltage across each component in the series circuit is:

$$U = IZ = \sqrt{(IR)^2 + \left(I\omega L - \frac{I}{\omega C}\right)^2} = \sqrt{U_R^2 + (U_L - U_C)^2} \quad (2-8-2)$$

If there are only resistance and inductance in series, then total voltage is:

$$U_{RL} = \sqrt{(IR)^2 + (I\omega L)^2} = \sqrt{U_R^2 + U_L^2}$$

So
$$L = \frac{1}{\omega}\sqrt{\left(\frac{U_{RL}}{I}\right)^2 - R^2} \qquad (2-8-3)$$

If there is not other series resistor in the circuit, then U_{RL} in (2-8-3) is voltage through L, and R is pure resistance of the coil L.

If there are only pure resistance and capacitor in the series circuit, then total voltage is:
$$U_{RC} = \sqrt{(IR)^2 + \left(\frac{I}{\omega C}\right)^2} = \sqrt{U_R^2 + U_C^2}$$

So
$$C = \frac{I}{\omega \sqrt{U_{RC}^2 - U_R^2}} \qquad (2-8-4)$$

in type (2-8-4), U_{RC} is total voltage of R and C in the series circuit.

2. Daylight lamp circuit, power Loss and power factor

Frequently-used daylight lamp circuit is shown in Fig. 2-8-2, where L is daylight lamp, D is ballast resistor, S is starter, and C is capacitor. Equivalence impedance of daylight lamp can be considered pure resistance, and suppose its resistance is R. and that ballast resistor is an inductance coil with an iron-core in it, its electric current in the circuit is associated with resistance r and inductance inductor L, so its ac circuit amount to resistance and inductance in series. With switch K_C not being pressed, switch on the ac power source, a series $R-L-r$ circuit is constituted.

Suppose the total voltage is U, voltage of daylight lamp is U_R, voltage of ballast resistor is U_{rL}, then from triangle vectorgraph of voltage, as shown in Fig. 2-8-3, we get
$$U_{rL}^2 = U^2 + U_R^2 - 2UU_R\cos\varphi \qquad (2-8-5)$$

in type (2-8-5), $\cos\varphi$ is called the power factor in daylight lamp circuit, and φ is the phase angle between the current I of load and the voltage U. So the power loss in the impedance is given by

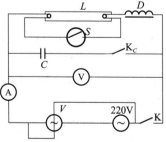

Fig. 2-8-2

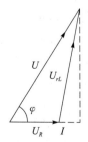

Fig. 2-8-3

$$P = IU\cos\varphi \qquad (2-8-6)$$

reactive power is:
$$Q = IU\sin\varphi = I^2 X_L \qquad (2-8-7)$$

Where X_L is inductance of the ballast resistor. Because of $X_L = \omega L$, so the inductance can be worked out from type (2-8-7)

$$L = \frac{U\sin\varphi}{I\omega} = \frac{U\sqrt{1-\cos^2\varphi}}{I\omega} \qquad (2-8-8)$$

When switch K_C is pressed, capacitor C is connected to the circuit, due to adding an electric current I_C whose phase angle leads 90° than voltage U, so the total current I' is the sum of vector of I_C and instead I with I', as shown in Fig. 2-8-4, namely $I' = I + I_C$, φ' is phase angle that I' lags behind voltage U, so the power loss in the circuit is

$$P' = I'U\cos\varphi' \qquad (2-8-9)$$

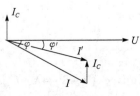

Fig. 2-8-4

As shown in Fig. 2-8-4, $I' < I$, $\varphi' < \varphi$, so the power factors $\cos\varphi' > \cos\varphi$. In the experiment, if supply voltage U, voltage of daylight lamp U_R, voltage of ballast resistor U_{rL}, total current I and I' are measured each time, $\cos\varphi$ and $\cos\varphi'$ can be calculated from type (2-8-5) and type (2-8-6), and moreover we can understand the meaning and method of increasing power factor.

【Procedure】

1. Measure inductance and capacitance

① Link components based on Fig. 2-8-1, and switch K_C is not be pressed at the moment, then check up the circuit.

② Short out circuits C and D of the capacitor, and switch on the power source. Regulate rheostat to increase voltage gradually, and keep electric current below the rating value of load. Measure electric current I and voltage U_rL, as well as known resistance, and we can get the inductance of L from type (2-8-3).

③ Short out circuits B and C of inductor L and cut-over resistance R, then switch the power source. Regulate rheostat to increase voltage gradually, keep voltage below the rating value of load. Measure electric current I and voltage U_{RC}, as well as known resistance, and we can get the capacitance C from type (2-8-4).

2. Determine power, and power factor of daylight lamp

① Interconnect links as shown in Fig. 2-8-2, after checking connect to the power source, press K down to make the daylight lamp shine.

② First, disconnect switch K_C, while do not cut-over capacitor C, and measure total voltage U, voltage of the daylight lamp U_R, voltage of ballast resistor U_{rL} and current I. Substitute them in (2-8-5), (2-8-6) and (2-8-8), and cipher out the power factor $\cos\varphi$, the power loss P and inductance of the ballast resistor L. And then measure the power of the circuit with wattmeter again, while compare the results of two methods and analyze the reasons.

③ Press K_C down, cut-over capacitor C, and measure total voltage U and current I'. Substitute them in type (2-8-9), and calculate the power factor $\cos\varphi'$ with paralleling capacitor.

【Questions】

(1) To connect capacitor in parallel for increasing power factor of daylight lamp, please

answer if it is good for a bigger value.

(2) Suppose the power loss of a circuit is P, how much capacitance should be connected in parallel to increase power factor from $\cos\varphi$ to $\cos\varphi'$ (frequency of power source is f, voltage is U)?

(3) Combine with experiment 6, how will you measure voltage, frequency, inductance and capacitance?

(Zhi Zhuang – zhi, Sun Bao – liang)

实验九　用线式电位差计测电池的电动势

【实验目的】

(1) 了解电位计的原理（补偿法）和滑线式电位计的构造。
(2) 用滑线式电位计测电池的电动势及其内阻。

【实验器材】

直流稳压电源、滑线板、标准电池、待测电池（1.5V 干电池一节）、检流计、滑线变阻器、电阻箱、保护电阻、万用表、开关、导线等。

【实验原理】

1. 补偿法原理

电压表可以测量电路各部分电压，但不能测量电源的电动势，因为电压表并联在电源的两端时（图 2-9-1），根据闭合电路的欧姆定律有 $V = E_X - Ir$，因而用电压表只能测出电池的端电压，要测量电池的电动势必须设法使待测电池中通过的电流 I 为零。将一个电动势可任意调节的电源 E_i 和待测电池 E_X 按图 2-9-2 的线路连接，调节 E_i 使检流计 G 指零，那么 E_i 和 E_X 的电动势一定是大小相等方向相反，称这时的电路达到了补偿。E_i 称为补偿电压，E_i 与 E_X 组成的回路称为补偿回路，如果 E_i 已知，则可求出 $E_X = E_i$，这种测量电动势的方法称为补偿法。用补偿法测电动势或电压时补偿电路中没有电流，所以不影响被测电路的状态，这是补偿法测量的最大优点。

图 2-9-1　路端电压与电源的电动势

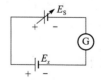

图 2-9-2　补偿电路图

2. 滑线式电位计工作原理

电位计就是根据补偿法原理构成的测量电动势和电势差的仪器。图 2-9-3 所示为实验室所用一种滑线式电位计的基本电路。

图中 a、b 为一均匀电阻丝，开关 K 合上后，电池 E 的两端接在 a、b 两点上供给稳定的工作电流（R 用来调节其电流的大小），将已知电动势为 E_s 的标准电池接于电路中后（K_2 合向 S 端），调节滑动接头 C 到某一位置 c_1，使检流计 G 中无电流通过（R_1 为保护电阻），因此标准电池的电动势 E_S 等于 ac_1 两点间的电势差，这一电势差就相当于图 2-9-2 中的补偿电压 E_i。如果 ac_1 段长度为 l_1，电阻为 r_1，通过 ab 线的电流为 I，则 $E_S = Ir_1$；使 K_2 合向 X 端，即以电动势为 E_X 的待测电池代替标准电池，仍如前调节 C 点，使检流计 G 指零。设此时 C 点位置在 c_2，ac_2 长度为 l_2，电阻为 r_2，则 $E_X = Ir_2$。

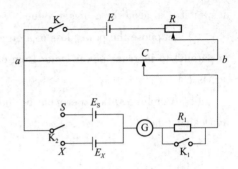

图 2-9-3 测电动势电路图

由以上两式得

$$\frac{E_X}{E_S} = \frac{r_2}{r_1}$$

由于均匀导线电阻 r_2 与 r_1 之比等于长度 l_2 与 l_1 之比，所以

$$\frac{E_X}{E_S} = \frac{l_2}{l_1}$$

即

$$E_X = \frac{l_2}{l_1} E_S \qquad (2-9-1)$$

上式中，标准电池电动势 E_S 已知，测得在同一工作电流下电位计处于补偿状态时的 l_1 和 l_2 代入上式，可求出待测电池的电动势 E_X。

3. 电池内阻的测定

导线的电阻一般是用惠斯通电桥测定的，但是电池由于有电动势的作用，不能用惠斯通电桥测电池的内阻。用电位计就可以测定电池的内阻。

将图 2-9-3 中 E_X 换成图 2-9-4 中的电阻为 R' 的电阻箱（已知）与待测电池 E_X 并联的电路，其余部分不变。当其与 K_2 接通时，合上 K_3 调节 C 的位置到 C' 使检流计 G 中无电流通过，设电池内阻为 r，R' 两端电势差为 E'，则

$$E' = I'R'$$

式中，I' 为 R' 与 E_X 所构成回路中的电流强度，显然

$$I' = \frac{E_X}{R' + r}$$

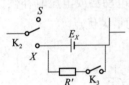

图 2-9-4 电位计测电阻电路图

因此，$E' = \dfrac{E_X}{R' + r} \cdot R'$

解此式得

$$r = \frac{E_X - E'}{E'} \cdot R' \qquad (2-9-2)$$

设 $ac_3 = l_3$，则 $\dfrac{E_X}{E'} = \dfrac{l_2}{l_3}$，此式两边各减 1，则有

$$\frac{E_X - E'}{E'} = \frac{l_2 - l_3}{l_3}$$

代入（2-9-2）式中得

$$r = \frac{l_2 - l_3}{l_3} \cdot R' \qquad (2-9-3)$$

所以，只要测出 l_2、l_3 之值，则电池内阻 r 就可由式（2-9-3）计算出来。

【实验步骤】

1. 测定待测电池的电动势

（1）按图 2-9-3 连接电路。接线时需断开所有开关，R 置于最大值，并特别注意工作电源 E 的正负极应与标准电池 E_S 和待测电池 E_X 的正负极相对，否则检流计 G 的指针总不会指零。（E = 6V）

（2）经指导教师检查后，接通电源，改变滑线变阻器 R 的滑动头，调节工作电流，此时，可在 ab 间以万用表测量电压 $U_{ab} \approx 2$ 伏即可，并在后面测量中，不再调节滑线变阻器，即保持 ab 中的电流为一定值。保护电阻 R_1 先置于最大值，以保护检流计 G。

（3）将 K_2 合向 S 端，然后接滑动接头 C，观察检流计指针的偏转，移动滑动接头 C，找到检流计大概不动的位置（粗调）。再逐渐减小保护电阻 R_1 的阻值，以至闭合 K_1 使阻值为零，找到检流计指针不动时 C 点准确位置 c_1，记录 ac_1 之长 l_1。

（4）使 R_1 为最大值，将 K_2 合向 X 端，按步骤 3 找出检流计 G 指零时点 C 的准确位置 c_2，记录 ac_2 之长 l_2。

（5）重复测量 l_1、l_2 各三次。由公式（2-9-1）计算待测电池电动势 E_X。

2. 测定电池的内阻（选作）

（1）按图 2-9-4 连好电路，使 R_1 为最大电阻，在电阻箱上取 $R' = 10\Omega$。

（2）将 K_2 接通，按上述步骤（3）找出检流计 G 指零时 C 点位置 c_3，记录 ac_3 之长 l_3。

（3）打开 K_1、K_2，改变电阻箱电阻 $R' = 20\Omega$，按上步骤再做一次，记录 ac_3' 之长 l_3'。

（4）将 l_3 及 l_3' 及对应 R' 值代入公式（3），求出电池内电阻，并求其平均值 $\bar{r}$。

【数据记录及处理】

1. 测定待测电池的电动势（表 2-9-1）

表 2-9-1 测定待测电池的电动势

项目 \ 次数	1	2	3	平均值	绝对误差	相对误差
l_1（cm）						
l_2（cm）						

$$\bar{E}_X = \frac{\bar{l}_2}{\bar{l}_1} \cdot E_S$$

$$\Delta E_X = \left(\frac{\Delta l_1}{\bar{e}_1} + \frac{\Delta l_2}{\bar{e}_2} \right) \cdot \bar{E}_X$$

测量结果：$E_X = \bar{E}_X \pm \Delta E_X$

2. 测定电池的内阻（自设表格）

【注意事项】

（1）使用电位计必须先接通工作回路（辅助回路），然后再接补偿回路。断电时必须先断补偿回路，再断开工作回路。

（2）实验中电动势 E 必须大于 E_S、E_X 的电动势，各电池的正极必须接在一起，否则不能达到平衡。

（3）寻找补偿点时，可先估计大概位置，移动滑动头 C 采用跃接法。在未达 G 中指针基本不动前不得减小保护电阻 R_1，以免损坏检流计。

（4）标准电池只能短时间通过 1μA 左右电流，否则将影响标准电池的精度或造成损坏，它只能用作电动势标准，而不能用它来供电。使用时应特别注意不可通过较大电流，不可较长时间通电，也不可用伏特计测量其两极间电势差。因标准电池是半流体的（内部有电解液），因此不得振动、倾斜或倒置。其电动势 $E_S = 1.018V$，精确值每支标准电池上都标出，温度不同时需校正，具体方法见实验二十二【实验步骤】（2）。

【预习要点】

为什么电动势 E 必须大于 E_S、E_X？为什么它们不能正极和负极相连？

【思考题】

（1）用滑线式电位计测量电池电动势实验中，如果发现检流计指针总向一个方向偏转而无法调到平衡，试分析此故障的产生有哪些可能的原因，并提出排除故障的方法。

（2）用电位计测电势差的方法有何优点？本实验的滑线式电位计还有什么引起误差的因素？

（李玉娟　孙宝良）

Experiment 9 Measurement of the Battery EMF with Wire Potentiometer

【Objects】

(1) To understand the principle (compensation method) of potentiometer and the structure of slide – wire potentiometer.

(2) To measure the electromotive force and internal resistance of a battery with slide – wire potentiometer.

【Apparatus】

Continuous stabilized voltage supply, slide wire board, normal element, element to measure (dry cell of 1.5V), galvanometer, slide – wire rheostat, resistance box, protective resistance, multimeter, switch, wire, etc.

【Theory】

1. The principle of compensation method

The voltmeter can measure the voltage of each part of the circuit except the electromotive force of the power source. When the voltmeter is parallel with the power source (Fig. 2 – 9 – 1), according to the Ohm law we know $V = E_X - Ir$, so the voltmeter can only measure the terminal voltage of the cell. If we want to measure the electromotive force of the power source, the current through the cell must be zero. Connect a power source E_i that can adjust at random with the power source to measure E_X, according to Fig. 2 – 9 – 2. Adjust E_i to make the displaying value of the galvanometer G become zero. Then the values of the electromotive force E_X and E_i must be equal, and the direction of them must be opposite. At that time we said the circuit has been compensated, and E_i is called offset voltage, while the circuit is called compensated circuit. If the value of E_i has been known, the value of E_X can be worked out. This method is compensation method. When we measure the electromotive force or voltage by this method, there is no current though the circuit, so the state of the circuit will not be affected which is the major advantage of the compensation method.

Fig. 2 – 9 – 1 Terminal voltage and the electromotive force of the power supply

Fig. 2 – 9 – 2 Compensation circuit

2. Slide – wire potentiometer principle

Potentiometer is an instrument to measure electromotive force and electric potention voltage based on compensation principle. The basic electric circuit of a commonly used potentiometer in lab is shown in Fig. 2 – 9 – 3. As shown in the picture, a and b is a uniformly distributed resistance. When switch K is turned on, the cell connects to a and b, and provides a constant current (R is used to adjust magnitude of current). When connecting a standard battery with a known electromotive force E_S into the circuit (K_2 is closed to end S), adjust slider end C to a certain position c_1, there will be no current flowing through galvanometer G (R_1 is the protection resistance). Thus, the electromotive force E_S of the standard battery equals to the potential between two ends of ac_1, and this potential is equal to the compensation voltage E_i shown in Fig. 2 – 9 – 2. If the length of ac_1 is l_1, its resistance is r_1, and the current in ab is I, thus $E_S = Ir_1$. Now close up K_2 to end X, and use the subjected battery with electromotive force E_X in place of the standard battery. Adjust slider C as before until galvanometer indicates zero. Assume that end C is now at position c_2, length of ac_2 is l_2, and its resistance is r_2, then $E_X = Ir_2$.

From the two equations above

$$\frac{E_X}{E_S} = \frac{r_2}{r_1}$$

Since the resistance is uniformly distributed, the ratio between resistance r_2 and r_1 is the same as the ratio between length l_2 and l_1. So

$$\frac{E_X}{E_S} = \frac{l_2}{l_1}$$

Which is the same as:

$$E_X = \frac{l_2}{l_1}E_S \qquad (2-9-1)$$

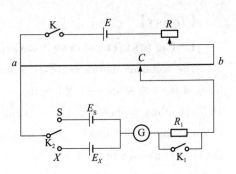

Fig. 2-9-3 Circuit diagram for measuring electromotive force

In the equation above, electromotive force E_S of the standard battery is known, and we measure the length of l_1 and l_2 when potentiometer is under compensation situation with the same working current. Substitute these into the equation above, and we can calculate the electromotive force E_X of the subjected battery.

3. The test of battery internal resistance

The resistance of wire is generally determined by the Wheatstone bridge, but because of electromotive force for the battery, the Wheatstone bridge can not be tested for the resistance of battery. Thus the potentiometer can test the resistance of battery.

Fig. 2-9-3 EX can be changed for resistance box R' (known) and EX parallel circuit batteries in Figure 2-9-4, and the rest is the same. When the K_2 is switched on, close K_3 regulation of C position so that no current through the galvanometer G. Supposed the internal resistance of battery is r, E' is the potential difference at both ends of R', so

$$E' = I'R'$$

I' is the strength of the current loop of R' and E_X, apparently

$$I' = \frac{E_X}{R' + r}$$

Thus, $E' = \frac{E_X}{R' + r}R'$

So
$$r = \frac{E_X - E'}{E'}R' \qquad (2-9-2)$$

Supposed $ac_3 = l_3$, so $\frac{E_X}{E'} = \frac{l_2}{l_3}$, both sides can subtract 1, so

$$\frac{E_X - E'}{E'} = \frac{l_2 - l_3}{l_3}$$

Substitution (2-9-2) formula in

$$r = \frac{l_2 - l_3}{l_3}R' \qquad (2-9-3)$$

So, as long as the values of l_2、l_3 can be measured, then the battery internal resistance r can be calculated by the formula 9-3.

【Procedure】

1. Measure the electromotive force of the cell

(1) Connect the circuit according to Fig. 2 – 9 – 3. Open all switches, adjust R to the max value, and the two poles of E must be opposite to that of E_S and E_X, otherwise the finger of the galvanometer can not point to zero ($E = 6V$).

(2) Request the teacher to inspect the circuit. Then switch on the power source, and remove the slick joint of R to make the voltage $U_{ab} \approx 2V$. Do not adjust the slide – wire rheostat again, namely keep the current of ab invariability. Adjust the protective resistance to the max value to protect the galvanometer G.

(3) Close K_2 to S. Then connect the slick joint C, observe the deflexion of the galvanometer, and move C to find the position that makes the indicating value of the galvanometer to zero (coarse adjustment). Diminish the value of the protective resistance R_1 gradually, till close K_1 to make the resistance be zero, find out the exact position c_1 that the finger of the galvanometer keep still, record the length l_1 of ac_1.

(4) Adjust R_1 to the max value, close K_2 to X, find out the place c_2 where the galvanometer finger points to zero, and record the length l_2 of ac_2.

(5) Repeatedly measure l_1 and l_2 three times. Work out the electromotive force E_X of the cell, according to equation (2 – 9 – 1).

2. Measure the internal resistance of the cell (alternative).

(1) As shown in Fig. 2 – 9 – 4 connection circuit, R_1 is the maximum resistance, $R' = 10$ ohm resistance box.

(2) K_2 is switched on, according to the above steps (3) to find the galvanometer G refer to zero C point of c_3, record the length l_3 of ac_3.

(3) Open K_1, K_2, change the resistance box $R' = 20$ ohm, according to the above steps to do it again, record the length l_3 of ac_3.

(4) Substitute l_3 and l_3' and the corresponding value of R' into formula (3), the internal resistance of battery and the average value can be calculated.

【Data and Calculations】

To measure the electromotive force of the cell is shown in Tab. 2 – 9 – 1.

Tab. 2 – 9 – 1 Measure the electromotive force of the cell

	1	2	3	average value	absolute error	relative error
l_1 (cm)						
l_2 (cm)						

$$\bar{E}_X = \frac{\bar{l}_2}{\bar{l}_1} \cdot E_S$$

$$\Delta E_X = \left(\frac{\Delta l_1}{l_1} + \frac{\Delta l_2}{l_2}\right) \cdot \bar{E}_X$$

The result: $E_X = \bar{E}_X \pm \Delta E_X$

【Note】

(1) Switch on the working circuit first, then switch on the compensating circuit. When turn off the power, break the compensable circuit first, then break the working circuit.

(2) The electromotive force E must be greater than that of E_S and E_X and the anodes of the cells must connect together, otherwise the circuit can not keep its balance.

(3) Account the place first when looking for the compensate point. Do not diminish the protective resistance R_1, before the finger G points to zero.

(4) The normal cell can only go through the current about $1\mu A$ for short time, otherwise its precision may be affected, and it can only be the normal electromotive force, but not the power supply. When using it, do not allow the current through it to be too strong or last too long time, nor to measure the electric potential difference of its two poles with voltmeter. Because the cell is semiliquid (some electrolyte in it), and it cannot be vibrated, leaned or converted. Its electromotive force $E_S = 1.018V$ (The exact value has been marked on every cell, and it needs to be corrected when the temperature changes. The specific method is in the experiment twenty – two [step])

【Preview Point】

why the electromotive force E must be greater than E_S、E_X? Why they cannot be connected to the cathode and anode?

【Questions】

(1) Why the electromotive force E must be lager than E_S and E_X? Why they cannot connect their positive poles with negative poles?

(2) What advantage does the method measuring the electric potential difference with potentiometer have? Which factors of the slide – wire potentiometer can cause error?

(Li Yu – juan, Zhao Yao)

实验十　透镜曲率半径的测量

【实验目的】

(1) 观察牛顿环上产生的干涉现象，进一步理解等厚干涉的形成及其特点。
(2) 掌握用牛顿环测量平凸透镜曲率半径的方法。
(3) 学会使用读数显微镜。

【实验器材】

牛顿环、读数显微镜、钠光灯等。

【实验原理】

牛顿环仪上透镜凸面 CD 与平玻璃板 AB 面的接触点为 O，它们之间形成一极薄的空气层，空气层厚度从接触点到边缘逐渐增加，且离接触点有相同半径的地方厚度相同。厚度相同的点在同一圆周上。

如图 2-10-1 所示，当用单色平行光垂直照射时，经空气层上、下表面反射的光具有一定的光程差，是两束相干光。它们相遇时将相互干涉，形成的干涉条纹是空气层的等厚各点的轨迹，这是一种等厚干涉。在反射方向观察，干涉条纹是以 O 为中心的明暗相间的同心圆环，其中心是暗斑。由于这种干涉条纹是牛顿最早发现的，故称牛顿环。如果在透射方向观察，则干涉条纹的强度分布正好与反射光的干涉条纹互补；中心由暗斑变成亮斑，原因是明条纹的地方变为暗条纹，暗条纹处变为明条纹，见图 2-10-2。设入射光波长为 λ，透镜的曲率半径为 R，空气层厚度为 e 的地方产生第 n 个暗环（从环中心 O 点数起），此暗环离 O 点半径为 r_n，由图 2-10-1 中的几何关系知

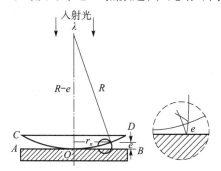

图 2-10-1　牛顿环仪结构图

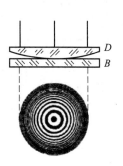

图 2-10-2　干涉图样

$$R^2 = (R - e)^2 + r_n^2 \qquad (2-10-1)$$

即

$$r_n^2 = 2Re - e^2$$

又

$$R \gg e$$

故可略去二阶小量 e^2，则有

$$r_n^2 = 2Re$$

或

$$e = \frac{r_n^2}{2R} \qquad (2-10-2)$$

由光路分析可知，第 n 级暗环所对应的两束相干光的光程差为

$$\delta = 2e + \frac{\lambda}{2} \qquad (2-10-3)$$

式中，$\frac{\lambda}{2}$ 是光线在 AB 表面反射时因半波损失而附加的光程差。把 (2-10-2) 式代入 (2-10-3) 式，则第 n 级暗环上各点处的光程差均为

$$\delta = \frac{r_n^2}{R} + \frac{\lambda}{2} \qquad (2-10-4)$$

由于光程差满足暗条纹的条件是

$$\delta = \frac{r_n^2}{R} + \frac{\lambda}{2} = (2n+1)\frac{\lambda}{2} \quad (2-10-5)$$

n 为条纹级次，于是得

$$r_n^2 = nR\lambda \quad (2-10-6)$$

或

$$\left(\frac{D_n}{2}\right)^2 = nR\lambda$$

式中，D_n 是第 n 级暗环的直径。当 $n=0$ 时，$r_n=0$，在理想情况下，牛顿环中心暗点应该是几何点，R、λ 一定时，n 越大，即环纹级次越高，两相邻暗环间距越小。条纹间距为

$$\Delta r = \sqrt{R\lambda}(\sqrt{n_{m+1}} - \sqrt{n_m})$$

如果入射光波长 λ 为已知，只要测出第 n 级暗环半径（或直径），就可由（2-10-6）式算出透镜曲率的半径。但是，由于透镜和玻璃板接触时有接触压力会引起弹性形变。使接触点不可能是一个几何点，而是一个圆面。有时接触处还附着灰尘，会引起附加程差，接近圆心处环纹就比较模糊。这样很难测准环纹的级次 n 和精确测出其半径 r_n，以致使条纹级数 n 和 r_n 与实际不符，这样会给测量带来较大的系统误差，为了消除上述误差，可取相隔 $(n-m)$ 环的两个暗环的直径差来计算：

第 n 级暗环：$\left(\dfrac{D_n}{2}\right)^2 = nR\lambda \quad (2-10-7)$

第 m 级暗环：$\left(\dfrac{D_m}{2}\right)^2 = mR\lambda \quad (2-10-8)$

（2-10-8）式 -（2-10-7）式：得 $R = \dfrac{D_m^2 - D_n^2}{4(m-n)\lambda}$，当波长已知时，测定第 n 级和第 m 级暗环的直径 D_n 和 D_m，由上式可以计算透镜的曲率半径 R；反之，若知道曲率半径，就可以计算波长 λ。

【实验步骤】

（一）调整测量装置

实验装置如图 2-10-3 所示，其主要部分为牛顿环，钠光灯及能精确测量微小间距的读数显微镜。

（1）调节牛顿环仪：牛顿环镜面如不干净需先用擦镜纸轻轻擦净，然后可调节三个紧固螺旋，使之用眼睛可直接见到干涉条纹，且使干涉条纹的圆形大致落在透镜中心。注意调节时要松紧得当，太松则条纹不稳定，太紧则变形大，会损伤透镜甚至可使透镜破裂。

（2）让读数显微镜读数指示在主尺中点附近，打开钠光灯并使之对准 45°玻璃片，将牛顿环置于镜下，调 45°玻璃片和钠光灯位置高低，使钠光灯射出来的光线照射到 45°玻璃片经反射后，接近垂直入射到牛顿环仪，再反射到读数显微镜。此时显微镜视场中亮度最大。

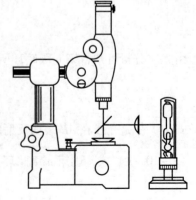

图 2-10-3 实验装置

(3) 先调显微镜目镜，看清里面的十字叉丝，然后用调焦旋钮对被测物进行调焦，方法是先下调镜筒使镜头接近被测物（牛顿环），要眼睛脱离目镜从侧面观察勿使镜头碰到牛顿环，然后用眼睛通过目镜观察，并使镜筒缓慢向上移动，直到从目镜中看到清晰的干涉条纹并且条纹与叉丝无视差（这样调节可避免镜头和待测物相碰）。

(4) 调节十字叉丝位置，旋转显微镜目镜筒，使十字叉丝中横丝与镜筒外面的主尺平行，然后缓慢移动牛顿环，使环纹中心与叉丝交点接近重合。

(5) 转动测微鼓轮，使叉丝从环纹中心向左、右移动的足够环数应大于要测的最大环级数，要求在这左右移动的范围内光照均匀，环纹清晰，叉丝横丝基本穿过直径与主尺平行，纵丝在两边移动中可与环纹相切（否则，应再细调45°玻璃片，聚焦及调牛顿环及钠光灯位置）。

（二）观察干涉条纹的分布特征

各级条纹粗细是否均匀，条纹间隔有无变化，牛顿环中心是亮斑还是暗斑，如何解释？

（三）测量牛顿环的直径

1. 测量时需考虑的问题

由于暗纹位置容易对准，所以对准暗纹测量，又由于接近中心的圆环宽度变化较大，不易测准，故测量时需尽量避开靠近中心的数环。同时为减小相对误差，级数差 $m-n$ 应适当取大些。此外，为减少误差需进行多次再取平均。

2. 测量方法

(1) m、n 值的选取：根据以上考虑，以取 5 组数据为例，例如级数 m 可依次取 50、49、48、47、49。级数 n 则可取 25、24、23、22、21。

(2) 测量各级暗环位置，将数据记录在表 2-10-1 内。

旋转测微鼓轮，使叉丝从环级中心往左移动，移出五十多环，然后反向旋转叉丝退回数环后，与第 50 条暗环相切，退回数环是为了消除因螺距间隙引起的空程差，记下对应读数（读数方法与螺旋测微器相同）。再旋转使纵丝继续右移，依次同各待测暗环相切（$m=50、49、48、47、46$；$n=25、24、23、22、21$），并记下读数填入表格。记录完中心左侧最后一个数据后，在继续右移叉丝过程中，要记准其离中心的环数 K，过中心后要使叉丝纵丝准确与右移第 K 级暗环相切，并记下数据。随后，使纵丝右移，依次同右侧各待测暗环相切，记下相应读数。

【数据记录及处理】

(1) 表 2-10-1 所示为参考数据表格，其中各环数为参考数，单色光波长 $\lambda = 589.3\text{nm}$。

表 2-10-1　数据记录

环的级数	m	50	49	48	47	46
环的位置（mm）	左					
	右					
环的直径（mm）	D_m					

续表

环的级数	n	25	24	23	22	21
环的位置（mm）	左					
	右					
环的直径（mm）	D_n					
D_m^2（mm²）						
D_n^2（mm²）						
$D_m^2 - D_n^2$（mm²）						

根据表 2-10-1 所测数据求出

$$\overline{D_m^2 - D_n^2} =$$
$$\Delta(D_m^2 - D_n^2) =$$

（2）测量结果及误差的计算

$$\overline{R} = \frac{\overline{D_m^2 - D_n^2}}{4(m-n)\lambda}$$

$$E = \frac{\Delta R}{\overline{R}} = \frac{\Delta(D_m^2 - D_n^2)}{D_m^2 - D_n^2} + \frac{\Delta(m-n)}{m-n}$$

$$\Delta R = E \cdot \overline{R} \qquad R = \overline{R} \pm \Delta R$$

式中，$\Delta(m-n) = \Delta m + \Delta n$。

Δm，Δn 是由于叉丝纵丝对准暗环纹中央所产生的对准误差，通过设此误差为条纹宽度的 1/10，故 $\Delta m = \Delta n = 0.1$，因而可以估计值 $\Delta(m-n) = \Delta m + \Delta n = 0.2$。

【注意事项】

（1）使用读数显微镜测一组数据时，只能从一个方向开始单向移动，即鼓轮应沿一个方向转动，中途不可倒转，以免引入螺距误差。

（2）仪器装置调好后，在测量过程中应避免碰动，特别要注意不可数错暗环数，否则要重新测量。

（3）实验完毕应将牛顿环上的三个紧固旋松开，以免长期受压力变形。

（4）钠光灯关闭后，必须稍等片刻才能重新打开。

【预习要点】

（1）透射光的牛顿环是如何形成的？如何观察？它与反射光的牛顿环在明暗上有何关系？为什么？

（2）在本实验中若遇下列情况，对实验结果是否有影响？为什么？

① 牛顿环中心是亮斑而非暗斑。

② 测直径 D 时，叉丝交点不通过圆环中心，因而测量的是弦而非直径。

【思考题】

（1）如显微镜目镜中视场不亮，应如何调节？

（2）在牛顿环实验中，假如平玻璃板上有微小的凸起，将导致牛顿环条纹发生畸变，试问该处牛顿环将局部内凹还是局部外凸？为什么？

（3）用白光照射时能看到牛顿环干涉条纹吗？此时条纹有何特征？

（4）用同样的实验方法，能否测定凹透镜的曲率半径？

【仪器介绍】

1. 牛顿环

牛顿环装置是一个曲率半径较大的平凸透镜和一块光学平玻璃所组成，通过紧固螺旋，使透镜的凸面中央部分与平玻璃接触，于是在透镜凸面和平玻璃间就形成一层空气薄膜，其厚度从接触点到边缘逐渐增加。当以平行单色光垂直入射时，入射光将在此薄膜上下表面反射，产生具有一定光程差的两束相干光。显然，它们的干涉图样是以接触点 O 为中心的一系列明暗相间的同心圆环，即牛顿环。

2. 读数显微镜

如图 2-10-4 所示，该仪器应放在牢固、平稳、无震动的工作台上，在 20℃±3℃ 条件下使用。被测工件放于台面玻璃 6 上，用弹簧压片 5 牢固压紧，并使工件的背面与台面全部接触。调整目镜

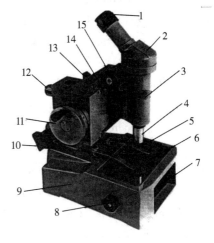

1. 目镜；2. 棱镜座；3. 镜筒；
4. 物镜；5. 弹簧压片；6. 台面玻璃；
7. 反光镜；8. 旋转手轮；9. 底座；
10. 立柱锁紧螺丝；11. 测微鼓轮；12. 横杆；
13. 横杆锁紧螺丝；14. 标尺；15. 调焦手轮

图 2-10-4 读数显微镜

1 使分划板刻线清晰可见，转动调焦手轮 15，从目镜观察使被测工件成像清晰。调整被测工件，利用测微丝杠移动瞄准显微镜，使被测部位的横向与显微镜的移动方向平行，即可读数。在标尺 14 上读取整数，在测微鼓轮 11 上读取小数，此二数之和即是此点的读数。如需改变观测位置，可将横杆锁紧螺丝 13 旋开，将观测系统转到所需的位置后，再用横杆锁紧螺丝 13 固定。

（支壮志，张 翼）

Experiment 10　Measurement of Radius of the Lens Curvature

【Objects】

（1）Observe the interference phenomenon produced by Newton's ring so as to further apprehend the formation and characteristics of equal thickness interference.

（2）Grasp the method of measuring the radius of curvature of planoconvex lens with Newton's ring.

(3) Learn to use reading microscope.

【Apparatus】

Newton's ring, a reading microscope, a sodium lamp.

【Theory】

The contact point of the plano-convex lens CD and flat glass plate AB is O, and an air gap is formed between them. The air-gap thickness increases gradually from the contact point to its edge, where the same thickness has equal radius, namely the same ring occurs at the same thickness.

As shown in Fig. 2 - 10 - 1, when monochromatic parallel light illuminates the lens, there will be an optical path length difference between the beam reflected from upper surface of air and the beam reflected from lower surfaces of air, and they are coherent beams. When they meet, they will interfere with each other. The interference pattern is a series track of the position at which the air-gap thickness is equal. And it is called the equal thickness interference. Observing from the reflection direction, we see a series of bright and dark rings centered on the contact point which is dark. It is called Newton's ring because it was found by Newton. If we observe from the transmitted light direction, we will see a complementary intensity corresponding to interference pattern of the reflected light, and the centre becomes bright. The position occurs with dark ring instead of initial bright ring and bright ring instead of initial dark ring, as shown in Fig. 2 - 10 - 2. We can assume that the incident light wavelength is λ, the radius of curvature is R, and the nth-order (count from the center O) dark ring occurs where the air-gap thickness is e, and the radius from this ring to the center O is r_n, from Fig. 2 - 10 - 1 we can conclude that

$$R^2 = (R - e)^2 + r_n^2 \qquad (2-10-1)$$

Then

$$r_n^2 = 2Re - e^2$$

and

$$R \gg e$$

Omit e^2, then

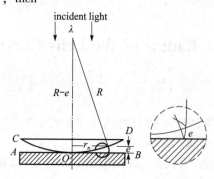

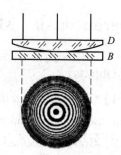

Fig. 2 - 10 - 1 Structure diagram of Newton ring apparatus

Fig. 2 - 10 - 2 Interference pattern

$$r_n^2 = 2Re$$

Or

$$e = \frac{r_n^2}{2R} \quad (2-10-2)$$

We know from the analysis of optical path that the optical path length difference of the two coherent beams corresponding to the nth order dark ring is:

$$\delta = 2e + \frac{\lambda}{2} \quad (2-10-3)$$

Where $\frac{\lambda}{2}$ is the extra optical path length for half-wave loss when beam is reflected at the surface AB, which is the boundary of a medium (glass) of higher refractive index. Substitute equation $(2-10-2)$ in equation $(2-10-3)$, and the optical path length difference of each point corresponding to the nth order dark ring is

$$\delta = \frac{r_n^2}{R} + \frac{\lambda}{2} \quad (2-10-4)$$

For cancellation

$$\delta = \frac{r_n^2}{R} + \frac{\lambda}{2} = (2n+1)\frac{\lambda}{2} \quad (2-10-5)$$

And n is the order of dark fringes, hence

$$r_n^2 = nR\lambda \quad (2-10-6)$$

Or

$$\left(\frac{D_n}{2}\right)^2 = nR\lambda$$

In the above equation, D_n is the diameter of the nth order. When $n=0$, $r_n=0$, under ideal condition, the centre of Newton's ring is geometrical point. When R and λ are given, the distance between the interfacing dark fringes will get less with n (the order of the dark fringe) increasing. The distance between the interfacing dark fringes is:

$$\Delta r = \sqrt{R\lambda}(\sqrt{n_{m+1}} - \sqrt{n_m})$$

If λ is given, we only need to measure the radius (diameter) of the nth dark ring, and the radius of curvature of the lens can be calculated. But when the lens contacts the glass plate, an elastic deformation will be caused because of the contact force. So the contact point can't be a geometrical point but a round plane. Sometimes the dust will attach to the contact point, it will cause an extra optical path length, and the fringe near the center will be obscure, then it will cause superior systematic error to the measurement. So we calculate the radius using of the difference of the two dark rings whose interval is $(n-m)$:

The nth order dark ring:

$$\left(\frac{D_n}{2}\right)^2 = nR\lambda \quad (2-10-7)$$

The mth order dark ring:

$$\left(\frac{D_m}{2}\right)^2 = mR\lambda \qquad (2-10-8)$$

Equation (2-10-8) minus equation (2-10-7), we can get $R = \dfrac{D_m^2 - D_n^2}{4(m-n)\lambda}$, when the wavelength is known, we can calculate the radius of curvature of the lens R according to the above equation by measuring the diameter of the nth order ring D_n and the diameter of the mth order ring D_m. Conversely we can figure out the wavelength λ when the radius of curvature is known.

【Procedure】

1. Adjust the measuring apparatus

The experiment instruments are arranged as shown in Fig. 2-10-3, it consists of the Newton's ring, the sodium lamp and the reading microscope.

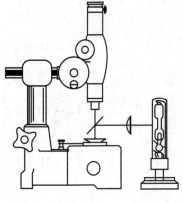

Fig. 2-10-3

① Adjust the Newton's ring: If the mirror of Newton ring is not clean, wipe it lightly with the lens paper, and then adjust the three screws to make the interference fringe can be seen directly by eyes, and make the interference fringe locate on the lens center. Notice that tighten the screws suitably, too lax will cause the fringe not to be steady; too close may cause the fringe to distort, which will damage the lens even make it broken.

② Make the reading of the reading microscope show around the center of the main ruler, turn on the sodium lamp then let it alignment to the 45° glass plate, and make the field of view brightness.

③ We must adjust the eyepiece so as to see the cross hairs, then focus on the object for measuring. First lower the microscope tube to approach the object to be measured (Newton ring), and the eyes should leave the eyepiece and observe from the flank and avoid eht lens contacting the Newton ring. Then view through the eyepiece and move up the microscope tube, until clear interference fringe occurs and there is no parallax with the cross hairs.

④ Adjust the cross hairs, and make the center of the fringe coincide with cross hairs. Rotate the eyepiece tubes of ye microscope, make the horizontal line of the cross hairs parallel to the main scale outside the microscope tubes. Then move the Newton's ring slowly.

⑤ Turn the micrometer screw, make the number of the rings which the cross hairs travel the rings from left to right to be larger than the order of the maximum ring to be measured. In the moving range, the illumination should be uniform, the fringes be clear; the horizontal line of cross hairs parallel to he main scale an cross through the fringer be clear; the horizontal line of cross hairs parallel to the main scale and cross through the diameter, the perpendicular line keep tangent with the fringe in moving process.

2. Observe the distribution characteristic of the interference pattern

For example: Are the width of fringes uniform? Do the intervals of fringes change? Is the spot on the center of zhe Newton's rings bright or dark? How to explain it?

3. Determine the diameter of the Newton's ring

① Take five groups of data as the example, the order m can be selected as 50, 49, 48, 47 and 49, the order n can be selected as 25, 24, 23, 22 and 21.

② Record the position of the different order for dark rings in Tab. 2 – 10 – 1.

【Data and Calculations】

(1) As shown in Tab. 2 – 10 – 1, the order is reference data. The wavelength of the monochromatic light is $\lambda = 589.3$ nm.

Tab. 2 – 10 – 1 Data of measurement

order of ring	m	50	49	48	47	46
position of the ring (mm)	left					
	right					
diameter (mm)	D_m					
order of ring	n	25	24	23	22	21
position of the ring (mm)	left					
	right					
diameter (mm)	D_n					
D_m^2 (mm²)						
D_n^2 (mm²)						
$D_m^2 - D_n^2$ (mm²)						

From the table above, we can get

$$\overline{D_m^2 - D_n^2} =$$
$$\Delta(D_m^2 - D_n^2) =$$

(2) The result and the error are

$$\bar{R} = \frac{\overline{D_m^2 - D_n^2}}{4(m-n)\lambda}$$

$$E = \frac{\Delta R}{\bar{R}} = \frac{\Delta(D_m^2 - D_n^2)}{D_m^2 - D_n^2} + \frac{\Delta(m-n)}{m-n}$$

$$\Delta R = E \cdot \bar{R}$$

$$R = \bar{R} \pm \Delta R$$

Where $\Delta(m-n) = \Delta m + \Delta n$, Δm and Δn are the errors which are caused by aiming the ordinate of the cross hairs at the center of the dark ring. Assume that the error equals 1/10 of the width of the ring, so $\Delta m = \Delta n = 0.1$, and we can estimate that $\Delta(m-n) = \Delta m + \Delta n =$

0.2.

【Note】

(1) When measuring a set of data with the reading microscope, it can be moved only in one direction. The micrometer screw should be rotated along one direction so that the screw-pitch error can be avoided.

(2) When the apparatus has been adjusted, it should not be changed, especially we could not count the wrong order of the dark ring in the measurement, otherwise you should measure it again.

(3) We should disentangle the three tightening screws on the Newton's ring after the experiment, for fear they will transmute for long time stressing.

(4) After the sodium lamp had been turned off, before it was turned on again we must wait for a minute.

【Preview】

(1) How to form the Newton's ring of the transmitted light? How to observe it? What relation is between the Newton's ring of the reflex and the transmitted light in being bright and dark, why?

(2) Will it influence the experimental result under the conditions below, why?

① There is no dark spot but the bright.

② When you are measuring the diameter D, the point of intersection of the cross hairs does not get across the center of the rings, so your measurement is not the diameter but the quadrature.

【Questions】

(1) How to adjust the microscope if its viewing field is not bright enough?

(2) Suppose that in the experiment of Newon's ring, there is ting prominece on the flat glass which may cause distortion of the ring. Under this situation, what will happen to Newton's ring, local dent or evagination? why?

(3) Whether interference fringe of the Newton's ring can be observed when the light is white? What characteristic of the fringe is there in the condition?

(4) Whether radius of curvature of a concave lens can be determined by using the same experimental method?

(Ma Jiao, Zhang Yi)

实验十一　利用劈尖干涉测量厚度

【实验目的】

(1) 观察劈尖上的等厚干涉条纹。

(2) 掌握利用劈尖干涉测量厚度的方法。

(3) 进一步熟悉读数显微镜的使用。

【实验器材】

劈尖、读数显微镜、钠光灯。

【实验原理】

将两块平面光学玻璃一端接触在一起，另一端夹一薄纸或细丝，则在两平面玻璃之间形成一楔形的空气薄膜，称空气劈尖。当一束单色平行光垂直入射时，经空气劈尖上下两表面反射的光是具有一定光程差的两束相干光，在劈尖上表面相遇，产生一组与棱边平行的明暗相间的干涉条纹，且条纹间距相等。如图 2 – 11 – 1 所示。凡空气层厚度 e 相同的地方，有级次相同的明条纹或暗条纹，故称等厚干涉。根据光程差公式和干涉条件，当

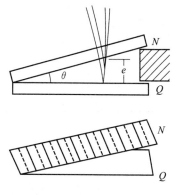

图 2 – 11 – 1　实验原理图

$$\delta = 2e_k + \frac{\lambda}{2} = (2k+1)\frac{\lambda}{2} \quad (k = 0,1,2,3\cdots) \quad (2-11-1)$$

得第 k 级暗条纹。劈棱处的厚度 $e_0 = 0$，对应级次 $k = 0$，即为零级暗条纹。

$$e_N = d = N \cdot \frac{\lambda}{2} \quad (2-11-2)$$

若待测薄片（或细丝的直径）出现在第 N 级暗条纹处，则待测薄片的厚度即为该处的膜厚。

实际测量时，因 N 值较大，不便于直接准确测量，可先测出第 k 级至第 $k+n$ 级干涉条纹的距离 l，测得劈棱至薄片（细丝）的距离（劈尖长度）为 L，则干涉条纹总数为

$$N = \frac{n}{l} \cdot L \quad (2-11-3)$$

将 N 代入式（2 – 11 – 2），得薄片厚度（或细丝的直径）为

$$d = \frac{n}{l} \cdot L \cdot \frac{\lambda}{2} \quad (2-11-4)$$

在测量中，只要将测量的任意 n 条暗条纹间距 l、劈尖长度 L 和单色光波长 λ，代入式（2 – 11 – 4），即可求出待测厚度 d。

【实验步骤】

(1) 把空气劈尖放在读数显微镜载物台上，调节 45°玻璃片，钠光灯位置及调节焦距（方法与牛顿环实验中相同），使从劈棱到薄片范围内都能清晰地看到劈尖等厚干涉条纹。

(2) 用读数显微镜测出 K 个暗纹间距的长度 S，为减小误差，K 可取大些（如 30 条）。再测出棱边到薄片处长度 L。将数据记录到表 2 – 11 – 1。

【数据记录及处理】

表 2-11-1　数据记录

次数		1	2	3	平均
棱长 L（mm）					
暗条纹位置	第 k 级暗条纹				
	第 $k+30$ 级暗条纹				

$$\overline{S} = |\overline{S_{k+30}} - \overline{S_k}|$$

$$\overline{d} = \overline{L} \cdot \frac{K}{\overline{S}} \cdot \frac{\lambda}{2} \tag{2-11-5}$$

由 (2-11-5) 式求出 $\overline{d}$。

【思考题】

在实验中若观察到的条纹发生畸变，且畸变处向接触棱处外凸，试解释其原因。

（张　翼）

Experiment 11　Determine Thickness by Interference of a Wedge-sharped Film

【Objects】

(1) Observe the pattern of equal thickness interference.
(2) Grasp the method of measuring thickness with a wedge.
(3) Further familiar with the use of reading microscope.

【Apparatus】

An air wedge, a reading microscope, a sodium lamp.

【Theory】

Two flat glass plates conneet with each other at one edge and are separated by a piece of paper or a filament at the other edge to form a wedge-shaped film of air, which is called an air-wedge. When the lens is illuminated by monochromatic parallel light from directly above, on the upper surface of glass patterns, which is a series of bright and dark fringe in equal interval and the fringes are parallel to the left-hand edge, as shown in Fig. 2-11-1, where the same thickness forms the same fringe, so it is called the equal thickness interference. In this case, according to the optical path length difference formula and interference condition, when

$$\delta = 2e_k + \frac{\lambda}{2} = (2k+1)\frac{\lambda}{2}(k=0,1,2,3\cdots)$$

$$(2-11-1)$$

Here k is the order number of the dark fringe. According to the interference conditions, $e_0 = 0$, corresponding to the order of the zero-order dark fringes, that is k =0.

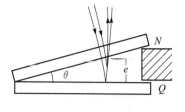

If the measured thickness of the sheet (or the diameter of the filament) to be appears at the position of N order dark fringes, the thickness of the sheet to be measured is the thickness of the film.

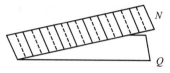

$$e_N = d = N \cdot \frac{\lambda}{2} \qquad (2-11-2)$$

Fig. 2-11-1

Because the value of N is too large to easy to measure directly. Firstly, the interval l from K to k + n can be measured, and the distance L (wedge length) from wedge edge to the measured sheet, then the total number of interference fringes is

$$N = \frac{n}{l} \cdot L \qquad (2-11-3)$$

Substituting N to (2-11-2), the measured thickness is

$$d = \frac{n}{l} \cdot L \cdot \frac{\lambda}{2} \qquad (2-11-4)$$

Actually, substituted the measured l (the interval from K to $k+n$), L (the length of wedge) and the known wavelength to (2-11-4), the unknown thickness d can be obtained.

【Procedure】

(1) Place the wedge on the stage of the reading microscope, then adjust the 45° glass plate, focal distance and the station of the sodium lamp (with the same method of the experiment of Newton's ring) to make the interference pattern clear from the entire wedge.

(2) Measure the length of the interval S of K dark fringes. In order to reduce the error, the value of K can be chose lager (such as 30). Then measure the length L, from one edge to the other edge. And record the measured data in Tab. 2-11-1.

【Data and Calculations】

Tab. 2-11-1 Data of measurement

times		1	2	3	average value
length of the arris (mm)					
the place of black stripe	the k order dark stripe				
	the $k+30$ order dark stripe				

$$\overline{S} = |\overline{S_{k+30}} - \overline{S_k}|$$
$$\overline{d} = \overline{L} \cdot \frac{K}{\overline{S}} \cdot \frac{\lambda}{2} \qquad (2-11-5)$$

Calculate the value of $\overline{d}$ according to $(2-11-5)$.

【Questions and Problem】

If the pattern observed distorts to the direction of the contacted edge, what reasons may cause it?

(Ma Jiao, Zhang Yi)

第三章　提高性实验

Chapter Three　Improving Experiment

实验十二　分光计的调节和使用

【实验目的】

（1）了解分光计的结构及各组成部分的作用。
（2）学习分光计的调节方法与正确使用。
（3）使用分光计测量三棱镜的顶角。

【实验器材】

JJY 型分光计、钠光灯、平面反射镜、三棱镜、照明放大镜等。

【实验原理】

本实验是利用自准直法测量三棱镜的顶角。其原理如图 3 - 12 - 1 所示，其中 AC 及 AB 是三棱镜的两个光学面，BC 为毛玻璃面，也是三棱镜的底面。用自准直法调节望远镜光轴使其既垂直于 AC 面又垂直于 AB 面，测出望远镜从 AC 面转到 AB 面所经过的角度 φ，则可根据

$$A = 180° - \varphi$$

求得三棱镜顶角 A。

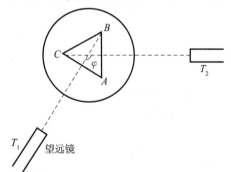

图 3 - 12 - 1　分光计测量三棱镜顶角的原理

【实验步骤】

1. 调节分光计

分光计结构见图 3 - 12 - 2。

调节分光计：调节前应对照仪器熟悉分光计各部件及各螺丝的作用，并先用眼睛估计各部件位置尽量合适。然后再对分光计进行调节，使平行光管发出平行光，望远镜聚焦无穷远，平行光管和望远镜的光轴均垂直于仪器中心轴。

（1）调节目镜焦距：调节手轮 11，使分划板位于目镜焦平面上。

（2）用自准直法调节望远镜使之聚焦无穷远：将平面镜如图 3 - 12 - 3 放置于平台上（为什么这样放?），慢慢转动平台，在望远镜中找到平面镜反射回来的亮十字像。松开螺丝 9，前后移动镜筒以改变物镜焦距，使亮十字像清晰。

（3）使望远镜光轴垂直于仪器中心轴：调节平台螺丝 6 中的 a、b 和望远镜高低调节螺丝 12，用渐近法使望远镜光轴垂直于仪器中心轴。

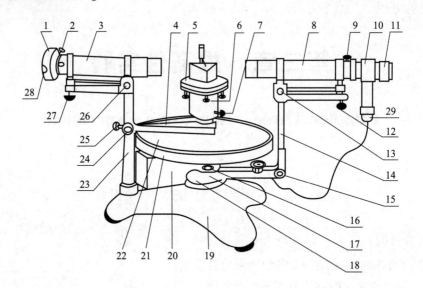

1. 狭缝装置；2. 狭缝装置锁紧螺丝；3. 平行光管部件；4. 制动架（1）；5. 载物平台；6. 载物台调平螺丝（3只）；7. 载物台锁紧螺丝；8. 望远镜部件；9. 目镜锁紧螺丝；10. 阿贝式自准直目镜；11. 目镜视度调节手轮；12. 望远镜光轴高低调节螺丝；13. 望远镜光轴水平调节螺丝；14. 支臂；15. 望远镜微调螺丝；16. 转座与度盘止动螺丝；17. 望远镜止动螺丝；18. 制动架（2）；19. 底座；20. 转座；21. 刻度盘；22. 游标盘；23. 立柱；24. 游标刻度盘微调螺丝；25. 游标盘止动螺丝；26. 平行光管轴水平调节螺丝；27. 平行光管光轴高低调节螺丝；28. 狭缝宽度调节手轮；29. 照明小灯

图 3-12-2 分光计结构

（4）调节平行光管：取下平面镜，将望远镜对准平行光管，利用已调好的望远镜为标准，调节平行光管产生平行光，且使平行光管光轴垂直仪器中心轴（参见仪器介绍部分）。

2. 测三棱镜顶角

（1）如图 3-12-4 所示将三棱镜放置在平台上（为什么如此放置？）。调节螺丝 6 中的 c 和 a，使望远镜光轴既能垂直于 AC 面，又能垂直于 AB 面。

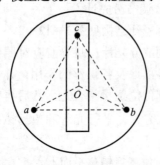   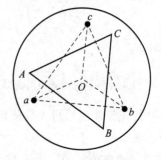

图 3-12-3 平面镜放置图示　　　图 3-12-4 三棱镜放置图示

（2）测量 φ 角 3 次，取其平均值，由此求出三棱镜的顶角 A。

利用望远镜自身产生平行光，用灯光照亮分划板，转动望远镜或平台，先使望远镜对准 AB 面使反射的亮十字像与分划板上方的十字线重合，固定望远镜和平台，记下刻度盘上两边的游标读数 Q_1 和 Q_2。然后再转动平台（望远镜固定或者转动望远镜将平台固定），使 AC 面正对望远镜，使反射亮十字像与划板上方十字线重合，固定平台（或望远镜），记下刻度盘游标读数 Q'_1 和 Q'_2（注意 Q_1 和 Q_2 不能颠倒），将所测数据记录于表 3-12-1。两次读数相减得 A 角的补角 φ，故 $A = 180° - \varphi$，可以证明：

$$\varphi = \frac{1}{2}(\varphi_1 + \varphi_2) = \frac{1}{2}[|Q_1 - Q'_1| + |Q_2 - Q'_2|]$$

【数据记录及处理】

分光计的准确度：_____

表 3-12-1　数据记录

位置 次数	T_1（AC 面正对望远镜）		T_2（AB 面正对望远镜）	
	Q_1	Q_2	Q'_1	Q'_2
1				
2				
3				
平均值				
绝对误差				

$$\overline{\varphi}_1 = \overline{Q}_1 - \overline{Q'_1}$$

$$\overline{\varphi}_2 = \overline{Q}_2 - \overline{Q'_2}$$

$$\overline{A} = 180° - \frac{1}{2}(\overline{\varphi}_1 - \overline{\varphi}_2)$$

$$\Delta A = \frac{1}{2}(\Delta Q_1 + \Delta Q_2 + \Delta Q'_1 + \Delta Q'_2)$$

【注意事项】

（1）严格遵守分光计的调节规则，不得改变先后次序。实验的先后次序也应按要求，不得变动。

（2）分光计上各光学透镜及三棱镜的光学平面不要用手摸碰，若有异物可用擦镜纸擦拭。

（3）狭缝不得合并，以免磨损刀口。

（4）望远镜调好以后，转动望远镜时不要搬动镜筒，应将望远镜与刻度盘固定，转动刻度盘来带动望远镜的转动。同样将平台与读数圆盘的内盘固定后，转动内盘来带动平台的转动以免使平台上的光学元件掉落。

（5）在测量读数时要注意：若使望远镜转动必须将读数圆盘的内盘固定（若转动平台即内盘，则必须将望远镜和外盘固定），读出的数据才是测量的准确值，否则读数圆盘的内、外盘将同时绕中心轴转动使测量数据不准确。

【预习要点】

（1）在调节分光计的过程中，调节望远镜的光轴与仪器的中心轴垂直是调节分光计的重要环节。实验中是怎样利用平面反射镜来调节望远镜光轴与中心轴垂直的？

（2）读数圆盘的游标的读数方法与我们以前学过的游标卡尺的读数方法有何异同之处？此游标为什么能读到1分？

【思考题】

（1）光栅平面不通过分光计的转轴时，对测量衍射角有无影响？

（2）什么是分光计的偏心差？如何消除此误差？

【仪器介绍】

1. 分光计结构

分光计是用来准确测量角度的光学仪器，如图3-12-2所示。在光学测量中需要测量的角度很多，如测量反射角、折射角、衍射角或最小偏向角等。分光计与衍射光栅、三棱镜配合可以观察衍射光谱、散射光谱，可以测量光波波长。分光计的结构与其他光学仪器如单色仪、摄谱仪等有很多相似的地方。

分光计由五个部分组成：平行光管、望远镜、载物平台、读数圆盘和底座。图3-12-2是其外形图。

（1）底座：底座19是整个分光计的支架。其中心有一垂直方向的转轴即仪器的中心轴。望远镜和刻度盘及游标盘均可围绕该轴转动。

（2）阿贝式自准直望远镜：阿贝式自准直望远镜10用来确定光线传播的方向。它由阿贝目镜b、消色差目镜c、全反射棱镜d、分划板e和消色差物镜f组成，见图3-12-5。物镜f装在镜筒的一端，阿贝目镜b装在镜筒的另一端套筒g中。松开螺丝9，在镜筒中移动或转动套筒，可对物镜进行调焦。在目镜套筒的侧面开有一小孔a，小孔旁装有一小灯泡，它发出的光束经棱镜全反射后照亮分划板上的小十字缝，并沿望远镜筒向外传播。调节手轮11改变目镜c和分划板e的相对位置，可对目镜调焦，从而看清分划板上的黑刻线，如图3-12-6所示。由于棱镜d在视场中挡掉一部分光线，故在分划板上呈现出它的阴影。

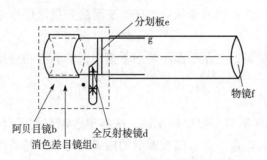
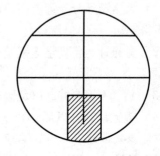

图3-12-5 阿贝式自准直望远镜内部结构图　　图3-12-6 分划板

调节螺丝12、13可改变望远镜的高低和水平方向的位置，使其光轴垂直于仪器中心轴。松开螺丝17，望远镜可绕中心轴转动；锁紧17后，可用螺丝15微调望远镜的水平位置。

（3）载物平台：载物平台5用以放置平面镜、三棱镜、光栅等光学元件。松开螺丝7，平台可绕中心轴自由转动，也可根据需要升高或降低。调节到所需位置后，再用螺丝7将之固定。螺丝6（由三只成正三角形放置的螺丝组成）用以调节平台水平，使之与中心轴垂直。

（4）读数圆盘：读数圆盘21、22，用来确定望远镜和载物平台的相对方位。锁紧螺丝16，松开螺丝17，望远镜可与外盘一起转动。外盘刻有720等分的刻线，每小格为半度；内盘为游标盘，通过螺丝7与平台相连。松开螺丝25，锁住17内盘可带动平台绕中心轴转动。锁紧25时，可用螺丝24微调内盘位置。在内盘相隔180°有两个对称游标A和B，各有30个分格（与外盘上29个分格的角度相等），故此游标可读到1分。读数按游标原理进行，例如图3-12-7所示的读数为116°+1′×12=116°12′，可持照明放大镜协助读数。

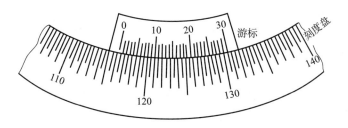

图3-12-7　读数圆盘示意图

为了克服由于内盘与外盘可能不是严格的同心所造成的偏心误差，每次测量必须分别读出A、B游标的读数，然后取其平均值。例：

	游标A	游标B
望远镜初始位置的读数：	330°45′	150°42′
望远镜转过θ角后读数：	90°47′	270°46′
差　　　数：	（360°+90°47′）-330°45′ =120°2′	270°46′-150°42′ =120°4′

则望远镜转过的角度θ：120°3′。

从此例可看到，望远镜转动时，游标A经过0°位置，要特别注意此时计算角度的方法。

（5）平行光管：平行光管用来获得平行光，由狭缝1和装在管的另一端的物镜组成。手松28用来调节狭缝宽度（0.02~2mm）。松开螺丝2可以调节狭缝和物镜的距离，并可转动狭缝。当狭缝正好位于物镜的焦平面时，从狭缝射进来的光经镜后成为平行光。平行光管与分光计的底座固定在一起，螺丝26、27分别用来调节平行光管的水平和高低位置。平行光管与望远镜之间的夹角可由读数圆盘读出。

2. 使用前调节

为使测量精确，必须满足以下三个条件：入射光和反射光是平行光，即望远镜要调焦无穷远；望远镜光轴既垂直于AC面又垂直于AB面；望远镜转动过程中所扫过的平面、三棱镜的主截面、平台平面和读数圆盘都相互平行。

（1）用自准直法使望远镜聚焦无穷远：自准直法是光学仪器调节中的一个重要方法，也是一些光学仪器进行测量的依据。当光点（物）S处在凸透镜的焦平面时，它发出的光线通过透镜后将成为一束平行光。

若用与主光轴垂直的平面镜将此光束反射回去，反射光仍为一束平行光。当它再次通过透镜后应会聚在与此平行光平行的副光轴和焦平面的交点 S' 上，S' 与 S 是关于主光轴对称的，其光路图如图 3-12-8 所示。

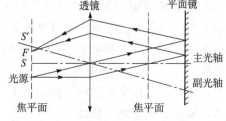

先调节目镜焦距，将分划板调至目镜的焦平面上，此时从目镜中能清楚地看到分划板上的"‡"形黑刻线。接着改变物镜焦距，使分划板同时又在物镜的焦平面上。那么透过十字缝的光线，经物镜出射后必为平行光。在载物平台上，放一面大致垂直于主光轴的平面镜，

图 3-12-8 自准直法光路图

利用该镜使平行光反射回望远镜中去，便从目镜中看到清晰的亮十字像，此时观察者若把头左右移动，发现亮十字像与黑刻线无相对位移即不存在视差，这时望远镜已聚焦于无穷远，即望远镜发出平行光也可接受于平行光。若有视差，说明目镜和物镜的焦平面还没完全重合，此时应该重复以上步骤，仔细调节两镜的焦距。

（2）使望远镜光轴垂直于 AB 面（或 AC 面）：在（1）的基础上，调节平面镜的倾斜度及望远镜的高低位置（实际上是通过螺丝 6 及 12 来实现的）。使十字像正好与十字缝处于对称的位置，如图 3-12-9 所示，则从自准直法原理可知，望远镜光轴已垂直于镜面，分别用棱镜的 AC 面（或 AB 面）取代平面镜，并做如上调节，就可使望远镜光轴垂直于 AC 面（或 AB 面）。

（3）使望远镜光轴垂直于仪器中心轴：仪器出厂时，读数圆盘已和仪器中心轴垂直，所以，只有当望远镜光轴及平台都垂直于中心轴时，望远镜转动时扫过的平面才能够同时与平台及读数圆盘平行。

在步骤（2）的基础上望远镜光轴已垂直于平面镜的一个镜面，但一般并没有和仪器中心轴垂直，此时将平台连同平面镜转过 180°，调节螺丝 6 及 12，使它再次符合图 3-12-9，这时望远镜光轴又垂直于另一镜面，如此反复，使望远镜光轴垂直于平面镜的任一镜面，那么望远镜光轴和平台一定同时垂直于平面镜的任一镜面，那么望远镜光轴和平台一定同时垂直于仪器中心轴。

图 3-12-9 调整十字像

如图 3-12-10 所示，望远镜光轴垂直其中一个镜面的情况只有图 3-12-10（1）、（2）中（a）所示的两种情况。当平台转过 180°后，显然只有图 3-12-10（1）符合实验要求。

调节时用二分之一渐近法。当望远镜光轴和平台相互垂直但不垂直于中心轴时，如图 3-12-11（a）实线所示。此时在目镜中可看到如图 3-12-9 所示的图像。若平台转过 180°，亮十字像必然有一个垂直位移，如图 3-12-11（a）中虚线所示，即反射像垂直位移（1）。目镜看到的是图 3-12-11（b），调节平台螺丝 6，此螺丝与观测者距离最近，使垂直位移减少一半，目镜中看到的如图 3-12-11（c）。实际上就是使平面镜平行于中心轴，即图 3-12-11（a）中垂直位移（2）；再调节望远镜高低螺丝 12，使位移完全消除。目镜中看到的仍然是图 3-12-11。实际上就是使望远镜光轴垂直于中心轴。

由于位移一半的估计往往不是很准确,所以需反复调节几次,直到无论平面镜转到哪一个镜面,望远镜光轴均与之垂直为止(注意:每次调节平台螺丝时,均调节与观测者距离最近的一个)。

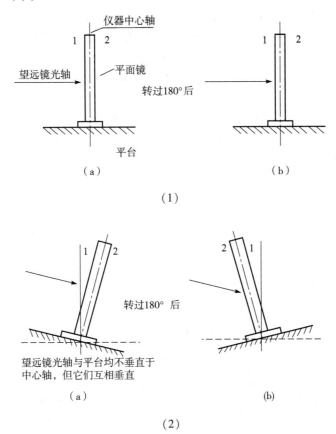

图 3-12-10　调整望远镜光轴垂直于仪器中心轴

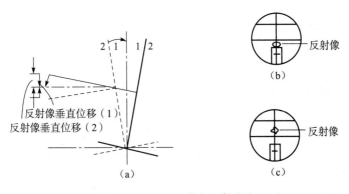

图 3-12-11　二分之一渐进法

在调试分光计中还需注意以下几点。

①本实验的重点是掌握分光计的调节方法。调节前必须先弄清分光计各螺丝的作用,直到能较熟练地调节好分光计后再进行测量。

②当望远镜目镜的焦距调好后，要用自准直法使望远镜聚焦无穷远（只改变物镜的焦距），此时不能再改变目镜的焦距。此后整个实验过程中不得再改变望远镜的任一焦距。

在用自准直法调节望远镜聚焦无穷远这一步中，当调好目镜焦距后，要细心寻找从平面镜反射回来的亮十字像，可先用眼睛观察，使望远镜光轴大致与平面镜的一个镜面垂直。再在望远镜附近对着平面镜寻找反射光，然后轻缓地左右转动平台，将反射光送回到望远镜中，此时可能在望远镜中看到一个绿色光斑（因为起初亮十字像未必一定落在目镜焦平面上），再慢慢前后移动镜筒 g，使之成清晰的像。注意，调节是否顺利，这一点是关键。

③平面镜按图 3-12-2 所示位置放在平台上，此时平面镜的表面与平台下调节螺丝 a、b 的连线相垂直（必须认准螺丝 a、b、c）。这样，只要调节螺丝 a 或 b，即可有效地改变平面镜的倾斜度。每次对一个镜面调节时，只需调节与观测者距离最近的螺丝。

④三棱镜应按图 3-12-11 放置在平台上。使三棱镜的底面 BC 与平台下调节水平螺丝 6 中的 a、b 连线垂直。对 AC 面调节螺丝 c，使之符合图 3-12-7；对 AB 面时调节螺丝 a，再使之符合图 3-12-7。如此反复，直到两个面均符合要求为止。注意，由于 AC 和 ao 平行，调节 a 不改变 AC 的法线方向。因此，当望远镜再对 AC 面时，亮十字像仍满足图 3-12-7。如果调节 c，则在 AB 面调好的同时，AC 面又移动了；如果调节 b，则对 AB 面根本不起作用。

⑤分光计调节好后，望远镜应取如图 3-12-5 所示位置。固定螺丝 25 使平台和游标盘一起转动，固定螺丝 16、17，使望远镜和外盘一起转动。

⑥当亮十字像与分划板上黑刻线水平位置相差很小时，就利用望远镜和游标盘的微调螺丝 15 和 24，使之精确对准后再读数（读数前必须认准 A、B 游标）。

⑦调节分光计前先掌握正确的调节步骤与方法，以免重复调节，延误时间。例如当已调好望远镜光轴垂直于仪器中心轴后，就不得再调望远镜了；调节三棱镜时应移动调平面镜时没有动过的螺丝 c。

3. 圆刻度盘偏心差的校正

用圆刻度盘测量角度时，为了消除圆刻度盘的偏心差，必须由相差为 180°的两个游标分别读数。我们知道，圆度盘是绕仪器主轴转动的，由于仪器制造时不容易做到圆度盘中心准确无误地与主轴重合，这就不可避免地会产生偏心差。圆度盘上的刻度均匀地刻在圆周上，当圆度盘中心与主轴重合时，由相差 180°的两个游标读出的转角刻度数值相等。而当圆度盘偏心时，由两个游标读出的转角刻度数值不再相等，如果只用一个游标读数就会出现系统误差。如图 3-12-12 所示，用 $\overparen{AB}$ 的

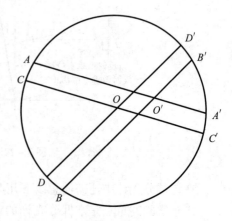

图 3-12-12　圆刻度盘的偏心差示意图

刻度读数，则偏大，用$\widehat{A'B'}$的刻度读数又偏小。由平面几何很容易证明

$$\frac{1}{2}(\widehat{AB} + \widehat{A'B'}) = \widehat{CD} = \widehat{C'D'}$$

（赵 喆 亓 霞）

Experiment 12 Adjustment and Use of Spectrometer

【Objects】

(1) To be familiar with the structure of a spectrometer and its basic parts.

(2) To know the test principle, and master the adjustment method of the spectrometer.

(3) To measure the apex angle of a grass prism with the spectrometer.

【Apparatus】

JJY – type spectrometer, sodium lamp, plane mirror, grass prism, magnifying glass.

【Theory】

In this experiment we will measure the apex angle of a grass prism with the autocollimating – method. Set the telescope to read the angle of the light that is reflected off each face of the prism. The difference between the scale readings at points T_1 and T_2, shown in Fig. 3 – 12 – 1, equals twice the apex angle. Calculate the value of A as follows (It should be close to sixty degrees.)

$$A = 180° - \varphi$$

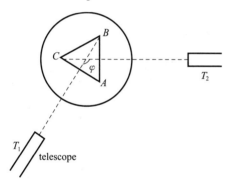

图 3 – 12 – 1 Principle of measuring prism top angle by spectrometer

【Procedure】

1. Adjust the spectrometer

Before making an adjustment, it is necessary to be familiar with the screws and every part of the spectrometer, and estimate their suitable positions judged by your eyes. Then adjust the spectrometer for following purposes: To make the collimator produce a collimated beam, and render the telescope to focus on a parallel light, both axises of the collimator and the telescope are perpendicular to the central axis of the spectrometer.

(1) Adjust the eyepiece: Adjusting the eyepiece11 Fig. 3 – 12 – 2 makes the image fall on cross hairs at the focusing ring.

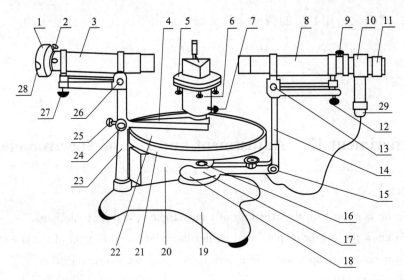

1. slit; 2. slit lock; 3. collimator; 4. the fixed frame (1); 5. load platform; 6. table leveling screws (three); 7. locking screw; 8. telescope; 9. eyepiece locking screw; 10. Abbe eyepiece; 11. eyepiece; 12. adjusting knurled knob; 13. telescope levelling screw; 14. arm; 15. fine adjusting; 16. dial locking screw; 17. telescope locking screw; 18. the fixed frame (2); 19. base; 20. turning base; 21. Scale plate; 22. Vernier plate; 23. column; 24. fine tuning screw; 25. lock screw of vernier dial; 26. level adjusting screw; 27. adjusting knurled knob; 28. slit adjuster; 29. light

Fig. 3 – 12 – 2 Spectrometer structure

(2) Adjust the telescope to make it focus on a distant object by autocollimating – method (See introduction to instruments 2): Place a reflector on the load platform just as shown in Fig. 3 – 12 – 3 (Why?), and rotate it slowly to find the cross image reflected from the reflector in the telescope. Loose the screw9, push the eyepiece in and out until you can see the crosswires sharply.

(3) Adjust the telescope to make its optical axis vertical to the center axis of the instrument: The 1/2 adjustment method is used to adjust the screw6 (a and b) underneath and the screw12, make the optical axis of the telescope vertical to the center axis of the instrument.

(4) Adjust the collimator: Remove the reflector, align the telescope with the collimator. Adjust the collimator to produce a parallel light taking the adjusted telescope as standard, and make the optical axis of the collimator vertical to the center axis of the instrument. (How to adjust?)

2. Measure the apex angle of the grass prism

(1) Place the grass prism on the load platform just as shown in Fig. 3 – 12 – 4 (Why place it like that?). Adjust screw6 (c and a) to make the optical axis of telescope vertical to not only the surface AC but also the surface AB.

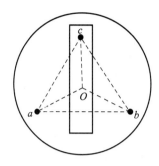 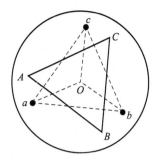

Fig. 3 – 12 – 3   Plane mirror placement diagram   Fig. 3 – 12 – 4   Prism placement diagram

(2) Measure the angle φ three times, take the average for the result, and then calculate the apex angle A of the prism.

A parallel light is produced by the telescope itself and the crosswire is lightened, and rotate the telescope or the load platform. then First, align the telescope with the surface AB to make the reflected image coincide with the crosswire. Fix the telescope and the load platform, then record the vernier readings of Q_1 and Q_2 on the degree plate. Then rotate the load platform (Fix the telescope or rotate the telescope while locking the table.) while making the surface AC face to the telescope, and make the reflected image coincide with the crosswire as well. Fix the telescope and the table, record the vernier readings Q_1' and Q_2' (Do not reverse them), then enter all the readings in Tab. 3 – 12 – 1. The difference of readings for two times is the auxiliary angle of angle φ, so $A = 180° - \varphi$, so we get

$$\varphi = \frac{1}{2}(\varphi_1 + \varphi_2) = \frac{1}{2}[\mid Q_1 - Q_1'\mid + \mid Q_2 - Q_2'\mid]$$

【Data and Calculations】

Tab. 3 – 12 – 1   Data of measurement of the spectrometer

position times	T_1 (AC face to the telescope)		T_2 (AB face to the telescope)	
	Q_1	Q_2	Q_1'	Q_2'
1				
2				
3				
average value				
absolute error				

$$\overline{\varphi}_1 = \overline{Q}_1 - \overline{Q_1'}$$

$$\overline{\varphi}_2 = \overline{Q}_2 - \overline{Q_2'}$$

$$\overline{A} = 180° - \frac{1}{2}(\overline{\varphi}_1 - \overline{\varphi}_2)$$

$$\Delta A = \frac{1}{2}(\Delta Q_1 + \Delta Q_2 + \Delta Q_1' + \Delta Q_2')$$

【Note】

(1) The particular skills and the precedence for adjusting the spectrometer must be obeyed strictly and do not allow to change.

(2) The polished sides of the prism and optical lens cannot be touched by hands. If there are foreign matters, you should clean it with lens papers.

(3) Do not close the slit avoid damaging the edges.

(4) After making one adjustment for the telescope at a time, try to adjust it again. Then the telescope would be fixed on the scale plate, and you can force the plate to rotate the telescope. Likewise after fixing the platform to the vernier plate, make it rotate to drive the platform to rotate for fear causing the optical component to drop from it.

(5) If you rotate the telescope, the vernier plate must be fixed (If you rotate the vernier plate, the telescope must be fixed.). Otherwise the measured data will not be accurate.

【Questions】

(1) In the process of adjusting the spectrometer, how to adjust the optical axis of the telescope vertical to the center axis by a reflector?

(2) What similarities and differences does the degree plate have comparing with the vernier calipers we have learned? Why can we read 1′ from the vernier.

(3) What is decentration error of the spectrometer? How to eliminate it?

【Introduction to Instrument】

1. The structure of the spectrometer

A spectrometer is an optical device for measuring angles precisely. Many kinds of angles can be measured in optical measurement such as reflection angle, refraction angle, diffraction angle and minimum deviation angle. And it also can be used to observe diffraction and diffusion spectrum and to measure wavelength of light using a diffraction grating or a prism. The structure of spectrometer has a lot of similarities with other optical instruments such as monochromator, spectrograph and so on. Spectrometer is composed of five parts: collimator, telescope, the stage, reading dial and base frame, as shown in Fig. 3 − 12 − 2.

(1) Base frame: Base frame is the basement of the spectrometer. There is a vertical rotation axis in the middle of the base frame. And telescope and the reading dial can move around the axis.

(2) Abbe autocollimator: Abbe autocollimator is used for determining the propagating direction of light. Its interior structure is shown in Fig. 3 − 12 − 5. where 1 is Abbe ocular (including achromatic ocular 2, total reflection prism 3 and reticule 4), and 5 is achromatic objective lens. The achromatic objective lens 5 is at the end of the lens tube and the Abbe ocular is at another end of the sleeve 6. The sleeve can be moved or rotate in the lens tube so as to adjust the focus of objective. There is a small hole on the side of the ocular sleeve and a small lamp near it. The light illuminates the opening sign " + " on the reticule after total reflection by prism, and

propages outside along the axis of telescope. Adjusting knob11 can change the relative position between ocular and reticule. The black sign "+" on the reticule can be seen clearly by focusing of the ocular. Fig. 3 – 12 – 6 is the front view of the reticule. Shadow is shown because of the prism.

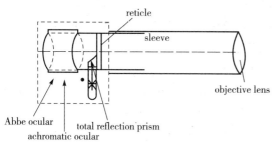

Fig. 3 – 12 – 5 Inner structure of Abbe type self – collimating telescope

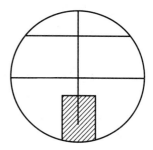

Fig. 3 – 12 – 6 Partition diagram

(3) The mounting table: The load platform5 is used to place optical elements such as: a reflector, a grass prism and a diffraction grating and so on. Loosen screw7, the table can rotate around the central axis freely and can be adjusted upper or lower. After being adjusted to the position needed, turn screw7 to fix it, and screws6 (three screws underneath) are adjusted to make the table horizontal and vertical to the central axis.

(4) Degree plate: The scale plate21 and vernier plate22 are used to adjust the position of the telescope relative to the load platform. Lock screw16 and loosen screw17, the telescope can rotate along with the scale plate. There are 720 equal intervals marked on the scale plate, and each of them is half degree. The vernier scales are connected to the scale plate with the screw17. Loosen screw25 and lock screw17, the load platform is driven by the vernier scales to rotate around the central axis. When locking screw25, the screw24 can make a fine adjustment for the vernier plate. There are two vernier scales A and B 180° diametrically – opposite to one another. Each vernier has 30 equal intervals which equal to 29 intervals of the scale plate, so the accuracy of the vernier is 1′. Its reading follows the theory of the vernier, as shown in Fig. 3 – 12 – 7, the reading is: $116° + 12 \times 1' = 116°12'$. A magnifying glass and a torch will be helpful for your reading.

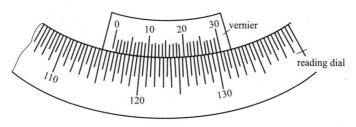

Fig. 3 – 12 – 7 Reading disk

To overcome the eccentricity of the inner and outer disks, which may not be strictly concentric, the cursor readings of A and B must be read out for each measurement and then averaged. For example:

	Cursor A	Cursor B
Reading of the initial position:	330°45′	150°42′
Reading after turns the angle :	90°47′	270°46′
The difference:	(90°47′ + 360°) − 330°45′ = 120°2′	270°46′ − 150°42′ = 120°4′

The angle of rotating of the telescope: = 120°2′。

As can be seen from this example, the cursor A passes through position 0 as the telescope rotates, with particular attention to the method of calculating the angle.

(5) Collimator: Collimator is used for producing a collimated light, it is consisted of a slit (1) and an objective lens setting at the other end of the tube. The width of the slit can be adjusted with the screw28 , its range is from 0.02mm to 2mm. The screw2 is used to fix the slit, loosen the screw2. While the distance between slit and the objective lens can be adjusted, and the slit can be rotated. The light emitting from the slit passes through the objective lens produces a parallel light when the slit locates on the focal plane of the objective lens. The collimator is fixed with the base frame. Its horizontal and vertical position can be adjusted by the screw26 and screw27 respectively. The angle between collimator and the telescope can be read out from the degree plate.

2. Adjustment before using

In order to make a precise measure, the spectrometer must be adjusted carefully: adjust the telescope to focus on an object at infinity, it means the incident light and the reflected light must be a parallel light, the optical axis of the telescope should be both normal to surface AB and surface AC, the optical axis of the telescope is vertical to the central axis of the spectrometer.

(1) Adjusting the focus of the telescope to bring a distant object into focus: Autocollimating – method is not only one very important method in adjusting optical instrument but also a standard for optical measurement.

The light from sourse S will be a parallel light after passing through the convex lens if S locates on the focal plane of the lens. The reflected light will be converged on the crossing point S' of focal plane and secondary optical axis which is parallel to the collimated light, S and S' is symmetrical about principal optical axis, as shown in Fig. 3 – 12 – 8.

To make the telescope focus distance by autocollimating – method: adjust the focal length of the ocular first, make the reticule locate on the focal plane, and at the moment the black scratch " ‡ " on reticule can be seen clearly.

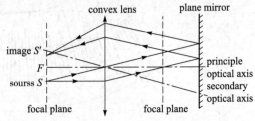

Fig. 3 – 12 – 8 Self – collimation path diagram

Then change the focal length of objective lens, make the reticule locate on the focal plane of objective lens, so transmitted light from the " + " slit will be collimated light.

Place a reflector on the platform, which is approximately vertical to the principle optical axis, then the parallel light will be reflected to the telescope, and a cross image will be seen from the ocular. At the moment, if you move your head back and forth (from left to right), the bright cross image has no relative displacement to the black scratch "‡", it means no "parallax" exists, then the telescope has focused distance, namely the telescope can produce a parallel light. If parallax exists, it shows the focal plane of the eyepiece does not coincide with that of objective lens, then you should repeat the steps above, and adjust the focal length carefully. Now the telescope has focused distance.

(2) Making the optical axis vertical to both AB and AC: Adjust the gradient of the reflector and the height of the telescope (by turning the screw 6 and12). Make the image of the crosswire appear at the upper cross center symmetric position of the " + " slit, as shown in Fig. 3 - 12 - 9, then the optical axis has been vertical to the mirror. Replace the mirror by the surface AC (or AB) and repeat the steps above, we can make the optical axis vertical to both AB and AC.

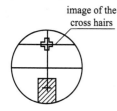

Fig. 3 - 12 - 9 Adjust the cross image

(3) Making the optical axis vertical to the central axis of the instrument: After focusing on the objective lens, a 1/2 adjustment method can be applied for adjusting the normal line of the plane mirror and the vertical position of the telescope, and make the "‡" image locate at the upper cross center and be symmetric with the " + " slit (as shown in Fig. 3 - 12 - 9). After the last step, the optical axis of telescope is normal to the mirror, but it may not be normal to it again for a turning of 180° of the mirror, as shown in Fig. 3 - 12 - 10. The two cases in Fig. 3 - 12 - 10 are all means the telescope is vertical to the reflector firstly, but only the case in Fig. 3 - 12 - 10 (a) means that the telescope is still vertical to the mirror, because it is vertical to the central axis of the instrument.

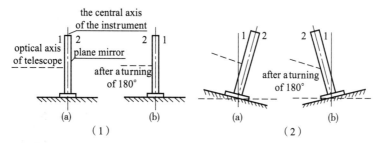

Fig. 3 - 12 - 10 Adjust the optical axis of the telescope perpendicular to the central axis of the instrument

Now, let's see the half asymptotic method. The continuous lines in Fig. 3 - 12 - 11 (a) show the case of the axis of telescope is parallel with the reflector of the stage but not vertical to the central axis of instrument. After a turning of 180° the bright reflective sign " + " will have a displacement (vertical difference), as the dotted line shown. The situation can be viewed through ocular and shown in Fig. 3 - 12 - 11 (b). Adjust the screw6 to make the vertical difference decrease by half, as shown in Fig. 3 - 16 - 11 (c). And then adjust the screw12 to make the bright "+" come back to the upper cross, as shown in Fig. 3-12-9. Because the half displacement estimated is not very accurate, so we need to repeat this adjusting for many times until the optical axis of telescope could be vertical to the central axis and the two planes of the reflector (Note: Rotate the nearest screw away the observer).

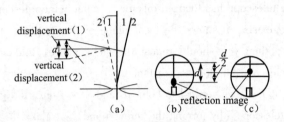

Fig. 3-12-11　Half asymptotic method

3. The decentration error of the vernier

The reading of vernier A and B must be read out respectively, and the average of them is taken in order to eliminate the decentration error because the main plate and the degree scale may not be concentric. The reason is shown as Fig. 3-12-12, we have

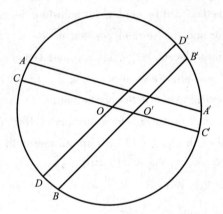

Fig. 3-12-12　Eccentric adjustment

$$\frac{1}{2}(\widehat{AB} + \widehat{A'B'}) = \widehat{CD} = \widehat{C'D'}$$

(Zhao Zhe, Su Ting-ting)

实验十三　用衍射光栅测定光波波长

【实验目的】

(1) 观察光波的衍射现象。
(2) 学会使用分光计与光栅测定光波波长的方法。

【实验器材】

分光计、透射式衍射光栅、钠光灯等。

【实验原理】

衍射光栅是透明玻璃片上刻有许多平行的等距离的刻痕而制成的。相邻刻痕之间的距离为 a,是透射光缝的宽度。两条相邻的透射光缝之间不能透射光波的距离为 b（即刻痕宽度）,$a+b$ 称为衍射光栅的光栅常数。

当波长为 λ 的单色光垂直照射到光栅平面时,由于光波的衍射作用,在光栅后面的光屏上形成明暗相间的衍射条纹。各级衍射亮条纹产生的条件由衍射公式确定,即

$$(a+b)\sin\varphi = k\lambda \quad (k=0,\pm 1,+2\cdots) \quad (3-13-1)$$

φ 为第 k 级亮条纹的衍射角,k 称为衍射级次。当 $k=0$ 时称为中央亮纹,衍射亮条纹的分布如图 3-13-1 所示,是以中央亮纹为中心对称排列的衍射图案。

若入射单色光的波长为已知,用分光计测量第 k 级亮条纹的衍射角 φ 就可能得到光栅的光栅常数 $a+b$,即

$$a+b = \frac{k\lambda}{\sin\varphi} \quad (3-13-2)$$

同样若光栅常数为已知,又可得到单色光波的波长 λ,即

$$\lambda = \frac{(a+b)\cdot\sin\varphi}{k} \quad (3-13-3)$$

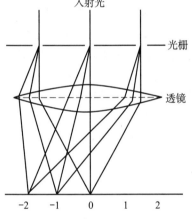

图 3-13-1 衍射光栅原理图

【实验步骤】

1. 分光计的调节

该实验在分光计上进行,要使实验满足于夫琅和费衍射条件和测量的准确,入射光应是平行光,衍射后应用望远镜观察和测量,所以对分光计的调节需要：望远镜聚焦于无穷远,望远镜光轴垂直于仪器中心转轴;平行光管产生平行光,其光轴垂直于转轴。具体调节方法见实验十二中分光计的调节和使用部分。

2. 光栅的调节

（1）调节光栅的平面（刻痕所在的平面）与仪器中心转轴平行,且光栅的平面与平行光管光轴和望远镜光轴垂直。调节方法：先把平行光管的狭缝对准钠光灯,把望远镜对准平行光管使入射的平行光的狭缝像聚焦于分划板中心十字线上与竖线重合,然后固定望远镜,关闭钠光灯（或用厚纸挡住灯光）。把光栅按图 3-13-2 的所示位置放在载物平台上用带铁片的螺丝夹稳光栅片。将载物平台与读数圆盘固定在一起。然后转动内盘使光栅的刻痕平面对准望远镜,观察从光栅面反射回来的亮十字像是否与分划板上方的十字线重合,否则调节载物台上螺丝 B_1 或 B_2,使反射亮十字像与分划板上方的十字线重合（注意望远镜不能调节,否则将重新调节分光计）,这时光栅平面与仪器转轴平行且垂直于平行光管和望远镜光轴。随即固定读数圆盘的内盘以免测量时发生转动。

（2）调节光栅刻痕与转轴平行：光栅平面虽然与转轴平行但平面上的刻痕可能不

平行于转轴。调节方法是：开启钠光灯照亮狭缝，这时在望远镜中可观察到衍射的中央亮纹，然后松动望远镜止动螺丝17，转动望远镜可观察到衍射光谱 $k=\pm 1$，$k=\pm 2$ 级亮条纹。若这些衍射亮条纹不是排列在同一水平线上，或各级衍射亮纹分布与分划板中心横线比较高低不整齐，表明光栅刻痕与仪器中心转轴不平行，应调节载物平台螺丝 B_3（图3-13-2），使各级衍射亮纹于同一水平线上。然后再检查一次光栅平面是否与仪器转轴平行，若有变化再进行调节。如此反复多次，直到两个都要符合条件为止。

3. 测量衍射光栅的各级谱线衍射角

（1）将望远镜对准平行光管，观察中央亮纹。使中央亮纹与分划板中心竖线重合于中心位置，固定望远镜，分别记录两个游标尺读数 Q_0 和 Q'_0。

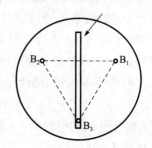

图3-13-2 调节光栅刻痕与转轴平行

（2）松动望远镜止动螺丝17，将望远镜向左转动可观察到第一级亮纹，使亮纹与分划板中心竖线重合。固定望远镜，分别记录两个游标尺读数为 $Q_{左1}$ 和 $Q'_{左1}$。继续向左转动望远镜，观测左侧第二级衍射光谱，使亮纹与分划板中心竖线重合，固定望远镜，分别记录两个游标尺读数为 $Q_{左2}$ 和 $Q'_{左2}$。

（3）将望远镜转回到光栅衍射光谱的中央亮纹位置，然后向右侧转动，分别测量右侧第一级、第二级衍射光谱，分别记录 $Q_{右1}$ 和 $Q'_{右1}$、$Q_{右2}$ 和 $Q'_{右2}$。

（4）重复上述测量过程，对每一测量量反复测量三次，将数据填入表3-13-1中。

【数据记录及处理】

1. 记录实验数据

光栅常数：$a+b = \dfrac{1}{300}$ mm 或 $a+b = \dfrac{1}{600}$ mm

分光计精确度：1′

表3-13-1 各光谱位置测量数据表

光谱级位置 次数	中央亮纹		一级光谱				二级光谱			
			左侧		右侧		左侧		右侧	
	Q_0	Q'_0	$Q_{左1}$	$Q'_{左1}$	$Q_{右1}$	$Q'_{右1}$	$Q_{左2}$	$Q'_{左2}$	$Q_{右2}$	$Q'_{右2}$
1										
2										
3										
平均值										
平均绝对误差										
衍射角	$\overline{\varphi}_{左1} = \dfrac{1}{2}(|\overline{Q_{左1}} - \overline{Q_0}| + |\overline{Q'_{左1}} - \overline{Q'_0}|) =$						$\overline{\varphi}_{右1} = \dfrac{1}{2}(|\overline{Q_{右1}} - \overline{Q_0}| + |\overline{Q'_{右1}} - \overline{Q'_0}|) =$			
	$\overline{\varphi}_{左2} = \dfrac{1}{2}(|\overline{Q_{左2}} - \overline{Q_0}| + |\overline{Q'_{左2}} - \overline{Q'_0}|) =$						$\overline{\varphi}_{右2} = \dfrac{1}{2}(|\overline{Q_{右2}} - \overline{Q_0}| + |\overline{Q'_{右2}} - \overline{Q'_0}|) =$			

2. 误差处理

根据衍射公式和公式（3-13-3），且 $\overline{\varphi_1} = \dfrac{\varphi_{左1}+\varphi_{右1}}{2}$，$\overline{\varphi_2} = \dfrac{\varphi_{左2}+\varphi_{右2}}{2}$

故待测单色光波波长为：

$$\lambda = \frac{1}{2}(a+b)\left[\sin\overline{\varphi_1} + \frac{1}{2}\sin\overline{\varphi_2}\right] =$$

$$\Delta\lambda = \frac{1}{2}(a+b)\left[\cos\overline{\varphi_1}\cdot\Delta\varphi_1 + \frac{1}{2}\cos\overline{\varphi_2}\cdot\Delta\varphi_2\right] =$$

（注意：计算 $\Delta\lambda$ 时，要将此式中的 $\Delta\varphi_1$、$\Delta\varphi_2$ 化为弧度。）

其中：

$$\Delta\varphi_1 = \frac{1}{4}(\Delta Q_{右1} + \Delta Q'_{右1} + \Delta Q_{左1} + \Delta Q'_{左1} + 2\Delta Q_0 + 2\Delta Q'_0)$$

$$\Delta\varphi_2 = \frac{1}{4}(\Delta Q_{右2} + \Delta Q'_{右2} + \Delta Q_{左2} + \Delta Q'_{左2} + 2\Delta Q_0 + 2\Delta Q'_0)$$

待测单色光波波长为：$\lambda = \overline{\lambda} \pm \Delta\lambda$

若待测光波波长为已知标准波长 λ_0（钠光标准波长为 589.3nm），则根据测量值与标准值的比较，计算实验结果的相对误差 E_r，即

$$E_r = [(\overline{\lambda} - \lambda_0)/\lambda_0] \times 100\%$$

待测波长的绝对误差：$\Delta\lambda = \overline{\lambda}\cdot E_r$

待测光波波长为：

$$\lambda = \overline{\lambda} \pm \Delta\lambda$$

【注意事项】

（1）光栅是一种精密、易碎的光学元件，使用时应小心谨慎，不允许用手触摸光栅刻痕表面，也不要用擦镜纸和棉花随意擦拭，以免损坏刻痕表面。

（2）分光计的调节应按要求调整好，以免造成测量上较大的误差。

（3）汞灯的紫外光很强，不可直视，以免灼伤眼睛。

【思考题】

（1）如果光栅平面与转轴平行，但刻痕与转轴不平行，那么整个衍射光谱有什么异常？对测量结果有什么影响？

（2）如以白光照射光栅，得到的衍射光谱是怎样按波长排列的？高级次衍射光谱的重合现象是怎样产生的？

（3）钠光的波长一般给定为 589.3nm，结合你的实际测量结果，分析误差产生的原因。

【仪器介绍】

1. 分光计的调节

分光计的调节要求同实验十二。

2. 垂直入射时光栅的调节

（1）光栅平面（刻痕所在平面）与仪器中心轴平行，且垂直于平行光管。调节的

方法是：把望远镜对准平行光管，使分划板的竖线与狭缝像的一边重合，横线在狭缝像的中央。固定望远镜（即固定螺丝 16、17、25）。

将光栅如图 3-13-2 所示放置在平台上，使光栅平面垂直平分 B_1、B_2 连线，光栅中心约在平行光管与望远镜的轴线上。松开螺丝 7，转动平台，以光栅面作为反射面，用自准直法调节光栅面与望远镜光轴垂直。然后固定螺丝 7。

（2）调节光栅使其刻痕与仪器中心轴平行，松开望远镜止动螺丝 17，左右转动望远镜，观察两边各级条纹。注意分划板上中心交点是否在各条谱线中央。如果不是，则调节 B_3 予以校正。调好后再回头检查光栅平面是否仍保持和中心轴平行，如果有改变，就要反复多次，直到两个要求都满足为止。

光栅如此放置的好处是：调节螺丝 B_1 与 B_2 就可以有效地改变光栅平面的倾斜度，使它垂直于望远镜光轴。注意此时望远镜已调好，不能再动；然后再调节螺丝 B_3 时，就能使光栅刻痕与仪器中心轴平行，而不会再影响光栅平面的倾斜度。注意此时不要再动 B_1 与 B_2。

（3）在调节望远镜光轴垂直于光栅平面时，从光栅反射回来的亮十字像往往有几个重影。这是因为光栅有好几个反射面（玻璃有两个面，复制的光栅还有胶膜面等），调节时以最清晰的像为准。

（4）因为狭缝有一定宽度，所以形成的谱线也有一定宽度。我们在测量时以谱线的一边为准。在读 0 级时要认准分划板的竖线与谱线的哪一边对齐，那么在测 ±1 级谱线时也应与这一边对齐。

（5）为了使分划板的竖线精确对准光谱线，最后必须使用望远镜微调螺丝 15 来对准。

（6）测量衍射角 φ 时应测出 ±1 级，然后取其平均值。应注意 +1 级与 -1 级的衍射角相差不能超过几分。否则垂直入射时应重新检查入射角 i 是否为零；斜入射时，重新检查入射线的位置是否有变动。

（赵　喆　李百芳）

Experiment 13　Determination of the Wavelength of Light Using a Diffraction Grating

【Objects】

(1) To examine line structure of visible spectra.

(2) To learn determining the wavelength of light with a spectrometer and a diffraction grating.

【Apparatus】

A spectrometer, a diffraction grating, a sodium lamp, etc.

【Theory】

A diffraction grating is a kind of optical device which consists of a large number of equally spaced, parallel slits or ridges. The transparent segments working as slits are a, and the opaque segments are b, and $(a+b)$ is called the grating constant of the diffraction grating.

When a monochromatic parallel light with wavelength λ is an incident normally upon a grating and an interference pattern forms on the screen located in the focal plane, from the grating equation

$$(a+b)\sin\varphi = k\lambda \quad (k = 0, \pm 1, +2\cdots) \quad (3-13-1)$$

Here $a+b$ is the distance between the two slits. φ is the angle between the incoming beam and the location where we observe the intensity (Fig. 3 - 13 - 1) and k is the order number of the interference. The $k=0$ peak is what you observe at 0 degree. As you go left or right, you will see colored stripes corresponding to $k = \pm 1$ for different wavelengths, and perhaps at $k = \pm 2$ at larger angles.

When the wavelengths of an incident light is known, through measuring the angle of diffraction for the bright line, from the following equation (3 - 13 - 1) we can get

$$a + b = \frac{k\lambda}{\sin\varphi} \quad (3-13-2)$$

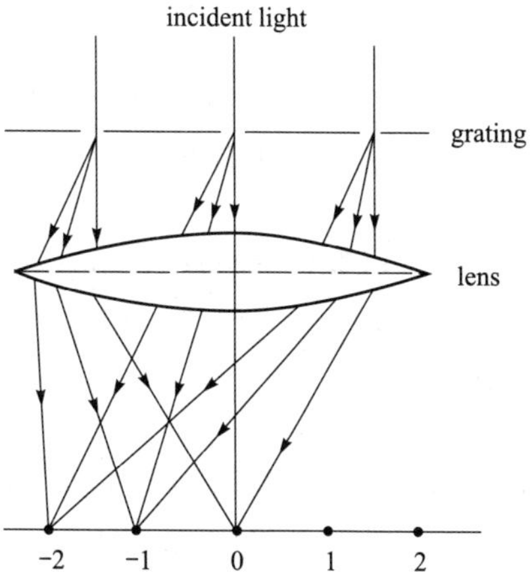

Fig. 3 - 13 - 1

In the same way, if the grating constant $(a+b)$ is known, and the wavelength λ of the incident light can be calculated by

$$\lambda = \frac{(a+b)\cdot\sin\varphi}{k} \quad (3-13-3)$$

【Procedure】

1. Adjust the spectrometer

A spectrometer is used in the experiment. To make the experiment meet condition of the Fraunhofer Diffraction and the accuracy of measurement, a beam of parallel light should be used as incident light, and its telescope is used for observing and measuring. So demands for adjusting are: the telescope fousing on distance, the optical axis of telescope is vertical to the center axis of the instrument, the collimator product a parallel light, its optical axis is vertical to the axis of the instrument. The skill of its adjustment can be found in Experiment 12.

2. Adjust the grating

(1) Adjust the grating surface (the flat of indent), make it parallel to the center axis of the instrument, and make it vertical to the optical axes of the collimator and telescope.

First, aim the collimator to the sodium lamp and the telescope to the collimator, make the image of the slit being coincidence with the central line on the reticle. And then, fix the telescope and close the sodium lamp. (Make the image of the slit being coincidence with the central line on the reticle). Put the grating on the load platform at the position shown in Fig. 3 - 13 - 2 and fix the grating with the special holder. Lock the platform and record the readings. Then rotate the platform aiming at the telescope and observe the " + " image reflected by diffraction grating to see whether it overlaps with the upper " + " on the reticule. If not, adjust the screw B_1 or B_2, then the plane of diffraction grating is parallel to the axis of the instrument and perpendicular to the axes of the collimator and telescope. At the same time, fix the degree plate not to move more during the measuring.

(2) Adjust the indent of grating make it parallel to rotational axis.

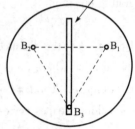

Fig. 3-13-2

The indent of grating may not be paralled to the axis eventhough the plane of the grating is parallel to the axis. Adjust method is to turn on the sodium lamp and light the slit. You can observe the central light fringe, and you can observe the diffraction spectrums by loosing the screw17 and rotating the telescope. These light lines are not at the same level. The grooves of the grating is not parallel to the central axis because different lines are not at the same level. Do it again until it is applied to the experiment requirement.

3. Measure the angle of different spectral lines

(1) Align the telescope with the collimator to observe the center bright line. Make the line coinciding with vertical center of the graduation board at the center. Fix the telescope, record the reading number Q_0 and Q_0' respectively.

(2) Loose the screw17 of the telescope, swing the telescope to left to observe the first order line. Make the line coinciding with vertical center of the graduation board at the center. Lock the telescope, record the reading number Q_{L1} and Q_{L1}' respectively. Go on swinging the telescope, observe the second order line, then record the reading number Q_{L2} and Q_{L2}'.

(3) Swing the telescope back to center line of the diffraction spectrum, then rotate it right sequentially, measure the first and the second order line, then record Q_{R1} and Q_{R1}'、Q_{R2} and Q_{R2}'.

(4) Make three measurements, and record the data in the following Tab. 3-13-1.

[Data and Calculations]

1. Record the experiment data in Tab. 3-13-1

Grating constant: $a+b = \dfrac{1}{300}$mm or $a+b = \dfrac{1}{600}$mm, spectrometer precision:

Tab. 3-13-1 Data of spectrum's measuring position

spectral lines times	center line		first order spectrum				second order spectrum			
			left side		right side		left side		right side	
	Q_0	Q_0'	Q_{L1}	Q_{L1}'	Q_{R1}	Q_{R1}'	Q_{L2}	Q_{L2}'	Q_{R2}	Q_{R2}'
1										
2										
3										

续表

spectral lines \ times	center line		first order spectrum				second order spectrum			
			left side		right side		left side		right side	
	Q_0	Q_0'	Q_{L1}	Q_{L1}'	Q_{R1}	Q_{R1}'	Q_{L2}	Q_{L2}'	Q_{R2}	Q_{R2}'
average value										
absolute error										
angle of diffraction	$\overline{\varphi}_{L1} = \frac{1}{2}(|\overline{Q_{L1}} - \overline{Q_0}| + |\overline{Q_{L1}'} - \overline{Q_0'}|) =$ $\overline{\varphi}_{L2} = \frac{1}{2}(|\overline{Q_{L2}} - \overline{Q_0}| + |\overline{Q_{L2}'} - \overline{Q_0'}|) =$						$\overline{\varphi}_{R1} = \frac{1}{2}(|\overline{Q_{R1}} - \overline{Q_0}| + |\overline{Q_{R1}'} - \overline{Q_0'}|) =$ $\overline{\varphi}_{R2} = \frac{1}{2}(|\overline{Q_{R2}} - \overline{Q_0}| + |\overline{Q_{R2}'} - \overline{Q_0'}|) =$			

2. Deviation treatment

According to diffraction formula and equation (3 – 13 – 3), and $\overline{\varphi}_1 = \frac{\overline{\varphi}_{L1} + \overline{\varphi}_{R1}}{2}$, $\overline{\varphi}_2 = \frac{\overline{\varphi}_{L2} + \overline{\varphi}_{R2}}{2}$

So the monochromatic wavelength to be measured is:

$$\lambda = \frac{1}{2}(a + b)\left[\sin\overline{\varphi}_1 + \frac{1}{2}\sin\overline{\varphi}_2\right] =$$

$$\Delta\lambda = \frac{1}{2}(a + b)\left[\cos\overline{\varphi}_1 \cdot \Delta\varphi_1 + \frac{1}{2}\cos\overline{\varphi}_2 \cdot \Delta\varphi_2\right] =$$

(Notice: when counting $\Delta\lambda$, change the unit of $\Delta\varphi_1$, $\Delta\varphi_2$ above into radian)

Among them:

$$\Delta\varphi_1 = \frac{1}{4}(\Delta Q_{R1} + \Delta Q_{R1}' + \Delta Q_{L1} + \Delta Q_{L1}' + 2\Delta Q_0 + 2\Delta Q_0')$$

$$\Delta\varphi_2 = \frac{1}{4}(\Delta Q_{R2} + \Delta Q_{R2}' + \Delta Q_{L2} + \Delta Q_{L2}' + 2\Delta Q_0 + 2\Delta Q_0')$$

So the measured wavelength is:

$$\lambda = \overline{\lambda} \pm \Delta\lambda$$

If the wavelength is known as standard λ_0 (589.3nm), depending on the contrast between measured value and standard value, work out the relative error E_r, namely

$$E_r = [(\overline{\lambda} - \lambda_0)/\lambda_0] \times 100\%$$

Then, the absolute error is: $\Delta\lambda = \overline{\lambda} \cdot E_r$

The result is:

$$\lambda = \overline{\lambda} \pm \Delta\lambda$$

【Note】

(1) Because grating is a precise and fragile optical element, care should be taken when using it. It is forbidden to touch the surface of grating indent with hands or clean it with lens paper or cotton for fear that indent surface is damaged.

(2) Regulate spectrometer according to instructions in order to avoid larger errors in measure.

(3) The ultraviolet ray of mercury vapour lamp is strong, so don't see it straightly in order to avoid burning eyes.

【Questions】

(1) If the grating plane runs parallel with the axis but indents not, how will the whole diffraction spectrum change? And whether it will do some effects to the measure results?

(2) If the grating is illuminated by a natural light, how does the diffraction spectrum range in order of wavelength? And how does the phenomena of diffraction spectrum's coincide occur?

(3) The wavelength of natrium light is generally 589.3nm. Combining you practical measure results, try to analyse the reason for errors.

(Zhao Zhe, Wang Xiao-fei)

实验十四　用阿贝折射计测定液体的折射率

【实验目的】

(1) 熟悉阿贝折射计的结构原理，学会正确使用方法。
(2) 学会用阿贝折射计测定液体折射率，并掌握由此确定液体浓度的方法。
(3) 熟悉实验数据的图示法和一元线性回归法。

【实验器材】

阿贝折射计、蒸馏水、无水乙醇、几种不同浓度的 NaCl 溶液、滴管、擦镜纸。

【实验原理】

我们知道，当一束光从一种介质射向另一种介质时遵守折射定律。如图 3-14-1 所示，n_1 和 n_2 分别表示第一种介质和第二种介质的折射率，i 表示入射角，r 表示折射角，由折射定律有

$$n_1 \sin i = n_2 \sin r \qquad (3-14-1)$$

若 $n_1 < n_2$，即光由光疏介质射向光密介质时，$i > r$；反之，若 $n_1 > n_2$，即光由光密介质射向光疏介质时，$i < r$。当 i 增大到某一值 i_0 时，此时的折射角恰等于 90°，则 i_0 称为临界角。在这种情况下，当光在光疏介质中的入射角在 0~90°范围内，即 0~90°范围均有光线时，在光密介质中只有临界角 i_0 内有光线，而在大于 i_0 范围为暗区，因而形成明暗分界线（用望远镜观察），如图 3-14-2 所示。

图 3-14-1　折射定律示意图　　图 3-14-2　明暗分界线示意图

根据光路的可逆性，若光线由光密介质射向光疏介质时，如果入射角为 i_0，则折

射角为90°，入射角继续增大，则光线将全部反射回光密介质中，这种现象称为全反射现象，而临界角 i_0 即为发生全反射时的特定入射角。

阿贝折射计就是根据上述全反射原理制成的，它是药物检验中常用的分析仪器。其原理如图3-14-3所示，它的主要部分是由两个同样的直角棱镜 ABC 和 DEF 所组成，两棱镜面夹待测液体薄层（设折射率为 n'）。棱镜 DEF 的 DF 面为磨砂表面，由普通光源发来的光由镜面 M 反射通过透光孔 P 进入棱镜 DEF，经折射后射到磨砂表面 DF，这使磨砂表面被照亮而成为发光面。由于磨砂面使光向各方向漫射，因此由 DF 面发出的漫射光线通过液层入射到棱镜 ABC 的 AC 面。因液层很薄，总可以有入射角非常接近90°的入射光，如图3-14-3所画出的光线 SO，称为掠射光线。设棱镜的折射率为 n，如果 $n>n'$，则 SO 射入 AC 面的折射角为棱镜对液体的临界角 i_0。光线 OR 在 BC 面的入射 i 及折射角 r 都由 i_0 而定。即出射线的位置也就由待测液体的折射率 n' 所决定。由于 SO 是所有入射 AC 面的光线中入射角最大的，故所有射入 AC 面的光线经两次折射后其出射线的方向只能在 Rl 的左边。若射入光线为单色光，则对准 Rl 方向的望远镜视野中，将看到一半明一半暗的图像，而 Rl 方向的光线所成的像就是这明暗的分界线，我们只要测定这分界线 Rl 的出射角 r 就可以求出待测液的折射率 n'。

图3-14-3中，设棱镜的棱角 $\angle ACB$ 为 φ，由三角形 ORC 可知：
$$i_0 + 90° = i + 90° + \varphi$$
即
$$i_0 = i + \varphi \quad (3-14-2)$$

对光线 SO 和 OR，由折射定律得
$$n'\sin 90° = n\sin i_0 \quad (3-14-3)$$

所以，
$$n' = n\sin i_0 = n\sin(i+\varphi) \quad (3-14-4)$$

在 BC 界面 R 点有
$$n\sin i = n_{空}\sin r = \sin r \quad (3-14-5)$$

故
$$\sin i = \frac{\sin r}{n}$$
$$\cos i = \sqrt{1-\sin^2 i} = \frac{1}{n}\sqrt{n^2-\sin^2 r}$$
$$(3-14-6)$$

将（3-14-6）式代入（3-14-4）式得
$$n' = \sin r\cos\varphi + \sin\varphi\sqrt{n^2-\sin^2 r}$$

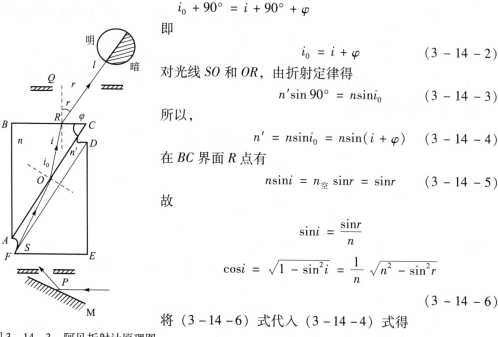

图3-14-3 阿贝折射计原理图

式中，棱镜的棱角 φ 和折射率 n 均为定值，因此 n' 由 r 决定，从折射计测得角 r，即可确定液体的折射率 n'。

实际上，阿贝折射计是经过换算校正的，可以从仪器刻度上直接读出待测液体的折射率。应用阿贝折射计来进行定性和定量分析的方法，称为折光分析法。通常在测量溶液浓度时，我们可用已知浓度的若干标准溶液在阿贝折射计上测出其折射率，从而求得该种溶液的折射率-浓度曲线，然后，测出待测溶液的折射率 n_x，再根据此标准曲线求出未知浓度 c_x。

【实验步骤】

（1）阿贝折射计外观如图 3-14-6 所示。打开反射镜，旋转目镜，使视野中的十字叉线成像清晰，然后合上反射镜。

（2）转动锁紧手轮 10，打开棱镜，用滴管将被测液体加在折射棱镜表面，合上棱镜并锁紧，打开遮光板，观察视场，要求液层无气泡。

（3）旋转手轮 15，在目镜视场中找到明暗分界线的位置，再旋转手轮 6，使分界线不带任何色彩（且不呈某一单色，如红或蓝色的边线），再微调手轮 15，使分界线于十字线中心，如图 3-14-4 所示，此时目镜视场下方显示出的与竖线对齐的刻度值即为该液体的折射率（上方标度为糖的百分含量），将实验数据记录在表 3-14-1 内。

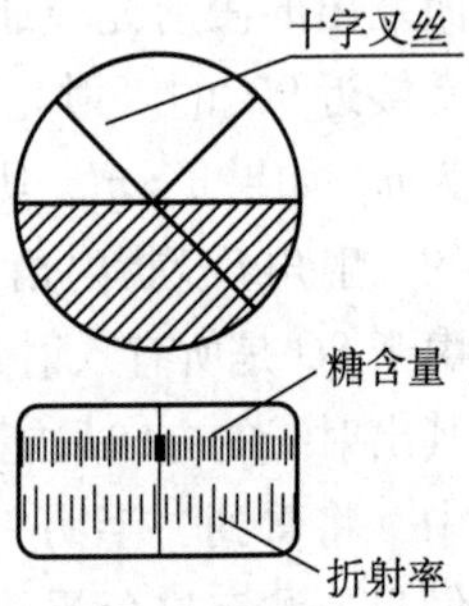

（4）将棱镜表面用擦镜纸擦干净，重复步骤（2）、（3），测定蒸馏水，7 种 NaCl 标准溶液的折射率，以浓度为横坐标，折射率为纵坐标，用坐标纸画出折射率-浓度关系曲线。

图 3-14-4 目镜视场示意图

（5）测未知浓度 NaCl 溶液的折射率，再由画出的折射率-浓度关系曲线确定未知浓度值。

（6）熟悉一元线性回归方法，设折射率-浓度的函数关系为 $y = a + bx$，由实验数据确定 a、b 值（与由曲线得到的值比较）并由方程确定未知浓度，试求相关系数。

【数据记录及处理】

表 3-14-1 乙醇和不同浓度 NaCl 的折射率

待测液体 次数	乙醇	NaCl 溶液							未知浓度
		0%	2%	4%	6%	8%	10%	12%	
第一次									
第二次									
第三次									
平均									

【注意事项】

（1）往棱镜上加待测液时，不得使滴管与棱镜表面接触。

（2）测完某液体，再测其他液体时，必须将棱镜擦洗干净，棱镜表面只能用擦镜纸擦洗，常用的清洗液有乙醇、乙醚、二甲苯等。糖类和易溶于水的盐类溶液应先用蒸馏水洗擦干净，再用有机溶剂洗涤，擦净并晾干再使用。

（3）测量有腐蚀性液体时，应尽量避免将被测液体与仪器金属部分接触。

（4）液体的折射率与温度有关，如测量同一种液体在不同温度下的折射率，可将温度计插入孔 15 内，通入恒温水，待温度稳定 10 分钟后方可测量。

（5）实验后，要用清洗液反复清洗，擦净棱镜，并晾干（15min）后方可合上棱镜。

【思考题】

(1) 阿贝折射计测定折射率的理论依据是什么？若待测物质的折射率大于折射棱镜的折射率，能否用阿贝折射计测定？为什么？

(2) 阿贝折射计中的进光棱镜起什么作用？

(3) 结合本实验思考如何测固体的折射率，并设计具体的实验步骤。

【仪器介绍】

1. 阿贝折射仪的光学系统

阿贝折射仪光学系统主要由两部分组成：望远镜系统与读数系统（图 3 – 14 – 5）。

望远镜系统：光线由反光镜 1 进入进光棱镜 2 及折射棱镜 3，被测定液体放在 2、3 之间，经阿米西棱镜 4 抵消由于折射棱镜及被测物体所产生的色散。由物镜 5

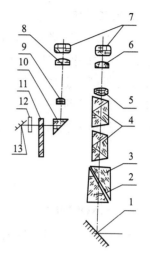

图 3 – 14 – 5　阿贝折射计的光学系统示意图

将明暗分界线成像于场镜 6 的平面上，经场镜 6 目镜 7 放大后成像于观察者眼中。

读数系统：光线由小反光镜 13 经过毛玻璃 12 和照明度盘 11，经转向棱镜 10 及物镜 9 将刻度成像于场镜 8 的平面上，经场镜 8 目镜 7 放大后成像于观察者眼中。

2. 阿贝折射仪的机械结构

阿贝折射仪的机械结构如图 3 – 14 – 6 所示。底座 14 是仪器之支承座，也是轴承座。13 为温度计座，因测量时的温度对折射率有影响，为了保证测定精度在必要时可加恒温器。

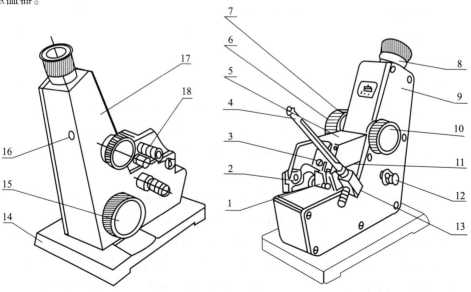

1. 反光镜；2. 转轴；3. 遮光板；4. 温度计；5. 进光棱镜座；6. 色散调节手轮；7. 色散值刻度圈；
8. 目镜；9. 盖板；10. 锁紧手轮；11. 折射棱镜座；12. 照明刻度聚光镜；13. 温度计座；
14. 支承座；15. 折射率刻度调节手轮；16. 校正螺钉孔；17. 壳；18. 恒温器接头

图 3 – 14 – 6　阿贝折射仪的机械结构

3. 准备工作

在开始测定前必须先用标准试样校对读数。将标准试样之抛光面上加 1~2 滴溴代萘，贴在折射棱镜之抛光面上，标准试样抛光一端应向上，以接受光线。当读数镜内指示于标准试样上之刻值时，观察望远镜内明暗分界线是否是在十字线中间、若有偏差则用附件校正扳手转动示值调节螺钉，使明暗分界线调整至中央（图 3-14-4）在以后测定过程中该螺钉不允许再动。此工作由实验室在准备实验时完成。

（张 翼 李百芳）

Experiment 14　Determination of Liquid Index of Refraction by Abbe Refractometer

【Objects】

(1) To familiar with the structure of Abbe refractometer, learn the correct operation of it.

(2) To learn to determine the liquid refractive index with Abbe refractometer, thereby confirm liquid concentration (refraction analytical method).

(3) To be familiar with graphic method of experiment data and the mono-linear-regression method.

【Apparatus】

Abbe refrectometer, distilled water, absolute alcohol, NaCl solution of different concentration, dropper, lens paper.

【Theory】

We know that when a ray of light is transmitted obliquely through the boundary between two materials of unlike index of refraction, the ray refracts the ray bends as shown in Fig. 3-14-1, and this phenomenon is called refraction. The way in which a ray refracts at an interface between materials with indices of refraction n_1 and n_2 is given by Snell' Law (refraction law)

$$n_1 \sin i = n_2 \sin r \qquad (3-14-1)$$

When $n_1 < n_2$, a ray of light passes from a material of lower index of refraction to one of higher index, because i must be larger than r (known from refraction law), when increase i to i_0, i_0 is called the critical angle. In this case, in the material of lower index of refraction, incident light exists from 0° to 90°, but refracted light exists only from 0° to i_0, the scope where the angle is larger than i_0 is dark, so the light and shade boundary is formed there (observing with telescope), as shown in Fig. 3-14-2.

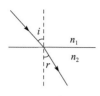
Fig. 3 – 14 – 1

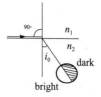
Fig. 3 – 14 – 2 Eyepiece field diagram

According to invertibility of optical path, when a ray of light passes from a material of higher index of refraction to one lower index, when angle of incidence is equal to i_0, then the angle of refraction is $90°$, if angle of incidence is larger than i_0, all the refracted ray is reflected, the phenomenon is called total internal reflection, and the critical angle i_0 is the particular angle of incidence when total internal reflection occurs.

Abbe refractometer is made according to the principle of total internal reflection. It is a frequently – used analysis apparatus in medicines examining, its structure is shown in Fig. 3 – 14 – 3, it consists of two same right angle prism ABC and DEF (as shown in Fig. 3 – 14 – 3).

Between the two prisms' surface is liquid which is measured (its refractive index is n'). Surface DF of the prism DEF is frosted glass. The light from common light source reflected by mirror M enters the prism DEF passing through nonopaque – hole P, strikes the frosted surface

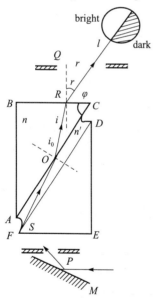
Fig. 3 – 14 – 3

DF, which makes the surface light, and becomes a radiative surface. Suppose the prism's refractive index is n, if $n > n'$ the angle of refraction which strikes surface AC is the critical angle i_0 of prism to liquid. The angle of incidence i and the angle of refraction r of the ray OR in the surface BC are both decided by i_0. That is the location of the emerging ray and it is also obtained by the refractive index n' of the measured liquid. If the incident light is monochromatic, as long as the angle of emergence of this parting line R measured, we can work out the refractive index n' of the liquid.

As shown in Fig. 3 – 14 – 3, suppose $\angle ACB$ is φ, from the triangle ORC we can know
$$i_0 + 90° = i + 90° + \varphi$$

Namely,
$$i_0 = i + \varphi \tag{3 – 14 – 2}$$

To the rays SO and OR, according to refraction law, we can know
$$n'\sin 90° = n\sin i_0 \tag{3 – 14 – 3}$$

So
$$n' = n\sin i_0 = n\sin(i + \varphi) \tag{3 – 14 – 4}$$

To point R in the BC interface

$$n\sin i = n_{空}\sin r = \sin r(n_{空} = 1) \quad (3-14-5)$$

So

$$\sin i = \frac{\sin r}{n}$$

$$\cos i = \sqrt{1 - \sin^2 i} = \frac{1}{n}\sqrt{n^2 - \sin^2 r} \quad (3-14-6)$$

Substitute $(3-14-6)$ to $(3-14-4)$, we get:

$$n' = \sin r\cos\varphi + \sin\varphi\sqrt{n^2 - \sin^2 r}$$

In the formula, the value of angle φ of prism and refractive index n are all settled, so n' is confirmed by r. Measure the angle r with refractometer, we can make sure the refractive index n' of the liquid.

Actually, Abbe refractometer has been corrected by through conversion, so we can directly read the refractive index of the liquid that needs to be measured from the instrument scale. It is named refraction alytical method that Abbe refractometer is applied to do qualitalive analysis and quantitative analysis with Abbe refractometer. Usually, while measuring the concentration of liquid, we can find out some standard solution with known concentrations and measure their refractive index with Abbe refractometer first, calculate refractive index - concentration curve of this kind of aqua after that. Then, find out refractive index of the aqua that needs to be measured by the concentration, calculate the concentration c_x according to this standard curve.

【Procedure】

(1) Shown as Fig. 3-14-5, Open reflector, rotate eyepiece and make the cross hairs' image clear in field of vision, then close the reflector.

(2) Turn the lock wheel 10, open the prism, drop measuring liquid on surface of refraction prism with dropper, close prisms and lock them, open light screen and observe the viewing field, it is desired no air bubbles appear in liquid layer.

(3) Rotate hand wheel 10, find the position of light and shade parting line in viewing field of the eyepiece, rotate hand wheel 6 again, make the colour of parting line eliminated (It does not present a certain colour, such as the line of blue or red), and fine adjusting handle 15 to align the intersection of the cross hairs with the parting line accurately, just as shown in Fig. 3-14-4. At the moment the lower scale value in viewing field, aligning at the upright line, is just refractive index of the measured solution (the upper scale value is the content of sugar), write down experimental data in Tab. 3-14-1.

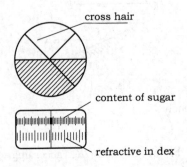

Fig. 3-14-4 Eyepiece field diagram

(4) Wipe up the prism using lens paper, repeat step (2) and step (3), measure the refractive indexes of distilled water, and seven kinds of standard solution of NaCl. Taking concentration as the abscissa and the refractive index as the ordinate, we can draw a refractive index - concentration relationship curve in the coordinate paper.

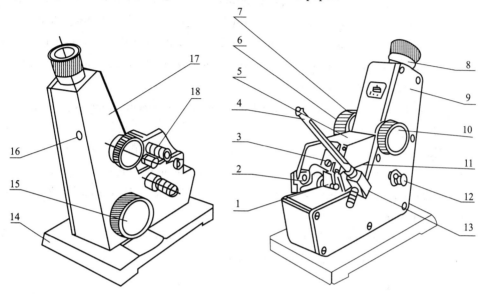

1. reflector; 2. rotor shaft; 3. light screen; 4. thermometer; 5. prism seat; 6. dispersion adjusting handle; 7. dispersion scale; 8. eyepiece; 9. cover board; 10. lock handle; 11. refraction prism seat; 12. illumination scale; 13. thermometer seat; 14. supporting seat; 15. refractive index scale adjusting handle; 16. correction screw hole; 17. cover; 18. attemperator joint

Fig. 3 - 14 - 5 Mechanical structure of Abbe refractometer

(5) Measure the refractive index of the NaCl solution while the concentration is unknown, then confirm the unknown concentration from the refractive index - concentration relationship curve.

(6) Acquaint with monobasic linear - regression method. Suppose the function relation of the refractive index - concentration is $y = a + bx$, and confirm the value of a and b by the experimental data (compared with the value got by curve), then we can work out the unknown concentration by the equation. Try to work out the related coefficient.

【Data and Calculations】

Tab. 14 - 1 Refractive index of alcohol and NaCl solution of different concentration

Measured Liquid Time	alcohol	NaCl solution							Unknown concentration
		0%	2%	4%	6%	8%	10%	12%	
First time									
Second time									
Third time									
Average									

【Note】

(1) Do not make dropper get in touch with the surface of prism, while putting the measured liquid on the prism.

(2) Scrub the prism after having measured one liquid. The surface of prism can use the lens paper to wipe only, and the mundificant used in common contain alcohol, diethyl ether, the xylol etc. After measuring carbohydrates or solution of the salt that is lyotropic in water, the prism should be scrubbed by stilled water first, then wiped by organic solvent. After that rub - up it and make it dry in the air.

(3) Avoid the liquid touching the metal parts of the instrument as far as possible, while measuring the amyctic liquid.

(4) The refractive index of the liquid has something to do with temperature. So if measure the refractive index of the same kind of liquid under different temperature, you can insert the thermometer into the bore 15, put in the homothermal water. Then measure it after being stabilized ten minutes.

(5) After the experiment, clean the prism with the mundificant, then rub - up it. Close the prism after drying it in the air about 15 minutes.

(Ma Jiao, Zhang Yi)

实验十五　利用旋光现象测定液体的旋光率和浓度

【实验目的】

(1) 观察线偏振光振动面的旋转现象。
(2) 了解旋光计的构造原理。
(3) 了解旋光性物质溶液浓度与旋光度的关系。
(4) 掌握用半荫式（或三荫式）旋光计测溶液的浓度和旋光率的方法。

【实验器材】

钠光灯、旋光计、待测溶液等。

【实验原理】

如图 3-15-1 所示，线偏振光通过某些物质的溶液后，特别是含有不对称碳原子物质的溶液，如包括糖溶液在内的许多有机物质（特别是药物）的溶液后，线偏振光的振动面将旋转一定的角度 φ，这种现象称为旋光现象，旋转的角度称为旋转角或旋光度（单位为度）。它与偏振光通过的溶液的长度 l（单位为 dm）和溶液中旋光性物质的浓度 c（单位为 g/cm³）成正比，即

$$\varphi = \alpha \cdot c \cdot l$$

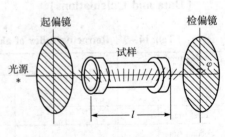

图 3-15-1　实验原理图

式中，α 称该物质的旋光率（又称为比旋光度），它表示偏振光通过单位长度、单位浓度的溶液后引起振动面旋转的角度。

某些晶体，如石英等也具有旋光性质。其旋转角 $\varphi = \alpha d$ ，式中 d 为平行光入射方向上晶体的厚度（单位为 mm）。

旋光物质依据使振动面旋转的方向又分为右旋和左旋两种。

实验表明，同一旋光物质的旋光率 α，在一定温度 t 下，与入射光波长 λ 有关，通常采用钠黄光的 D 线（$\lambda = 589.3$nm）来测定旋光度，故旋光率就以 $[\alpha]_D^t$ 来表示。因此上式可写成

$$\varphi = [\alpha]_D^t \cdot c \cdot l \qquad (3-15-1)$$

若已知待测旋光性溶液的浓度 c 和溶液长度 l，则由旋光计测其旋光度 φ，就可计算出该物质的旋光率 $[\alpha]_D^t$；反之，若测出待测溶液的旋光度 φ，查表知其旋光率 $[\alpha]_D^t$ 时，就可算出它的浓度。

测量旋光性溶液的浓度时，可采用比较法。

对于浓度不同的同一种旋光性溶液，可以利用浓度已知的溶液来测定出未知的浓度，不必知道它的旋光率 $[\alpha]_D^t$。原理如下：

设 c_0、l_0、φ_0 与 c_X、l_X、φ_X 分别为已知和待测溶液的浓度、长度、旋光度，则

$$\varphi_0 = [\alpha]_D^t \cdot c_0 \cdot l_0$$
$$\varphi_X = [\alpha]_D^t \cdot c_X \cdot l_X$$

二式相除，得

$$\frac{\varphi_X}{\varphi_0} = \frac{c_X \cdot l_X}{c_0 \cdot l_0}$$

则有

$$c_X = \frac{\varphi_X \cdot l_0}{\varphi_0 \cdot l_X} c_0 \qquad (3-15-2)$$

已知 c_0，测得 φ_0、φ_X、l_0、l_X 后，由（3-15-2）式即可求出待测浓度 c_X。

待测溶液的浓度还可由 $\varphi - c$ 曲线查出。其方法是：在旋光性溶液中长度和温度不变时，依次改变溶液的浓度 c，测出相应的旋光度 φ，然后画出 $\varphi - c$ 曲线（旋光曲线），则得到一条直线，其斜率为 $[\alpha]_D^t \cdot l$。根据这条直线可以从测得溶液的旋光度查出该溶液的浓度。并从直线的斜率计算出旋光率 $[\alpha]_D^t$。

另外，亦可由最小二乘法求 $[\alpha]_D^t$ 及 c_X。设 φ/l 与 c 之间的经验公式为

$$\varphi/l = bc + a \qquad (3-15-3)$$

根据最小二乘法原理（参见实验绪论第二节的六、统计最佳直线），则有

$$[\alpha]_D^t = b = \frac{\overline{c \cdot \varphi/l} - \bar{c} \cdot \overline{\varphi/l}}{\overline{c^2} - \bar{c}^2} \qquad (3-15-4)$$

$$a = \overline{\varphi/l} - b\bar{c}$$

$$r = \frac{\overline{c \cdot \varphi/l} - \bar{c} \cdot \overline{\varphi/l}}{\sqrt{(\overline{c^2} - \bar{c}^2)[\overline{(\varphi/l)^2} - (\overline{\varphi/l})^2]}} \qquad (3-15-5)$$

将未知浓度溶液的旋光度值代入上面已经确定的经验公式后，即可求出 c_X。

在此，我们要注意温度对于旋光率的影响。对于大多数物质，用 $\lambda = 589.3\mathrm{nm}$（钠光）测定，当温度每升高或降低 1℃时，旋光率约减小或增加 0.3%。对于要求较高的测定工作，最好能在 20℃±2℃的条件下进行。

【实验步骤】

1. 旋光计的调节和使用

（1）将仪器接于 220V 交流电源，开启开关，约 5 分钟后钠光灯发光正常就可开始工作。

（2）调节旋光计的目镜，使能看清视场中两（或三）部分的分界线（即调节目镜的焦距）。

（3）转动手轮（检偏镜），观察并熟悉视场明暗变化的规律。检查零点（暗均匀视场）位置，记下刻度盘上相应的读数——零点校正值。

（4）将溶剂（如蒸馏水）注满测试管，然后放进旋光计的暗盒中，检验溶剂是否具有旋光现象。

2. 测定旋光性溶液的旋光率和浓度

（1）记录旋光物质的温度和光波波长。

（2）将纯净的待测物质（如葡萄糖）事先配制成不同百分浓度的溶液，分别注入同一长度的测试管中。测不同浓度的旋光度各 3 次，将测量数据记录在表 3-15-1 中。

（3）根据数据处理结果，在计算机上作 $\varphi - c$ 曲线，并由曲线求未知浓度 c_X。

（4）以 $0.06\mathrm{g/cm^3}$ 浓度为标准液，用比较法求 c_X [代入式（3-15-2）]。

（5）应用最小二乘法求 $[\alpha]_D^t$ [代入式（3-15-4）] 和相关系数 r [代入式（3-15-5）]，并由经验公式 [式（3-15-3）] 求 c_X。

【数据记录及处理】

$\lambda =$ ，$T =$ ，零点校正值：$\varphi_{左校} =$ $\varphi_{右校} =$

表 3-15-1 数据记录

浓度 c		H_2O	0.02	0.04	0.06	0.06	0.08	0.10	X
管长 l (dm)		1	1	1	1	2	1	1	1
$\varphi_左$	1								
	2								
	3								
	$\overline{\varphi_左}$								
$\varphi_右$	1								
	2								
	3								
	$\overline{\varphi_右}$								
$\varphi_{左修}$									
$\varphi_{右修}$									
$\varphi = \dfrac{\varphi_{左修} + \varphi_{右修}}{2}$									
φ/l									

其中

$$\varphi_{左修} = \overline{\varphi}_{左} - \varphi_{左校}$$
$$\varphi_{右修} = \overline{\varphi}_{右} - \varphi_{右校}$$

【注意事项】

溶液应尽量装满测试管，残存的气泡应调整到测试管中凸出部分，使其不影响光线通过溶液的长度。

【预习要点】

（1）旋光度与哪些因素有关？在实验中，我们考虑了哪些因素？忽略了哪些因素？

（2）进行零点校正的目的是什么？做空白实验（测试管中只装溶剂）的目的是什么？在什么情况下可以省去空白实验？零点校正能不能省去？

【思考题】

（1）什么是偏心差？采用双游标读数为什么能消除偏心差？

（2）在某次测量中，视野中出现明暗一致（暗均匀）的视场，读数为右旋10°，也可说为左旋170°，那么你怎样判断取哪个读数为对？也就是如何判断是左旋物质还是右旋物质？

【仪器介绍】

旋光计是测量物质旋光度的仪器，通过对旋光度的测定可确定物质的浓度、纯度、含量等，可供一般的成分分析之用。广泛应用于石油、化学、制药、制糖、香料及食品等工业。

旋光计的主要结构如图3–15–2所示。其中度盘和检偏镜固为一体，借助手轮的转动同步旋转。

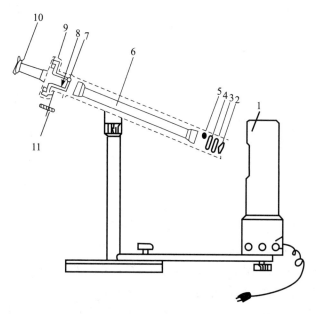

1. 光源；2. 汇聚透镜；3. 滤色片；4. 起偏镜；5. 半波片；6. 测试管；
7. 检偏镜；8. 望远镜物镜；9. 刻度盘；10. 望远镜目镜；11. 度盘转动手轮

图3–15–2 旋光计示意图

旋光计的测量原理：首先在不放入测试管的情况下，旋转检偏镜，使其偏振化方向与起偏镜的偏振化方向相互正交，这时在望远镜目镜中看到的视场最暗，然后装入有旋光性溶液的测试管，视场变稍亮。调整转动检偏镜，使视场重新达到最暗，则在上面两次过程中检偏镜所旋转过的角度即为被测溶液的旋光度。

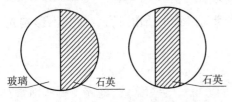

图 3-15-3　石英片的两种安装方法

因为人的眼睛难以确切地判断视场是否最暗，故实际中多采用半荫板，用比较视场中相邻两光束的强度是否相同来确定检偏镜的方位。半荫板是一个半圆形的玻璃片与半圆形石英片（半玻片）胶合成的透光片，如图 3-15-3 所示。当从起偏镜得到的偏振光通过半荫板时，透过玻璃半圆部分的光的振动方向保持不变，而透过石英半圆部分的线偏振光的振动方向则旋转了某个角度 θ（θ 为入射线偏振光振动方向与石英半玻片光轴间夹角的 2 倍），因此，通过半荫板的这束线偏振光就变成振动方向不同（两振动面间夹角为 θ）的左右两半了。

如果以 OP 和 OA 分别表示起偏镜和检偏镜的偏振化方向；又以 OP' 表示透过石英片的偏振光的振动方向，并将 OP 与 OA 的夹角记为 β，OP' 和 OA 的夹角记作 β'；再以 A_P 和 A_P' 分别表示通过玻璃片和石英片的偏振光在检偏镜偏振化方向上的分量；则由图 3-15-4 可知，当转动检偏镜时 A_P 和 A_P' 的大小将发生变化，反映在从目镜中见到的视场上将出现亮暗的交替变化（图 3-15-4 的下半部）。图中列出了四种显著不同的情形。

在图 3-15-4（a）情况下，$\beta' > \beta$，$A_P > A_P'$，故当通过检偏镜观察时，跟石英片对应的部分为暗区，跟玻璃片（即起偏镜）对应的部分为亮区，视场被分为清晰的两（或三）部分。当 $\beta' = 90°$ 时，亮暗的反差最大。

在图 3-15-4（b）情况下，$\beta = \beta'$，$A_P = A_P'$，故通过检偏镜观察时，视场中两（或三）部分界线消失，亮度均匀，较亮。

在图 3-15-4（c）情况下，$\beta > \beta'$，$A_P > A_P'$，视场又分为两（或三）部分，与石英片对应的部分为亮区，与玻璃片对应的部分为暗区。当 $\beta = 90°$ 时，亮暗反差最大。

在图 3-15-4（d）情况下，$\beta = \beta'$，$A_P = A_P'$，视场中两（或三）部分界线消失，亮度均匀，最亮。

由此可见，当检偏镜每转过 180°，则视场中就会出现以某一状态开始的，上述四个显著态的一个连续变化。

由于在亮度不太大的情况下，人眼辨别亮度微小差别的能力较强，所以常取 3-15-4（b）所示的视场为参考视场，并将不放测试管时，此时检偏镜的位置取作刻度盘的零点（如果仪器作好校准，这个读数应是 0°）。记录下此时的读数——零点校正值（属仪器的系统误差）。

当旋光计中放入测试管后，两束线偏振光均通过测试管，它们的振动面转过相同

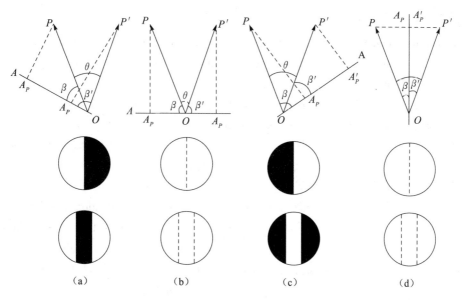

图 3-15-4 转动检偏镜时，目镜中视场的亮暗图

的角度 φ，并保持两振动面间夹角 θ 不变，视场中两半圆亮度出现差异，旋转检偏镜使视场仍旧回到图 3-15-4（b）所示的暗均匀状态，这时刻度盘上的读数与零点校正值读数之差就是被测溶液的旋光度 φ。注意零点校正值本身有正、负之分。

有些仪器中不采用半荫式，而采用三荫式，即把石英片做成条状，位于三荫板中间部分，如图 3-15-3 所示，其原理与半荫板完全相同，只不过比较的是条状部分与左右两部分之间界限消失的情况。

另外，本仪器采用双游标对称读数（即在度盘上设有两个读数窗），以消除度盘偏心差。游标窗前方装有两块 4 倍的放大镜，供读数时用。（有的型号仪器采用光学游标跳线对准的读数装置，具有对线方便、读数准确、测量精度高等优点，只是读数之前需调焦）

（李玉娟 李百芳）

Experiment 15 Determine Specific Rotation and Concentration of Liquid Utilizing Optical Phenomena

【Objects】

(1) To observe the rotation phenomenon of the vibration surface of linear polarized light.

(2) To understand the structure principle of polarimeter.

(3) To understand the relation between the concentration of optically active solution and the optical rotation.

(4) To master the method of measuring the concentration and specific rotation of solution with polarimeter.

【Apparatus】

Sodium lamp, polarimeter, solution to measure, etc.

【Theory】

As shown in Fig. 3 – 15 – 1, after the linear polarized light passes through the solution of some materials (especially the solution which contains unsymmetrical carbon, such as solution of some organic matter (especially medicine)), its vibration plane will revolve a certain angle φ, which is called the optical phenomenon. The rotation angle (degree in unit) calls the angle of rotation or optical rotation. It has a direct proportion with the length l (unit for decimeter) of the solution passed and the concentration of optically active solution c (unit for g/cm^3), namely

$$\varphi = \alpha \cdot c \cdot l$$

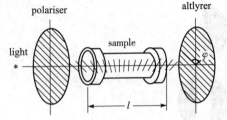

Fig. 3 – 15 – 1 Polarization principle

Where α is called specific rotation (or specific rotatory power) of the material. It represents the angle that the vibration plane rotated, after the polarized light goes through the solution of the unit concentration and unit length.

The optically active substance is divided into the dextro – rotation and laevo – rotation, according to the direction rotating vibration surface.

The experiment expresses that the specific rotation α of the same optically active substance, has something to do with the wavelength λ of the incident light, under the certain temperature t, and it usually adopts the D line ($\lambda = 589.3$ nm.) of natrium light to measure the optical rotation, so the specific rotation can use $[\alpha]_D^t$ to express. Thereby the equation can be expressed by:

$$\varphi = [\alpha]_D^t \cdot c \cdot l \qquad (3-15-1)$$

If we have already known the concentration c of the measured optically active solution and the length l, and its angle of rotation φ can be measured by the polarimeter, the specific rotation $[\alpha]_D^t$ of that material can be worked out; Where as we have detected the optical rotation φ that needs to be measured, we can check the table to know its specific rotation, thereby, we can calculate its concentration.

$$c_X = \frac{\varphi_X \cdot l_0}{\varphi_0 \cdot l_X} c_0 \qquad (3-15-2)$$

Known c_0, after measuring φ_0、φ_X、l_0、l_X, the (3 – 15 – 2) type can calculate the concentration of c_X.

The concentration of test solution can be found by $\varphi - c$ curve. The method is: when the length and temperature in the active solution are the same, the solution concentration c change in turn, measuring the degree of rotation φ accordingly, and then draw the curve of $\varphi - c$ (rotation curve), then get a straight line and the slope $[\alpha]_D^t \cdot l$. According to this line, it can be measured from the solution of optical rotation detected the concentration of the solution. And from the slope of the line, the rotation rate $[\alpha]_D^{tc}$ can be calculated.

In addition, by the least – squares method $[\alpha]_D^t$ and c_X can be obtained. Supposed the empirical formula for φ/l and c

$$\varphi/l = bc + a \qquad (3-15-3)$$

According to the principle of least-square method (see the introduction section second 6. statistical optimal linear), is

$$[\alpha]_D^t = b = \frac{\overline{c \cdot \varphi/l} - \bar{c} \cdot \overline{\varphi/l}}{\overline{c^2} - \bar{c}^2} \qquad (3-15-4)$$

$$a = \overline{\varphi/l} - b\bar{c}$$

$$r = \frac{\overline{c \cdot \varphi/l} - \bar{c} \cdot \overline{\varphi/l}}{\sqrt{(\overline{c^2} - \bar{c}^2)[\overline{(\varphi/l)^2} - (\overline{\varphi/l})^2]}} \qquad (3-15-5)$$

The optical rotation values of unknown concentration solution have been determined in the above experience formula, c_X can be calculated out.

Here, we must pay attention to the temperature effect on the optical rotation. For most materials, $\lambda = 589.3$nm (sodium light) determination, when the temperature of each increase or decrease in rotation rate is about 1 ℃, the decrease or increase of the optical rotation is 0.3%. For the determination of demanding work, the best condition is in 20 ℃ ± 2 ℃.

【Procedure】

1. The polarimeter's adjustment and application

(1) Connect instrument with the 220V A.C. power, turn on the switch, we can begin to work when the sodium lamp shines normally after about five minutes.

(2) Adjust eyepiece of the polarimeter and make the parting line of the viewing field clear (by adjusting the eyepiece's focal distance).

(3) According to the measured data, make a $\varphi - c$ relationship curve on the computer, and obtain the unknown concentration according to the curve.

(4) Fill the test tube with solvent (such as the distilled water), then put it into the dark chamber of the polarimeter, examine whether solvent has the optical activity or not.

2. Determine the specific rotation and concentrations of optically active liquor

(1) Record the temperature and the wavelength of the optically active substance.

(2) Compound the pure unknown substance (such as dextrose) into different proportions of concentration solvent, infuse them into the test tubes with the same type respectively. Test the optical rotation five times respectively corresponding to each concentration.

(3) According to the measured data, make a $\varphi - c$ relationship curve on thecomputer, and obtain the unknown concentration according to the curve.

(4) Take 0.06 g/cm³ concentrations as the standard liquid, work out c_X with the relative method.

(5) Work out $[\alpha]_D^t$ and c_X accordingly by the use of least square method.

【Data and Calculations】

$\lambda =$, $T =$, zero-point adjusted value: $\varphi_{\text{left error}} =$
$\varphi_{\text{right error}} =$

Tab. 3-15-1 Data of measurement

concentration		H_2O	0.02	0.04	0.06	0.06	0.08	0.10	X
length of tube l (dm)		1	1	1	1	2	1	1	1
φ_L	1								
	2								
	3								
$\overline{\varphi_L}$									
φ_R	1								
	2								
	3								
$\overline{\varphi_R}$									
$\varphi_{\text{left correction}}$									
$\varphi_{\text{right correction}}$									
$\varphi = \dfrac{\varphi_{\text{left correction}} + \varphi_{\text{right correction}}}{2}$									
φ/l									

Among them

$$\varphi_{\text{left correction}} = \overline{\varphi}_{\text{left}} - \varphi_{\text{right error}}$$
$$\varphi_{\text{right correction}} = \overline{\varphi}_{\text{right}} - \varphi_{\text{right error}}$$

【Note】

The test tube should be filled up with the solution, and the remaining bubbles should be adjusted to the bulgy part of the test tube, so that the length that the ray passes the solution will not be affected.

【Questions】

(1) Which factors have something to do with the optical rotation? Which factors did we consider in the experiment? Which factors did we neglect?

(2) What is the purpose of the zero correction? What is the purpose of the blank experiment (the test tube was only filled with the solvent)? Under what condition can we omit the blank test? Can the zero-point adjustment be omitted?

(3) In a certain measurement, there appears the dark and even in viewing field, the number that appears is a dextro-rotation $10°$, and it also can be expressed: the laevo-rotation $170°$, then how do you judge which reading number should you take? So as to how to judge the unknown substance is laevo-rotation or dextro-rotation?

【Introduction to Instrument】

Fig. 3-15-2 shows that main structure of polarimeter. As shown in the picture, dial scale is fastened to the polarization analyzer as one piece, and they revolve together when rotating the hand-wheel.

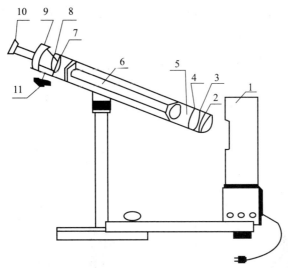

1. light source; 2. convergent lens; 3. colourfilter; 4. polarizer; 5. half-wave plate; 6. test tube; 7. polarization analyzer; 8. objective lens of telescope; 9. dial scale; 10. eye lens of telescope; 11. rotating dial hand wheel

Fig. 3-15-2　Picture of polarimeter

Here is how it works: first, rotate analyzer without the sample tube until minimum light is seen through the eyepiece of the telescope, and this is when the polarization direction of analyzer is orthogonal to that of the polarizer. Now place in the sample tube containing optically active solution, the sight will be brighter inside the telescope because more light has come through. Rotate analyzer again to let minimum light come through. The angle of the analyzer between these two tests is the optical rotation of the subjected solution.

Because it is very difficult for a human eye to accurately determine if the current sight is the darkest, thus an half-shade plate is often used in practical terms. It is used to determine analyzer direction by deciding if the intensities of the two beams of

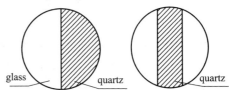

Fig. 3-15-3　Two methods for setting quartz plate

neighboring light are the same. Semitone polarization plate is a transparent plate made of a semicircle glass plate and a semicircle quartz plate (half-plate) by gluing them together. As shown in Fig. 3-15-3, after polarized light from the polarizer passing through the semitone plate, vibration direction maintains on the glass side, but it will revolve an angle θ on the quartz plate side. (θ is two times of the angle between the polarization direction of incoming light and that of the optical axis of quartz half-plate). Thus, one beam of polarized light is divided into two beams of light with different vibration direction (the angle between these two oscillation surfaces is θ). OP and OA are used to represent the polarization direction of polarizer and analyzer, while OP' represents that of the light passing through quartz plate. The angle between OP and OA is β, and the angle between OP' and OA is β'. A_P and A_P' are the components of light on the polarization direction of analyzer for the beams of light coming from glass and quartz respectively. As shown in Fig. 3 – 15 – 4, when rotating the analyzer, A_P and A_P' will change, and user will see the change of brightness when looking through eyepiece (see the bottom half of Fig. 3 – 15 – 4).

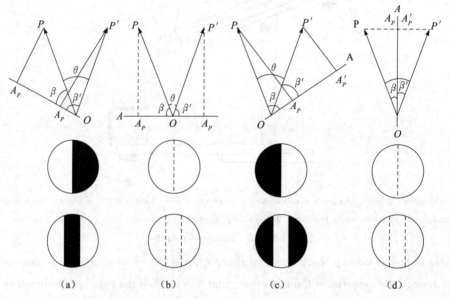

Fig. 3 – 15 – 4 Picture for brightness in eyepiece when rotating the analyzer

When it is not too bright, human eyes can detect very small difference in brightness. So we often use the sight shown in Fig. 3 – 15 – 4(b) as the reference sight. Without sample tube, we use current analyzer position as zero position, and current readings on the dial is zero position correctness value.

When sample tube is placed into polarimeter, both of those beams of polarized light will pass through the sample tube, and their vibration surface will both rotate an angle of φ. The angle between the two vibration surfaces is unchanged (θ). The brightness of the two semicircles will be different. Rotate the analyzer until the brightness of the two semicircles are the same as shown in Fig. 3 – 15 – 4(b), the difference of dial reading between current position and zero position correctness reading is the optical rotation φ of subjected solution.

(Li Yu – juan, Su Ting – ting)

实验十六　用模拟法描绘静电场

【实验目的】

(1) 了解用模拟法描绘和研究静电场的分布状况。

(2) 掌握测绘柱形电极和平行板电极间的电场分布。

（3）了解模拟法应用的条件和方法。
（4）加深对电场强度及电势等基本概念的理解。

【实验器材】

水槽式静电场模拟仪（包括水槽、探针）、同轴电极、平行板电极、游标卡尺、绘图纸等。

【实验原理】

1. 恒稳电流场模拟静电场

电场强度和电势是表征电场特征的两个基本物理量，为了形象地表示静电场，常采用电场线（电力线）和等势面来描绘静电场。电场线与等势面相互正交，因此有了等势面的分布图形就可大致画出电场线的分布图，反之亦然。我们要测出某任意带电体的静电场分布，由于其形状一般来说比较复杂，用理论计算其电场分布非常困难。若仪表探测头放入静电场，因感应电荷使被测场原有分布状态发生畸变，无法直接测绘出真实的静电场。为了解决上述问题，本实验采用一种间接测量的办法，仿造一个与待测静电场分布完全一样的电流场模型（称为模拟场），使它的分布和静电场的分布完全一样，当用探针去探测该模拟场时，它不受干扰，因此可以间接测出被模拟的静电场。

一般情况下，要进行模型模拟，模拟量和被模拟量两者在数学形式上要相同，在一定的初始条件和边界条件下，数学方程的特解相同，这样才可以进行模拟。由电磁学理论可知，电解质（或水）中稳恒电流的电流场与电介质（或真空）中的静电场具有相似性，都是有源场和保守场，都可以引入电势，两个场的电势都符合拉普拉斯方程。

在相同的边界条件下，这两个方程的特解相同，即这两种场的电势分布相似。实验中只要两种场带电体的形状和大小、相对位置以及边界条件一样，就可以用电流场来研究和测绘静电场的分布。模拟场的成立需要注意以下条件：①稳恒电流场中的导电质分布必须相应于静电场中的介质分布。②如果产生静电场的带电体表面是等位面，则产生电流场的电极表面也应是等位面。为此，可采用良导体做成电流场的电极，而用电阻率远大于电极电阻率的不良导体（如石墨粉、自来水或稀硫酸铜溶液等）充当导电质。③电流场中的电极形状及分布，要与静电场中带电导体的形状及分布相似。

2. 长同轴圆柱面电极间静电场的模拟

图 3–16–1 所示为一个同轴圆柱电极，内电极半径为 a，外电极半径为 b，内电极电势为 V_a，外电极电势 $V_b = 0$，在两极间距轴心 r 处的电势为：

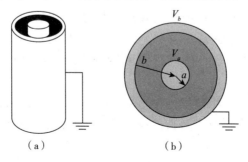

图 3–16–1　同轴圆柱面电极

$$V_r = V_a - \int_a^r E \cdot dr \qquad (3-16-1)$$

由高斯定理知半径为 r 的圆柱面上的电场强度为：

$$E = \frac{\lambda}{2\pi\varepsilon r} \quad (a \leqslant r \leqslant b) \qquad (3-16-2)$$

式中，λ 是圆柱面单位长度上的电量；ε 是两极间介电常数。由两式可得：

$$V_r = V_a - \int_a^r \frac{\lambda}{2\pi\varepsilon r} \cdot dr = V_a - \frac{\lambda}{2\pi\varepsilon} \cdot \ln\left(\frac{r}{a}\right) \qquad (3-16-3)$$

当 $r = b$ 时，

$$V_b = V_a - \frac{\lambda}{2\pi\varepsilon} \cdot \ln\left(\frac{b}{a}\right) = 0 \qquad (3-16-4)$$

则有

$$\lambda = \frac{2\pi\varepsilon V_a}{\ln(b/a)} \qquad (3-16-5)$$

代入 (3-16-3) 式，得到：

$$V_r = V_a - \frac{\lambda}{2\pi\varepsilon} \cdot \ln\left(\frac{r}{a}\right) = V_a - \frac{\ln(r/a)}{\ln(b/a)} = V_a \cdot \frac{\ln(b/r)}{\ln(b/a)} \qquad (3-16-6)$$

此式即为同轴圆柱电极间静电场中的电势分布公式。

若在同轴圆柱电极间充填均匀不良导体，在该电极间将形成稳定的电流场。同上道理，为了计算电流场的电位分布，先计算两极间的电阻，然后计算电流，最后得出两电极间任意两点的电位差。也可推导出稳定电流场中的电势分布公式为

$$V_r = V_a \cdot \frac{\ln(b/r)}{\ln(b/a)} \qquad (3-16-7)$$

以上分析说明恒定电流场与静电场的电势分布函数是相同的，可用尺寸相同、边界条件相同的稳恒电流场来模拟静电场。

检测电流场中各等势点时，为了不影响电流线的分布，探测支路不能从电流场中分流电流，因此必须使用高内阻电压表或平衡电桥法进行测绘。但直流电压长时间加在电极上会使电极产生"极化作用"而影响电流场的分布，若把直流电压换成交流电压则可消除这种影响。当电极接上交流电压时，产生交流电场的瞬时值是随时间变化的，但交流电压的有效值与直流电压是等效的，所以在交流电场中用交流电压表测量有效值的等势线与直流电场中测量同值的等势线，其效果和位置完全相同。由静电场理论我们知道，场强 E 在数值上等于电势梯度，方向指向电位降落的方向。考虑到 E 是矢量，而电势 U 是标量，从实验测量来讲，测定电势比测定场强容易实现，所以可先测绘等势线，然后根据电场线与等势线正交，画出电场线。

实验中把连接电源的两个电极放在不良导体如稀薄溶液（或水）中，在溶液中将产生电流场。电流场中有许多电势彼此相等的点，测出这些电势相等的点，描绘成面就是等势面。

【实验步骤】

（1）按照图3－16－2连接线路，电源可取静电场描绘仪信号源、其他交流电源或直流电源，经滑线变阻器 R 分压为实验所需要的两电极之间的电压值。V 表可用交流毫伏表（晶体管毫伏表）、万用表或数字万用表。下面分别测绘各电极电场中的等电位点。

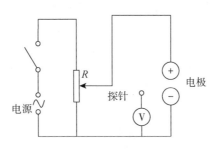

图3－16－2　线路示意图

（2）用水准仪调平水槽架底座。在水槽内注入一定量的水，在水槽架上层压好绘图用白纸，用于记录测绘点，探针置于水槽外。

（3）开启电源前，输出电压调节旋钮应逆时针旋到最小。关闭电源前，也应如此操作，以避免冲击电流过大损坏仪器。正确连接线路，接通电源，数字电压表置"输出"档，调节电压至10V 左右（粗调）接通电源。

（4）将电表转换开关拨向"探测"，让探针接触同轴圆柱中心电极，调节电压至10V（细调）。若 中心电极的电压显示为0V，则改变电源电压输出极性。

（5）确定同轴电极的轴心和边界。用探针沿外电极内、外侧分别取3 个和1 个记录点，用于确定电极的圆心和外电极的厚度。

（6）将探针置于水中，在两极间慢慢移动，依次测出电压分别为7.0V、5.0V、3.0V、1.0V 的等势线，每条等势线均匀测4 个点，测绘时沿径向移动，能较快确定测绘点。

（7）关闭电源，换记录纸，用平行板电极换下同轴圆柱电极重复上述步骤，分别沿7.5V、5.0V、2.5V 三条等势线各记录至少10 个测量点（内少外多），打出确定电极位置的点。在记录等势点时，除了两极板之间的区域外，两端延伸以描出边缘效应。

（8）在平行板电极测量纸上用不同符号标注出各等势线上的测量点和等势线数值，画出电 极，绘出实验等势线和电场线。

（9）关闭电源，将水倒掉，擦干电极并将其倒扣放置，避免电极氧化生锈，整理仪器。

【数据记录及处理】

（1）计算同轴圆柱面电极静电场等势线理论值，记录如下。

$r_{7.0\text{ v}}=$ _____ cm，$r_{5.0\text{ v}}=$ _____ cm，$r_{3.0\text{ v}}=$ _____ cm，$r_{1.0\text{ v}}=$ _____ cm

（2）将同轴圆柱电极模拟场等势线的电压分布记录在表3－16－1。

表 3-16-1 模拟场等势线电压分布测量半径

等势线测量点	r_1	r_2	r_3	r_4	r_5	平均值
V_1 (7.0V)						
V_2 (5.0V)						
V_3 (3.0V)						
V_4 (1.0V)						

将上述测量值的平均值与理论值比较，计算用模拟法描绘静电场各等势面的相对误差。

（3）绘制同轴圆柱电极电场分布图。在同轴圆柱电极记录纸上用几何方法确定圆心，画出内、外电极，用不同符号标注出各等势线上的测量点和等势线数值，绘出理论等势线（公式计算）和电场线，如图 3-16-3 所示。

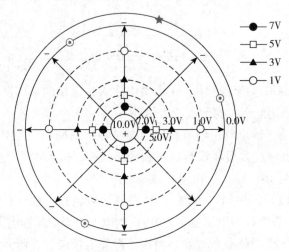

图 3-16-3 同轴圆柱电极电场

（4）绘制平行板电极电场分布图。沿等势线做测量点时，不可只局限于电极两端之间的区域内，一定要向外延伸扩展。实验中的平行板电极是有限长的，在电极两端电场存在边缘效应，所以在测量中沿等势线先在中间做两个测量点，其次在电极两端各做两个测量点，然后向两端外延再各做两个测量点，最后，再沿两个电极板的四个角各做一个测量点，确定电极的位置，如图 3-16-4 所示。

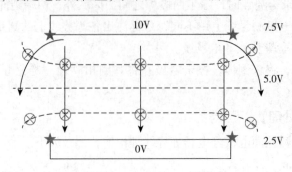

图 3-16-4 平行板电极电场分布图

【注意事项】

（1）模拟场除满足与被测场有相似的数学方程和边界条件外，还要求水槽电极放置时要端正水平，水槽中装入的水不可太少，也不可漫过电极，以避免等势线失真。

（2）开启电源前，输出电压调节旋钮应逆时针旋到最小。关闭电源前，也应如此操作，以避免冲击电流过大损坏仪器。

（3）使用同步探针时，应轻移轻放，避免变形而导致上、下探针不在同一条轴线上；在上层记录纸上打点时，不要用力过猛，轻轻按即可，以免移动电极带来误差。

（4）描绘电场线应始于高电势电极的外表面，终止于低电势电极的内表面，且处处与等势线垂直。电场线的密度反映了电场强度的大小。

（5）做实验时，要确定圆心，要确定电极位置，要描出两极板之间的区域外向外延伸的边缘效应。

【预习要点】

（1）为什么静电场不容易直接测量出来，需要引入模拟场？

（2）可以用恒稳电流场模拟静电场的前提是什么？

【思考题】

（1）试从你测绘的等位线和电力线分布图，分析何处的电场强度较强，何处的电场强度较弱。

（2）从对长直同轴圆柱面的等势线的定量分析看，测得的等势线半径和理论值相比是偏大还是偏小？有哪些可能的原因导致这样的结果？

【仪器介绍】

静电场模拟仪由电极架、电极（水槽电极）、同步探针等组成，还有配套的静电场模拟仪电源。

1. 静电场模拟仪

静电场模拟仪示意图如图 3-16-5 所示。仪器的下层用于放置水槽电极，上层用于安放坐标纸，P 是测量探针，用于在水中测量等势点，P′是记录探针，可将 P 在水中测得的各电势点同步记录在坐标纸上（打出印迹）。由于 P、P′是固定在同一探针架上的，所以两者绘出的图形完全相同。

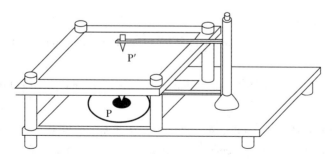

图 3-16-5 静电场模拟仪示意图

2. 水槽电极

电极的外形如图 3-16-6 所示。

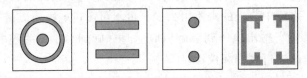

图 3-16-6　不同种类的水槽电极

3. 同步探针

同步探针由装在探针座上的两根同样长短的弹性簧片及装在簧片末端的两根细而圆滑的钢针组成。如图 3-16-7 所示，下探针深入水槽自来水中，用来探测水中电流场各处的电势数值，上探针略向上翘起，两探针处于同一直线上，当探针座在电极架下层右边的平板上自由移动时，上、下探针探出等势点后，用手指轻轻按下上探针上的揿钮，上探针尖就在坐标纸上打出相应的等势点。

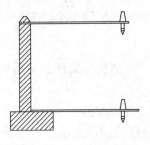

图 3-16-7　同步探针

静电场模拟仪一般配合专用测试电源使用。支架上层会配有磁性压条，固定坐标纸，避免在绘图过程中移动。

（孙　言）

Experiment 16　Simulation Method for Depicting Electrostatic Field

【Objects】

(1) Study the principle and the method for measuring and depicting the distribution of electrostatic fields.

(2) Master the electric field distribution between the coaxialcylindrical electrodes and the parallel plate electrodes.

(3) Understand the conditions and methods of simulation application.

(4) Understanding of basic concepts such as electric field intensity and potential.

【Apparatus】

Flume electrostatic fieldplotter (including flume, probe), cylindrical electrodes, parallel plate electrodes, vernier caliper, drawing paper, etc.

【Theory】

1. Simulation of electrostatic field by constant current field

The electric fieldintensity and potential are two basic physical quantities which characterize the electric field. To visually represent the electrostatic field, electric field lines (power lines) and equipotential surfaces are usually used to depict the electrostatic field. The electric field lines and the equipotential surfaces are orthogonal to each other, so that the distribution pattern of the equipotential surfaces can roughly plot the distribution of the electric field lines, and vice versa. When we want to measure the electrostatic field distribution of an arbitrary charged body, it is very difficult to theoretically calculate its electric field distribution because its shape is generally complicated. If the meter's probe is placed in an electrostatic field, the original distribution state of the measured field will be distorted due to the induced charge, and the actual electrostatic field cannot be directly mapped. In order to solve the above problems, this experiment uses an indirect measurement method to imitate a current field model (called a simulated field) that is exactly the same as the electrostatic field to be measured, so that its distribution and the distribution of the electrostatic field are exactly the same. When the probe detects the simulated field, it is undisturbed, so the simulated electrostatic field can be indirectly measured.

Generally, to carry out model simulation, both thereal and the simulated states have the same mathematical form. Under certain initial conditions and boundary conditions, the special solutions of the mathematical equations are the same, so that the simulation can be performed. It can be known from the electromagnetic theory that the current field of the steady current in the electrolyte (or water) is similar to the electrostatic field in the dielectric (or vacuum). The two kinds of fields can both introduce potential. The potentials are all Laplace equations.

Under the same boundary conditions, the special solutions of the two equations are the same, namely, the potential distributions of the two fields are similar. In the experiment, as long as the shape, size, relative position and boundary conditions of the charged bodies in the two fields are the same, the current field can be used to study and depict the distribution of the electrostatic field. Therefore, the establishment of the simulation field needs to pay attention to the following conditions: ①The distribution of the conductivity in the steady current field must correspond to the distribution of the medium in the electrostatic field. ②If the surface of the charged body that generates the electrostatic field is an equipotential surface, then the electrode surface of the current field should also be an equipotential surface. For this reason, a good conductor can be used as the electrode of the current field, and a poor conductor (such as graphite powder, tap water or dilute copper sulfate solution) having a resistivity much larger than the resistivity of the electrode can be used as the conductive material. ③The shape and distribution of the electrodes in the current field are similar to the shape and distribution of the charged conductors in the electrostatic field.

2. Simulation of electrostatic field between long coaxial cylindrical electrodes

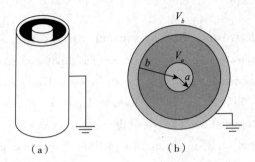

Fig. 3 - 16 - 1   Long coaxial cylindrical electrodes

As shown in Fig. 3 - 16 - 1, a coaxial cylindrical electrode has an inner electrode radius a, an outer electrode radius b, the inner electrode potential V_a, and the outer electrode potential $V_b = 0$. The potential at the axial distance r of the two pole pitch is:

$$V_r = V_a - \int_a^r E \cdot dr \quad (3-16-1)$$

According to Gauss theorem, the electric field intensity on the cylindrical surface with radius r is as follows:

$$E = \frac{\lambda}{2\pi\varepsilon r} \quad (a \leqslant r \leqslant b) \quad (3-16-2)$$

Where λ is the amount of electricity per unit length of the cylindrical surface, and ε is the dielectric constant between the two poles.

$$V_r = V_a - \int_a^r \frac{\lambda}{2\pi\varepsilon r} \cdot dr = V_a - \frac{\lambda}{2\pi\varepsilon} \cdot \ln\left(\frac{r}{a}\right) \quad (3-16-3)$$

When $r = b$,

$$V_b = V_a - \frac{\lambda}{2\pi\varepsilon} \cdot \ln\left(\frac{b}{a}\right) = 0 \quad (3-16-4)$$

Then

$$\lambda = \frac{2\pi\varepsilon V_a}{\ln(b/a)} \quad (3-16-5)$$

Substituting formula (16 - 3), we get

$$V_r = V_a - \frac{\lambda}{2\pi\varepsilon} \cdot \ln\left(\frac{r}{a}\right) = V_a - \cdot \frac{\ln(r/a)}{\ln(b/a)} = V_a \cdot \frac{\ln(b/r)}{\ln(b/a)} \quad (3-16-6)$$

The formula (3 - 16 - 6) is the potential distribution formula in the electrostatic field between coaxial cylindrical electrodes.

If the coaxial cylindrical electrodes are filled with uniform and bad conductors, a stable current field will be formed between the electrodes. Similarly, in order to calculate the potential distribution of the current field, the resistance between the two electrodes is first calculated, then the current is calculated, and finally the potential difference between any two points between the two electrodes is obtained. The above formula can also be derived as a formula for the potential distribution in a stable current field.

$$V_r = V_a \cdot \frac{\ln(b/r)}{\ln(b/a)} \qquad (3-16-7)$$

The above analysis (3 – 16 – 7) shows that the constant current field and the electrostatic field have the same potential distribution function, and the static current field can be simulated by the same constant current field with the same size and the same boundary conditions.

When detecting the equipotential points in the current field, in order not to affect the distribution of the current lines, the detection branch cannot take current from the current field, so it is necessary to use a high internal resistance voltmeter or a balanced bridge method for mapping. However, if the DC voltage is applied to the electrode for a long time, the electrode will have a "polarization effect" and affect the distribution of the current field. If the DC voltage is replaced by an AC voltage, the effect can be eliminated. When the electrode is connected to the AC voltage, the instantaneous value of the AC electric field changes with time, but the effective value of the AC voltage is equivalent to the DC voltage, so the equipotential line of the effective value is measured with an AC voltmeter in the AC electric field. The equipotential lines that measure the same value in the DC electric field have the same effect and position. From the theory of electrostatic field, we know that the fieldintensity E is numerically equal to the potential gradient, and the direction points to the direction in which the potential falls. Considering that $\boldsymbol{E}$ is a vector and the potential U is a scalar, from the experimental measurement, the measured potential is easier to measure than the measured field intensity, so the equipotential line can be mapped first, and then the electric field line is drawn according to the electric field line and the equipotential line.

In the experiment, the two electrodes connected to the power supply were placed in a poor conductor such as a dilute solution (or water), and a current field was generated in the solution. In the current field, there are many points where the potentials are equal to each other, and the points where the potentials are equal are measured, and the surface is depicted as the equipotential surface.

【Procedure】

(1) Connect the line according to Fig. 3 – 16 – 2. The power supply which can take the electrostatic field plotter signal source, other AC power or DC power supply, is divided by the sliding rheostat R into the voltage between the two electrodes required for the experiment. The V meter can be used with an AC millivoltmeter (transistor millivoltmeter), a multimeter, or a digital multimeter.

(2) Level the sink base with a level. Inject a certain amount of water into the flume, and laminate the drawing white paper on the water tank rack to record the mapping point, and place the probe outside the water tank.

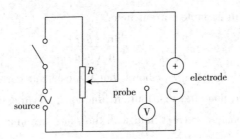

Fig. 3 – 16 – 2 Schematic circuit diagram

(3) Before turning on the power, the output voltage adjustment knob should be turned to the minimum. This should also be done before turning off the power to avoid damage to the instrument due to excessive inrush current. Connect the line correctly, turn on the power, set the "output" file on the digital voltmeter, and adjust the voltage to about 10V (coarse adjustment) before turn on the power.

(4) Turn the meter switch to "Probe" and let the probe touch the center of the coaxial cylinder and adjust the voltage to 10V (fine adjustment). If the voltage of the center electrode is displayed as 0V, the polarity of the power supply voltageneeds to change.

(5) Determine the axis and boundary of the coaxial electrode. Using the probe to take three and one recording point along the inside and the outside of the outer electrode, respectively, for determining the center of the electrode and the thickness of the outer electrode.

(6) Place the probe in the water and slowly move between the two poles. Then measure the equipotential lines with voltages of 7.0V, 5.0V, 3.0V, and 1.0V, and measure 4 points uniformly for each equipotential line. When moving in the radial direction, the mapping point can be determined relatively quickly.

(7) Turn off the power, change the recording paper, and replace the coaxial cylindrical electrode with the parallel plate electrode. Repeat the above steps to record at least 10 measuring points along the three equal potential lines of 7.5V, 5.0V, and 2.5V. , and click out the points that determine the position of the electrode. When recording the equipotential point, in addition to the area between the two plates, the ends extend to trace the edge effect;

(8) Mark the measurement points and equipotential lines with different symbols on the electrode measurement paper, draw the electrodes, and plot the experimental equipotential lines and electric field lines.

(9) Turn off the power, pour off the water, wipe the electrode and place it upside down to avoid oxidation and rusting of the electrode, and organize the instrument.

【Data and Calculations】

(1) Calculate the theoretical value of the electrostatic field equipotential line of the coaxial cylindrical electrode and fill in below.

$r_{7.0 v}$ = _____ cm , $r_{5.0 v}$ = _____ cm, $r_{3.0 v}$ = _____ cm, $r_{1.0 v}$ = _____ cm

(2) Fill the voltage distribution of the simulated field equipotential line of coaxial cylindrical electrode in Tab. 3 – 16 – 1.

Tab. 3 – 16 – 1 Simulated field equipotential voltage distribution measurement radius

Measuring points	r_1	r_2	r_3	r_4	r_5	mean
V_1 (7.0V)						
V_2 (5.0V)						
V_3 (3.0V)						
V_4 (1.0V)						

By comparing the average value of the measured value with the theoretical value, the relative error of each equipotential surface of the electrostatic field with the simulation method are obtained.

(3) Drawing electric field distribution of coaxial cylindrical electrode. On the drawinging paper, the center of the circle is determined by geometric method, draw the inner and outer electrodes, the measuring points and the values of the equipotential lines are marked with different symbols, and the theoretical equipotential lines (formula calculation) and electric field lines are drawn as Fig. 3 – 16 – 3.

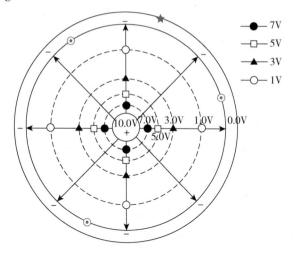

Fig. 3 – 16 – 3 Electric field distribution of coaxial cylindrical electrode

(4) Draw the electric field distribution diagram of the parallel plate electrode. As shown as Fig. 3 – 16 – 4, when drawing measurement points along the equipotential lines, it should not be limited to the area between the two ends of the electrodes, and should be extended outward. The length of parallel plate electrode in the experiment is finite, and there is an edge effect on the electric field at both ends of the electrode. Therefore, two measurement points in the middle along the equipotential line in the measurement are first drawn, and two measurement points at each end of the electrode are drawn next, and then two measurement points are respectively extended to both ends. Finally, a measurement point is drawn at each of the four corners of the two electrode plates to determine the position of the electrode.

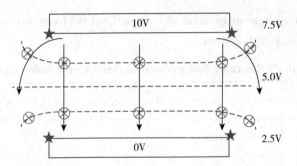

Fig. 3 – 16 – 4 Electric field distribution of parallel plate electrode

【Note】

(1) In addition to satisfying the mathematical equations and boundary conditions similar to the measured field, the simulated field also requires the electrodes in the flume to be placed horizontally. The water in the flume should not be too little, nor should it be diffused through the electrodes, so as to avoid the distortion of the equipotential line.

(2) Before turning on the power supply, the output voltage should be turned to the minimum. This should also be done before the power supply is turned off, which can avoid damage to the instrument due to excessive impulse current.

(3) When using synchronous probes, they should be moved lightly to avoid deformation so that the upper and lower probes are not on the same axis; when dotting on the upper recording paper, they should not be pushed too hard, so as to avoid moving the electrodes and causing errors.

(4) The depiction of the electric field line should start at the outer surface of the high potential electrode and end at the inner surface of the low potential electrode. The electric field line should be perpendicular to the isopotential line everywhere. The density of electric field line reflects the intensity of electric field.

(5) In the experiment, the center of the circle and the position of the electrode should be determined, and the edge effect of the outer extension of the area between the two plates should be described.

【Preview】

(1) Why is it not easy to measure the electrostatic field directly and need to introduce a simulated field?

(2) What is the precondition for simulating electrostatic field with steady current field?

【Questions】

(1) From the isoline and power line distribution maps you plotted, try to analyze where the electric field intensity is stronger and where the electric field is weak.

(2) From the quantitative analysis of the equipotential line of the long coaxial cylinder, is the measured radius of the equipotential line larger or smaller than the theoretical value?

What are the possible reasons for this?

【Introduction to instrument】

The electrostatic field plotter is composed of an electrode holder, an electrode (flume electrode), a synchronous probe, and a matching power supply for the electrostatic field simulator.

1. The electrostatic field plotter

The schematic diagram of the electrostatic fieldplotter is shown in Fig. 3 – 16 – 5. The lower layer of the instrument is used to place the flume electrode, the upper layer is used to place the drawing paper, the P is the measuring probe for measuring the equipotential point in the water, and P′ is the recording probe, which can be used to record the potential points measured in the water simultaneously on the coordinate paper. Since P and P′ are fixed on the same probe holder, the figures drawn by the two probes are identical.

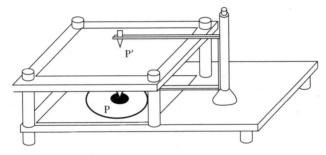

Fig. 3 – 16 – 5 Schematic diagram of electrostatic field plotter

2. Flume electrodes

The shape of the electrode is shown as Fig. 3 – 16 – 6.

Fig. 3 – 16 – 6 Different kinds of flume electrodes

3. Synchronous probe

The synchronous probe consists of two elastic reeds of the same length mounted on the probebase and two thin and smooth steel needles mounted at the end of the reed, as shown in Fig. 3 – 16 – 7. The lower probe penetrates into the tap water in the flume to detect the potential values of the current field in the water. The upper probe is slightly tilted up, and the two probes are in the same straight line. When the probe base moves freely on the plate on the right side of the lower layer of the electrode holder, the upper and lower probes can detect the equal potential points. Then gently press the button on the upper probe with your finger, and the upper probes point punches the corresponding equipotential points on the coordinate paper.

Electrostatic fieldplotter is usually used with special test power supply. The upper layer of the bracket will be equipped with magnetic battens to fix drawing paper to avoid moving during the drawing process.

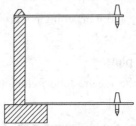

Fig. 3 – 16 – 7　Synchronous probe

(Sun Yan)

实验十七　核磁共振

近年来随着实验技术的发展，核磁共振（简称 NMR）实验方法在各个科学领域内得到广泛的应用，已制成各种测试设备，成为进行物理和化学研究的标准实验方法之一。NMR 实验技术不但是直接测定原子核磁矩和研究核结构的标准方法而且通过对液体或固体样品核磁共振谱线的研究，可以获得物质结构方面的知识；在基本实验测量方面，核磁共振还是精确测量磁场的标准方法之一，其准确度一般可达 ±0.001% 以上，NMR 实验已成为国内外高等院校物理实验课必做的实验之一。

【实验目的】

（1）了解核磁共振现象及其实验原理。

（2）掌握精确测定磁场的实验方法。

（3）掌握测定核的旋磁比和 g 因子的方法。

【实验器材】

（1）掺入硫酸铜的水溶液样品，可以获得强的质子共振。

（2）一个聚四氟乙烯棒，可以获得一个弱的氟核共振。

（3）掺有三氯化铁和硫酸铜的重水溶液（选做题用），可以获得一个弱的氘核共振。

【实验原理】

大家知道用经典力学理论描述微观粒子的运动是有很大近似性的，微观粒子服从经典规律是有一定极限的。根据量子理论具有动量矩 P_I 和磁矩 μ_I 的核，应满足下列公式

$$P_I = \sqrt{I(I+1)}\hbar$$

式中，I 是表示核性质的自旋量子数。对于不同的核，I 是表示核性质的自旋量子数。对于不同的核，I 可以取下列诸值之一：

$$0, \frac{1}{2}, 1, \cdots$$

$$\mu_I = \gamma \sqrt{I(I+1)}\hbar$$

γ 即为核的旋磁比，并用另一量子数 m_I（磁量子数）表示核磁矩在某一方向（如沿磁场 Z 的方向）上的分量，m_I 可取下列值：

$$m_I = I, I-1, I-2, \cdots -I+2, -I+1, -I$$

则

$$\mu_z m_I \gamma \hbar = m_I g \mu_N$$

一般将核磁矩投影的最大值称为核磁矩，即

$$\mu = I g \mu_N$$

在有外磁场 $\boldsymbol{B}_0$ 存在时，旋进的核获得附加的能量 E_m，它满足

$$E_m = -\mu_z \boldsymbol{B}_0 = -m_I \gamma \hbar B_0$$

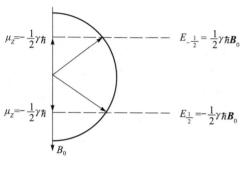

图 3-17-1　能极图

此时核的每一能级将分裂为 $2I+1$ 个次能级，每一个能级与磁场矩在空间一定取向对应，核磁矩可以处在次能级所表征的任一状态中，考虑最简单的情况，对氢核而言，$I = \frac{1}{2}$，仅有两个次能级存在，对应于 $m = -\frac{1}{2}$ 和 $m = +\frac{1}{2}$，画出以 $\hbar$ 为单位的能级图，见图 3-17-1。

因此邻近次能级的能量差，对应 $\Delta m = \pm 1$，有

$$\Delta E = E_{-\frac{1}{2}} - E_{+\frac{1}{2}} = \gamma \hbar B_0 = \omega_0 \hbar$$

上式表示若这个核再吸收一个附加的频率 $\omega = \omega_0$ 的电磁场量子能时，有可能发生能级跃迁，事实上，只要在放入恒定磁场 $\boldsymbol{B}_0$ 内的核上，附加一个垂直于 $\boldsymbol{B}_0$ 平面内旋转的，频率为 ω 的弱场 $\boldsymbol{B}_1$，若 $\omega = \omega_0$，则次能级间的跃迁便成为可能。注意到此时从 $m_{-\frac{1}{2}} \to m_{+\frac{1}{2}}$ 和 $m_{+\frac{1}{2}} \to m_{-\frac{1}{2}}$ 的跃迁可能性是相同的，即受激吸收与受激辐射的概率相等。在核磁共振中，由于磁能级之间的差别很小，自发辐射完全可忽略，这是与光频能级跃迁不同之处。为了更形象地了解磁场 $\boldsymbol{B}_1$ 的作用，我们返回到经典力学的模型上来。如果除了外加磁场 $\boldsymbol{B}_0$ 外，在 $X-Y$ 平面上还加上旋转磁场 $\boldsymbol{B}_1$，其频率也为 ω_0，为了方便地研究磁矩 μ 的运动，我们在以 $\boldsymbol{B}_0$ 为轴角速度为 ω_0 的旋转坐标系中来研究这一运动，如图 3-17-2，这样一来，磁矩 μ 在该坐标系中保持不动。现在加一射频场 $\boldsymbol{B}_1$，因为它亦将以角频率 ω_0 旋转，在旋转坐标系中，$\boldsymbol{B}_1$ 是以恒定场出现，因此磁矩 μ 在力矩 $\mu \times \boldsymbol{B}_1$ 的作用下，将开始绕 $\boldsymbol{B}_1$ 运动。在图 3-17-2（a）中 $\boldsymbol{B}_1$ 对 μ 产生的力矩 $\mu \times \boldsymbol{B}_1$ 使 μ 和磁场 $\boldsymbol{B}_0$ 之间的夹角 θ 加大，我们知道 μ 在磁场 $\boldsymbol{B}_0$ 中的能量为

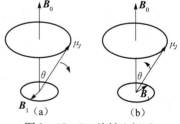

图 3-17-2　旋转坐标系

$$E = -\mu \cdot \boldsymbol{B}_0 = |\mu| \cdot |B_0| \cdot \cos\theta$$

因此 θ 增大，意味着系统的能量增加。在图 3-17-2（b）中，B_1 对 μ 产生的力矩 $\mu \times B_1$ 使 μ 和 B_0 之间的夹角 θ 减小，这意味着系统能量减少。系统的这个能量变化可借助于外电路加以探测。

综上所述，核在一定恒磁场和一定旋转场的作用下，当旋转场的圆频率等于拉莫尔旋进频率时，会引起核的磁量子态发生跃迁，这种现象就称之为核磁共振。

【实验装置】

实验装置方框图见图 3-17-3。均匀

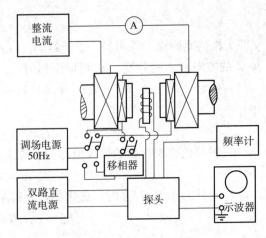

图 3-17-3 实验装置

的恒定磁场是由稳压电源供电的电磁铁产生，其强度由磁极距离及磁化电流调节。电磁铁极头直径约为 10cm，极隙约为 2cm，极头上装有一对扁平的调制场线圈，提供一个可调的均匀调制磁场，其最大强度约为 20Gs，用移相器可以调整示波器水平输入与调制场之间的相位。探头是一振荡器，由它在谐振圈中产生射频磁场。圆柱形样品放在谐振线圈内，并应注意调准其轴线与恒定磁场方向垂直。射频频率用数字频率计测量，共振吸收信号用示波器检测。

如前所述，核磁共振产生的必需条件是提供一个角频率等于核拉莫尔旋进频率（$\omega = \omega_0$）的旋转磁场，而在图 3-17-4 中可以看见，在含样品的射频线圈内实际建立的是一个线性振动的磁场，但从振动理论可推论出它相当于一个旋转磁场的存在。从图 3-17-4可以看出，一个线性简谐运动等效于两个旋转方向相反的旋转运动，和拉莫尔

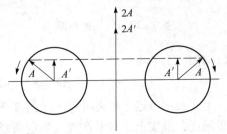

图 3-17-4 旋转磁场示意图

旋进方向相反的旋转运动，和拉莫尔旋进方向相同的旋转磁场，从而产生共振，引起跃迁；另一成分因为旋转方向相反，不能引起共振旋转磁场的作用，不会对样品产生效果。

为了观察共振信号，必须应用调场技术，对吸收的范围予以扫描，即在恒定磁场 B_0 方向上，加上一个低频（为方便起见，我们选用 50Hz）调制磁场 B'，此时样品所在的实际磁场为 $B_0 + B'$。由于调制场的振荡不大，总磁场方向保持不变，只是大小按调制频率发生周期性变化，对应的拉莫尔频率 ω_0 也发生周期性变化。由于局部场等原因的存在，实际的共振吸收发生在磁场的一定范围内，所以只有在射频频率 ω 所对应的共振吸收范围 B_0' 被 $B_0 + B'$ 扫过的时间内才产生核磁共振（图 3-17-5），可观察到共振吸收信

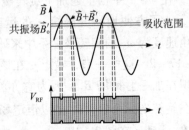

图 3-17-5 共振吸收范围示意图

号；而在其他时刻不会发生共振吸收，也没有共振信号出现，若$|B'_0-B_0|$小于调制场幅度$|B'_{max}|$时，则磁场曲线在一周内必与B_0在两处相交，它们都满足共振条件，此时将出现间隔不均等的共振信号，如图3-17-6（a）所示。如果改变恒定磁场B_0的大小或变化射频频率ω，则将使信号的相对位置发生变化，出现信号"相对走动"现象，这是因为此时的共振不是电磁铁的恒定磁场值，若出现间隔相等的共振信号，见图3-17-6（b），其位置与调制场B'的辐值无关，并随B'辐度的减少信号变宽，图3-17-6（c）此时表明与射频频率对应的共振场恰好是电磁铁的稳定磁场B_0，应有核磁共振关系式。

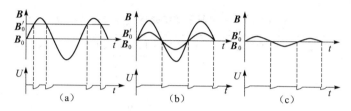

图3-17-6 对应于射频频率的共振场

在示波器荧光屏上出现的核磁场共振信号通常有如图3-17-7的形状，除有一个高峰外，还包含称为弛豫尾巴的一系列辐度衰减的小峰。NMR实验技术水平的重要标志之一是讯号噪音比，简称讯噪比，它除和探测仪器的性能有关外，还和样品的体积V、核的旋磁比γ以及磁场B_0等量有关系。

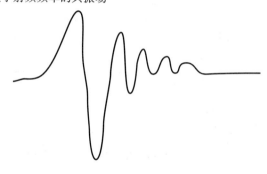

图3-17-7 核磁场共振信号形状

【实验内容】

（1）观察质子的核磁场共振。变化恒定磁场、射频频率、注意共振吸收讯号的位置和幅度的变化，并对实验结果作简略分析。

精确测定两个以上不同的磁场值，其计算公式如下：
$$B_0 = 2.3487355 \times 10^{12} f$$

B_0为被测磁场值，以"特斯拉（T）"为单位；f为射频频率，以"赫兹（Hz）"为单位。

（2）观察氟的核磁共振现象，测出其旋磁比，计算g因子和磁矩并给出值做比较，对结果误差做简略估计和分析。

应用含氟核的样品，通过调节磁场首先观察^{19}F的共振信号，测出对应的共振频率f，然后再测出磁场B_0，便可由下式算出
$$\gamma = 2\pi \frac{f}{B_0}$$

g因子的计算可由下式得出
$$g = \frac{\gamma \hbar}{\mu_N}$$

式中，$\hbar = \dfrac{h}{2\pi}$；h 为普朗克常数，等于 6.626×10^{-34} J·s；μ_N 为核磁子，真值为 5.051×10^{-27} J/T。用 $\mu = Ig\mu_N$ 计算核磁矩。

（3）测定氢核（^{1}H）的横向弛豫时间 T_2，T_2 称为横向弛豫时间，它是反映当原子核的平衡状态被破坏后，沿外磁场垂直的方向上总磁化强度分量恢复到平衡状态快慢的物理量，其数量级在 $10^{-5} \sim 10^2$ s 之间。（详见梅振岳主编《原子核物理学》）

（4）（选做）观察氟核磁共振现象。测出氟核的旋磁比及其磁矩。参考表 3-17-1。

表 3-17-1　一些原子核核磁场共振的有关参数

元素名	电荷数	质量数	共振频率 $\times 10^6$（Hz/10^4T）	磁矩（μ_N）	自旋量子数 I
H	1	1	42.57	2.792745	1/2
Li	3	7	16.55	3.257	3/2
B	5	11	13.66	2.687	3/2
C	6	13	10.71	0.702	1/2
F	9	19	40.06	2.628	1/2
Al	13	27	11.09	3.641	5/2
P	15	31	17.24	1.131	1/2

（支壮志　孙宝良）

Experiment 17　Nuclear Magnetic Resonance

Foreword：In recent years, along with the development of experimental technology, the experimental technique of the nuclear magnetic resonance (called NMR for short) has a widespread application in each different fields. It has been made into various kinds of testing facilities, and became one of standard experimental techniques carried on physics and chemistry research. The NMR experiment technology is not only the standard method of immediate determination of atomic nucleus magnetic moment and the research on nuclear structure, but also a method through research on nuclear magnetic resonance spectral line of the liquid or solid sample, we may obtain knowledge on material structure. As to the basic measurement, nuclear magnetic resonance is one of the standard methods of magnetic field precision measurement, and its accuracy may reach generally above $\pm 0.001\%$, therefore NMR has become one of the essential and required physical experiments in domestic and foreign institutions.

【Objects】

(1) To understand nuclear magnetic resonance phenomenon and its experimental principle.

(2) To grasp the the experimental technique for accurate determination of magnetic field.

(3) To master the method of determinating nucleus gyromagnetic ratio and g factor.

【Apparatus】

The aqueous solution sample mixed with cupric sulfate is to obtain strong proton resonancing. A polytef stick is to obtain a weak fluorine nucleus resonating. Deuterated solution mixed with iron trichloride and cupric sulfate (used for being selected) is to obtain a weak triton resonating.

【Theory】

It is known that the description of the microscopic particle movement with the classical mechanics theory has its great approximate, and the microscopic particle obeys the classical rule in a certain range. According to the quantum theory, nucleus with the moment of momentum and the magnetic moment should correspond with the following formula

$$P_I = \sqrt{I(I+1)}\hbar$$

I is nucleus spinning quantum number. Regarding different nucleuses, and it may adopt one of following various values

$$0, \frac{1}{2}, 1, \cdots,$$

$$\mu_I = \gamma \sqrt{I(I+1)}\hbar$$

Namely γ is the nucleus magnetogyric ratio. Another quantum number m_I (magnetic quantum number) is the nuclear magnetic moment heft in some direction (for example along the magnetic field Z direction), which may take the following value:

$$m_I = I, I-1, I-2, \cdots, -I+2, -I+1, -I$$

then

$$\mu_Z = m_I \hbar \gamma = m_I g \mu_N$$

In general, the maximum value of the nuclear magnetic moment projection is called the magnetic moment, namely

$$\mu = I g \mu_N$$

When an external magnetic field exists, the precession nucleus will obtain an additional energy E_m, it satisfies

$$E_m = -\mu_Z B_0 = -m_I \gamma \hbar B_0$$

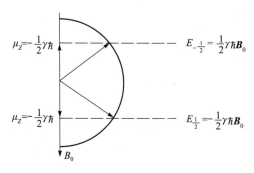

Fig. 3-17-1 Energy level diagram

At this time each nucleus energy level is split into $2I+1$ second-energy levels, each energy level is corresponding to magnetic field moment with a certain spatial orientation. The nuclear magnetic moment may occupy any inferior energy state of levels. For the simplest case, to hydrogen nucleus, $I=1/2$, only two second-energy levels exist, corresponding to $m = -1/2$ and $m = +1/2$. In Fig. 17-1, energy – level diagram with unit $\hbar$ is shown.

Therefore the neighborhood energy level difference is corresponding to $\Delta m = \pm 1$, and $\Delta E = E_{-\frac{1}{2}} - E_{+\frac{1}{2}} = \gamma \hbar B_0 = \omega_0 \hbar$. The above formula shows if the nucleus absorbs an additional electromagnetic field quantum energy with $\omega = \omega_0$ again, a transition will be produced. In fact, so long as a weak rotating magnetic field $\boldsymbol{B}_1$ is added on the constant magnetic field $\boldsymbol{B}_0$, which is vertical to the $\boldsymbol{B}_0$ plane, the energy level transition becomes possible. Notes: possibilities for jumping of $m_{-\frac{1}{2}} \to m_{+\frac{1}{2}}$ and that of $m_{+\frac{1}{2}} \to m_{-\frac{1}{2}}$ are equal, namely the odds of stirred absorption and stirred radiation are equivalent. In nuclear magnetic resonance, spontaneous radiation may be neglected due to the minor magnetism energy level difference, which is different to the jumping of optical frequency energy. To understand magnetic field ($\boldsymbol{B}_1$) function more vividly, it is necessary to return to the classical mechanics model.

If there is an exception of the additional magnetic field $\boldsymbol{B}_0$, a rotary field $\boldsymbol{B}_1$ is also added in $X - Y$ plane, whose frequency is also ω_0. In order to conveniently study the movement of magnetic moment μ, the rotating coordinate system is adopted, whose angular speed is ω_0, and $\boldsymbol{B}_0$ is axis as Fig. 3 – 17 – 2. Then, magnetic moment μ maintains motionless in this coordinate system. Now add a radio frequency field $\boldsymbol{B}_1$, for it revolves at the angular frequency ω_0. In this rotating coordinate system, $\boldsymbol{B}_1$ is the constant field, therefore the magnetic moment μ starts to encircle $\boldsymbol{B}_1$ under the function of $\mu \times \boldsymbol{B}_1$. In Fig. 3 – 17 – 2 (a),

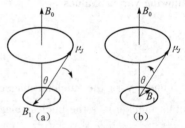

Fig. 3 – 17 – 2 Rotating coordinate system

moment of force $\mu \times \boldsymbol{B}_1$ makes the θ angle enlarge, which is between μ and $\boldsymbol{B}_0$. We know that the energy of μ in the magnetic field $\boldsymbol{B}_0$ is

$$E = -\mu \cdot \boldsymbol{B}_0 = -|\mu| \cdot |\boldsymbol{B}_0| \cdot \cos\theta$$

Therefore system energy will increase with increasing θ. In Fig. 3 – 17 – 2 (b), moment of force $\mu \times \boldsymbol{B}_1$ makes the θ angle decrease, which means the decrease of system energy. The change of system energy can be detected by usage of external circuit.

In summary, the magnetism quantum state of nucleus will jump in the function of certain permanent magnetic field and certain rotating field, only when the circular frequency of rotating field is equal to that of Larmor precession, this phenomenon is called nuclear magnetic resonance.

【Experimental Installation】

The block diagram of the experimental installation is shown in Fig. 3 – 17 – 3. The even constant magnetic field is produced by the electromagnetsupplied by stabilized voltage supply, and its intensity is regulated by the distance of magnetic poles and the magnetization current. The diameter of electromagnet pole is approximately 10cm, the gap of poles is about 2cm. A pair of flat modulation field coils are fixed on the poles. They provide even modulation magnetic field, whose maximum is 20 Gauss.

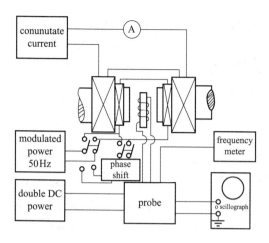

Fig. 3 – 17 – 3　Experimental device

The phase between oscilloscope level input and modulation field may be adjusted by phase shifter. Probe head is an oscillator, which produces the radio frequency magnetic field. The cylindrical sample is placed in the resonance inductor, whose axis is vertical to the constant magnetic field direction. The radio frequency is measured by the digital frequency meter, and the resonance absorption signal is determined by oscilloscope respectively.

As mentioned above, it is necessary for the nuclear magnetic resonance to provide a rotary magnetic field, whose angular frequency is equal to that of Larmor precession. It can be seen in Fig. 3 – 17 – 4 that a linear vibration magnetic field is generated in radio – frequency coil, which is equal to a rotary magnetic field.

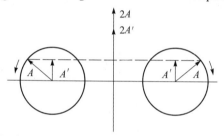

Fig. 3 – 17 – 4　Diagram of rotating magnetic field

From the Fig. 3 – 17 – 4, a linear simple harmonic motion is equivalent with two opposite rotary motions: one direction is same to Larmor precession direction, the other is opposite. So it leads to resonate and jumping. The latter can't cause the effect on the sample for its opposite direction to Larmor precession direction.

In order to observe the resonates signal, adjusting field technology must be applied to scan in the absorption scope, namely, low frequency B' (for convenience, 50 Hz is adopted) is added on the constant magnetic field B_0. Actual magnetic field is $B_0 + B'$. Due to little vibration of the modulation field, the total magnetic field's direction is invariable, whose size changes periodically with the modulation frequency correspondingly, and the corresponding Larmor frequency ω_0 also has the periodic variation. As a result of existence of the local field, the actual resonance absorption occurs in a certain magnetic field scope. Only in the resonance absorption scope corresponding to radio frequency ω, the nuclear magnetic resonance is generated in the scanning time, in which B_0' is scanned by $B_0 + B'$ (as shown in Fig. 3 – 17 – 5). The resonance absorption can't generate in other times, and the resonated signal does't appear, too. If $|B_0' - B_0|$ is smaller than $|B'_{max}|$, the magnetic field curve must intersect with B_0 at two places in a period. They all satisfy the resonates condition, at this time the resonate signal with different

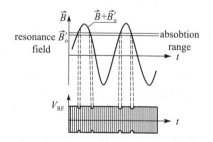

Fig. 3 – 17 – 5　Diagram of resonance absorption range

intervals will appear, as shown in Fig. 3 – 17 – 6(a).

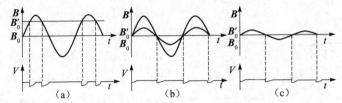

Fig. 3 – 17 – 6 Resonance fields corresponding to radio frequency frequencies

If the size of constant magnetic field B_0 or the radio frequency frequency is changed, the relative position of signal will change, and the phenomenon "walk relatively" occurs. The reason is resonate magnetic field value is not equal to electro – magnet's constant magnetic field value. If the resonate signal appears with the same interval, as shown in Fig. 3 – 17 – 6 (b), whose position has nothing to do with amplitude of the modulation field B', and it will be widened with decreasing amplitude of B', as shown in Fig. 3 – 17 – 6(c). It shows resonate field corresponded to radio frequency is just the electro – magnet stable magnetic field, which should satisfy the nuclear magnetic resonance relationship.

On the oscilloscope fluorescent screen, the resonance signal of nuclear magnetic field usually takes the shape, which is shown in Fig. 3 – 17 – 7. Besides a peak, it also contains a series of small peak with amplitude decay which are called the relaxation tail. One of the important symbols for NMR experiment techniques is the ratio of signal and noise, which is called the signal – noise ratio for short. It is related not only with detecting instrument property, but also with the sample volume V, the nucleus magnetogyric ratio γ and the magnetic field B_0, etc.

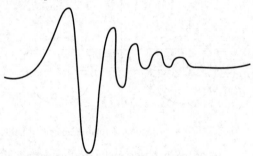

Fig. 3 – 17 – 7 Shape of nuclear magnetic resonance signal

【Procedure】

(1) The observation of nuclear magnetic resonance of the proton. Change the constant magnetic field and frequency of the radio frequency, pay attention to the position of resonance absorption signal and the changing amplitude, and make a abbreviated analysis of the experiment result. Determinate accurately more than two different magnetic fields values, its formula is as follows:

$$B_0 = 2.3487355 \times 10^{12} f$$

B_0 is the measured magnetic field value, take "T" as unit; f is the frequency of radio frequency, take "Hz" as unit.

(2) Observe nuclear magnetic resonance phenomenon of fluorine, determine its magnetogyric ratio and calculate the g factor and the magnetic moment, then make a comparison with

the given value, and make a brief estimate and analysis of the result error.

Using the sample containing fluorine nucleus, observe the resonance signal first through adjusting the magnetic field, determine the corresponding resonance frequency f, and determine the magnetic field B_0 again, then the value may be figured out by the following formula

$$\gamma = 2\pi \frac{f}{B_0}$$

The g factor may be calculated by next formula.

$$g = \frac{\gamma \hbar}{\mu_N}$$

Among them, $\hbar = \frac{h}{2\pi}$, and h is Planck constant whose value is $6.626 \times 10^{-34} J \cdot s$, μ_N is nuclear magneton, its true value is the $5.051 \times 10^{-27} J/T$. Use formula $\mu = I g \mu_N$ to calculate nuclear magnetic moment.

(3) Determine the transverse relaxation time of hydrogen nucleus (^1H) T_2, which is called transverse relaxation time. It is a physical quantity which the total magnetization component restores to the state of equilibrium (after being destroyed) fast or slowly along in the direction vertical to outside magnetic field. Its order of magnitude is between $10^{-5}s$ and $10^2 s$.

(4) (Alterative) Observe the deuteron magnetic resonance phenomenon, and determine the deuteron's magnetogyric ratio and its magnetic moment (Tab. 3 - 17 - 1).

Tab. 3 - 17 - 1 Parameters of nuclear magnetic resonance

element	charge number	mass number	resonance frequency $\times 10^6$ (Hz/T)	magnetic moment (μ_N)	spin quantum number I
H	1	1	42.57	2.792745	1/2
Li	3	7	16.55	3.257	3/2
B	5	11	13.66	2.687	3/2
C	6	13	10.71	0.702	1/2
F	9	19	40.06	2.628	1/2
Al	13	27	11.09	3.641	5/2
P	15	31	17.24	1.131	1/2

(Zhi Zhuang – zhi)

实验十八 激光全息照相

全息照相（全息术）英文为 holography，原由希腊文 halos（全部）和 graphein（记录）组成，即"完全记录"之意。全息照相是以光的干涉和衍射等物理光学规律为基础，它不但能同时记录物体光波的全部信息（振幅和相位），又能完全再现被摄物光波的全部信息，所以全息照相形象逼真，富有立体感。波前再现的理论，早在1948年由英国物理学家盖伯提出并完善，从而奠定了三维照相的基础，1971年，由于在全息术方面的贡献，他获得了诺贝尔物理学奖。20 世纪 60 年代初高强度相干光——激光的问世，使全息照相迅速发展，其应用领域不断扩大，成为近代光学领域的重要分支。它在无损检测、精密测量、遥感技术、生物医学及军事方面有非常广泛的应用。近些年风靡世界的激光全息模压图片，以其绚丽多彩的图像和高技术含量，在防伪技术和精美包装方面起着巨大的作用，产生了良好的效益。

【实验目的】

(1) 学习拍摄静态全息照片的技术和再现观察的方法。
(2) 了解全息照相技术的主要特点。

【实验器材】

全息台、氦氖激光器、曝光定时器、分束镜（透射 95%）、全反射镜（2 个）、扩束镜（100 倍和 40 倍）、拍摄物。

【实验原理】

物体上各点发出的光（或反射的光）是电磁波，借助于它们的频率、振幅和相位信息的不同，人可以区别物体的颜色、明暗、形状和远近。普通照相用透镜将物体成像在感光底片平面上，曝光后，它仅记录了物体表面光强（光振动振幅的平方）的分布，却无法记录光振动的相位，因此它得到的是物体的一个平面像。而全息照相能够把光波的全部信息——振幅和相位全部记录下来，并能再现物体的立体像。

1. 全息照相的记录原理

全息照相的实验原理光路图见图 3-18-1。图中氦氖激光器射出的激光束通过分束镜 S 分成两束，一束经反射镜 M_1 反射，再由扩束镜 L_1 使光束扩大后照射到被摄物体 D，经物体表面反射（或透射）后照射到感光底片 H（全息干板）上，这部分光称为物光（O 光）。另一束光经 M_2 反射，经 L_2 扩束后，直接投射到感光底片 H 上，这部分光称为参考光（R 光）。两束光到达底片上的每一点都有确定的相位关系。由于激光的高度相干性，两束光在底片上迭加，形成稳定的干涉图样并记录下来。

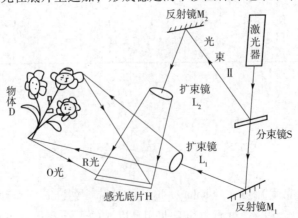

图 3-18-1 全息照相的实验原理光路图

为简单起见，我们分析物体上某一物点 C 的情况。假设参考光为垂直底片表面的平面波，如果感光片对物点 C 所张的立体角充分小，从物点发出的球面波在感光片上任一小区域，如图 3-18-2 所示，可以简化为平面波加以处理，如图 3-18-3 所示。在这个小区域内，物光和参考光的干涉可简化为两束平行光的干涉，见图 3-18-4。可以证明，它们形成的条纹间距为

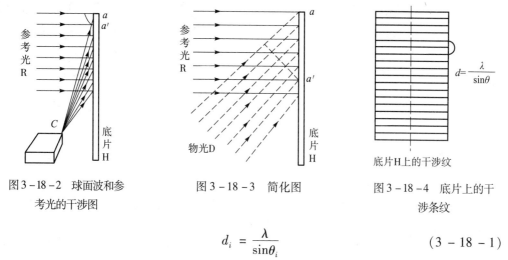

图 3-18-2 球面波和参考光的干涉图　　图 3-18-3 简化图　　图 3-18-4 底片上的干涉条纹

$$d_i = \frac{\lambda}{\sin\theta_i} \quad (3-18-1)$$

式（3-18-1）中，λ 为相干光的波长，θ_i 为物光与参考光之间的夹角。

干涉图像中亮条纹和暗条纹之间明暗程度的差异，取决于两束光波的强度（振幅的平方）。由于同一物点发出的物光在感光板上不同区域与参考光的夹角不同，不同物点发出的物光在感光板上同一区域的光强以及与参考光的夹角也不同，因此其干涉条纹的浓黑程度、疏密和走向也各不相同。

总的物光波可以看成由无数物点发出光波的总和，因此在全息感光板平面上形成的是无数组浓黑程度、疏密走向情况各不相同的干涉条纹的组合。曝光以后，经过显影和定影等底片处理过程，物光波全部信息的干涉图像就被记录下来了。

2. 全息照相的再现原理

全息照相在感光板上的记录的不是被摄物体的直观形象，而是无数组干涉条纹的复杂组合。其中每一组干涉条纹有如一组复杂的光栅，因此当我们观察全息照相记录的物像时，必须采用一定的再现手段，即必须用与原来参考光完全相同的光束去照射（这光束称为再现光）。再现观察时的光路如图 3-18-5 所示。

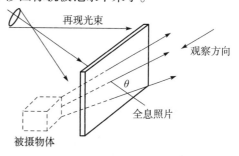

图 3-18-5 全息照相的再现原理图

在再现光束的照射下，全息照片相当于一块透过率不均匀的障碍物，再现光通过它时就会发生衍射，如同一幅极为复杂的光栅衍射一样。以全息照片上某一小区域 ab 为例，为简单起见，把再现光看成一束平行光，且垂直照射在全息照片上，再现光将发生衍射。其中 +1 级衍射光是发散光，在各原物点处成一虚像。-1 级衍射光是会聚光，各会聚点在各原物点对称的位置上，如图 3-18-6 所示。

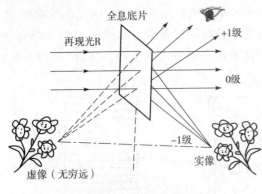

按光栅衍射原理，这时衍射角满足

$$\sin\theta_i = \frac{\lambda}{d_i}$$

这样，一幅复杂而又极不规则的光栅的集合就产生了衍射图像。其中+1级衍射光形成一个与原物完全对称的虚像，称为虚像；-1级衍射光形成一个实像，称为赝像。0级光仍按再现光原方向传播。迎着+1级衍射方向观察，在原物体的位置上，就能看到与原物形象完全一样的立体像。

图 3-18-6 全息照片

3. 全息照相的特点

（1）全息照片所再现出的被摄物体像具有完全逼真的三维立体感。当人移动眼睛从不同角度观察时，就好像面对原物一样，可看到它的不同侧面，在某个角度被摄物遮住的另一物体，也可以从另一角度看到。

（2）由于全息底片上任一小区域都以不同的物光倾角记录了来自整个物体各点的光信息，因此，一块打碎的全息照片，我们只需取出任一小碎片，就能再现出完整的被摄物体立体像。

（3）同一张全息感光板可以进行多次重复曝光、重复记录。只要稍微改变感光板的方位（如转动一个角度），或改变参考光束的入射方向，就可在同一感光板进行多次重复曝光记录，并能互不干扰地再现各自的图像。如果全息过程光路各部件都严格保持不动，只使被摄物体在外力作用下发生微小的位移或形变，并在形变前后重复使感光板曝光，则再现时物体变形的前、后两次记录的物光波将同时再现，并形成反映物体形态变化特征的干涉条纹。这就是全息干涉计量的基础。

（4）若用不同的波长的激光照射全息照片，再现像可以得到放大或缩小。根据公式，再现光的波长大于原参考光时，像被放大，反之缩小。

（5）全息照片再现时物光波是再现光束的一部分，因此，再现光束越强，再现出的物像就越亮。实验指出，亮暗的调节可达10^3倍。

【实验装置】

为了实现光波的全息记录，静态全息照相必须具备下列三个基本实验条件。

1. 相干性好的光源

氦氖激光器具有较好的相干性，它输出激光束的波长为632.8nm。若谱线宽度$\Delta\lambda=0.0020$nm，则相干长度$L_m=\frac{\lambda^2}{\Delta\lambda}=20$（cm）。有了相干性较好的光源，实验中还必须注意以下事项。

（1）尽量减少物光和参考光的光程差。实验中必须妥善安排好光路，使它的光程差控制在数厘米之内。

（2）参考光和物光的光强比一般选在2:1~10:1之间，因此需要挑选较为合适的分光板和衰减片。参考光和物光的光强可以用光电池配以灵敏电流计进行测量。由于在光通量不太大的范围，光电池的光电流与入射光通量成正比，因此，在此条件下，当入射光为参考光与物光时，其分别在光电池中产生的光电流之比，即为参考光与物光的光强比。

2. 高分辨率的记录介质

感光板记录的干涉条纹一般都是非常密集的。以图 3-18-2 为例，如果 $\theta = 30°$，$\lambda = 632.8\text{nm}$，则形成的干涉条纹间距 $d = \dfrac{\lambda}{\sin\theta} = 1.3 \times 10^{-3}$（mm），亦即每毫米将记录近千条条纹。随着夹角的增大，条纹间距将进一步减小。而普通照相感光片的分辨率仅约每毫米 100 条左右。因此全息照相需要采用高分辨率的介质——全息感光片进行。这种感光片的分辨率可大于每毫米 1000 条以上，但感光灵敏度不高，所需曝光时间一般比普通照相感光片要长（45s~1min）。用于氦氖激光的全息干板对红光最敏感，全息照相的全部操作可在暗绿色灯光下进行。

3. 良好的减震装置

由于干涉条纹密集，故曝光记录时必须有一个非常稳定的条件，轻微的振动或扰动只要使光程差发生波长数量级的变化，条纹即会模糊不清。为此全息实验室一般都选在远离震源的地方，全息照相光路各元件全都布置在全息防震工作台上，被摄物体、各光学元件和全息感光片全都严格固定。同时，拍摄时还需防止实验室内有过大的气流流动。

【实验步骤】

1. 拍摄静物的全息照片

（1）按照图 3-18-1 在全息台上布置光路，符合下列要求。

①物光路和参考光路大致等光程（计算物光和参考光光程，应从何处算起？）。为方便起见，扩束镜 L_1 和 L_2 可暂勿放入光路，感光板亦可先以其他玻璃板代替。

②放入扩束镜 L_1 和 L_2（尽量充分利用激光光能，在物光路尤其如此），被摄物和感光板位置分别受到物光束及参考光束的均匀照明。严格防止扩束后的物光束直接照射到感光板位置。

③使参考光和物光的光强比在合适的范围。

（2）曝光

①由光强情况选定曝光时间（由实验室给出）。

②遮住激光束，装感光片，感光片乳胶面应向着激光束。

③静止数分钟，然后曝光。曝光过程中严防震动，各光学元件应严格固定，除暗绿灯光外，无其他杂散光干扰。

（3）感光板的显影、定影：显影采用 D-19 显影液，定影采用 F-5 定影液。处理过程与普通感光板相同，但仍可在暗绿灯光下进行。

（4）经冲洗、甩干后，即可准备观察再现。

2. 观察全息照片的再现物像

（1）取与原参考光方向尽量一致的再现光照明全息照片（感光乳胶面仍向着激光束）观察再现虚像，体会再现像的立体性（从哪些现象可以说明你所观察到的再现像是立体像？），比较再现虚像的大小、位置与原物的情况。

（2）通过小孔观察再现虚像，并改变小孔覆盖在全息片上的位置进行观察并写出观察结果。

（3）观察再现实像：未扩束的激光直接射到全息照片的玻璃基面（乳胶面的背面）上，选取适当的角度，再用平毛玻璃观察屏来接收再现实像。改变激光束的入射点，观察视差特性。改变屏的位置，观察实像大小及清晰程度的变化，只有像质最佳的位置才是实像的位置。

（4）将再现光换成钠光灯或汞灯，观察并记录再现物像的变化。

（赵喆 马骄）

Experiment 18　Laser Hologram Photography

Hologram is based on the rules of physical optics such as interference and diffraction. It not only can record the whole information of the lightwave (amplitude and phase), but also can reappear it. So the hologram is realistic and stereo. Wavefront reappearance theory has been provided by English physical scientist Gebor in 1948. This theory is a foundation of hologram, and in 1971 Dr. Dennis Gabor was awarded the Nobel Prize in physics for his work in holography. The appearance of laser which is a kind of high strength coherent light makes the hologram develop rapidly. And its application area is becoming wider and wider, it has been an important branch of optical field. It can be wildly used in nondestructive inspection, precision measure, remote sensing technology, biomedicine, military affairs, etc.

【Objects】

(1) Learn to film the hologram and the method of reappearance.

(2) To realize the main features of hologram.

【Apparatus】

Holographic desk, He-Ne laser, exposure time controller, splitter (transmission 95%), mirror (two), lens (100 times and 40 times), object.

【Theory】

Light emitted (or reflected) by every point of an object is electromagnetic wave. People can distinguish the object's colour, light and shade, figure, and distance by its frequency, amplitude and phase. For normal photograph, the image of the object emerges on sensitization plate by lens. After being exposed, it records the information of the intensity distribution for the object surface (square of amplitude), but the vibration phase of light cannot be recorded. So the image is planar. Howevers hologram can record the whole information of light (amplitude and phase) and reappear it.

1. The Recording principle of Hologram

The optical path of the hologram is shown in Fig. 3-18-1. where the laser beam from He-

Ne laser is split into two parts when passing through splitter S, one part is reflected by mirror M_1, then is expanded by lens L_1 to illuminate the object D. After being reflected by D, it illuminates the photographic film H (photographic plate), this part is called object light (O light). The other part is reflected by mirror M_2, after being expanded by lens L_2, it illuminates directly the photographic film H, and this part is called reference light (R light). Every point, on which light arrives the photographic plate, has a certain phase. As a result of high coherence of laser light, an interference pattern is produced on the film by the two beams.

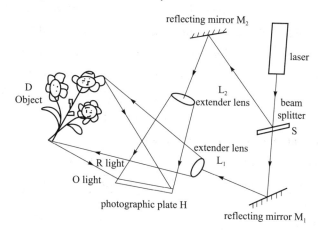

Fig. 3 – 18 – 1　Optical path of experimental principle for holography

We can analyze the case of a certain point C of the object. Suppose the reference light is a plane wave that is vertical to holographic plate. If the point angle that the holographic plate opens to the object point is small enough, the spherical wave emitted from the object point can be assumed as plane wave on the area of holographic plate shown as Fig. 3 – 18 – 2. In Fig. 3 – 18 – 3 and Fig. 3 – 18 – 4, the interference of object light and the reference light can be predigested to the interference of parallel light. It can be proved that the separation distance of fringe

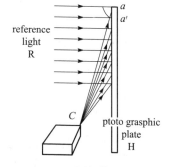

Fig. 3 – 18 – 2　Spherical wave and reference beam interference figure

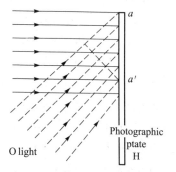

Fig. 3 – 18 – 3　Simplified optical path diagram

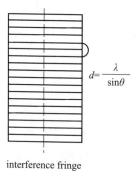

Fig. 3 – 18 – 4　Interference fringe of film

$$d_i = \frac{\lambda}{\sin\theta_i} \qquad (3-18-1)$$

In the formula: λ—wavelength of coherent light; θ_i—the angle of object light and the reference light.

The luminance difference between bright fringe and dark fringe is decided by the intensities of the two beams. In the different areas of holographic plate, it has different angles, which is between the reference light and the object light from the same point of the object. In the same area, the angle of the reference light and the light emitted from different object points is different, too. So the luminance, spacing and trend of them are different.

Total object light can be regard as the summation of light emitted from incalculable object points. So the figure is the combination of incalculable interference fringe with different luminance, spacing and trend. After being exposed, through the process of developing and fusing, the interference pattern containing all information of object points is recorded.

2. The reappearance principle of hologram

The image recorded on the sensitization plate is not the directly perceived figure of the object, but the combination of incalculable interference fringe. Each set interference fringes is like a complex grating. So when we observe the image recorded by hologram, we must adopt certain reappearance method, namely, it is shone on this plate with the same light (the light is called reappearance light). The optical path is shown in Fig. 3-18-5.

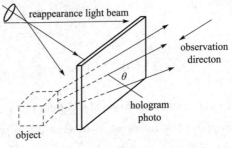

Fig. 3-18-5　Principle of holographic reconstruction

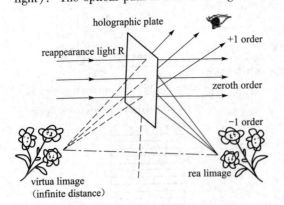

Fig. 3-18-6　Hologram

When the hologram is shone with the reappearance light, it can be regard as a fraise with heterogeneous transmissivity. So the reappearance light will be diffracted just as complex grating diffraction. For simplicity, we discuss a little area of the photograph, and assume the reappearance light is a parallel beam that is vertical to the photograph, then it will diffract. The positive first order diffracted ray is emanative light, it can engender a virtual image on the pristine position. The negative first order diffracted ray is converging light, the position of the convergence is symmetry to the pristine position (as shown in Fig. 3-18-6).

According to grating diffraction principle, the angle of diffraction suffice

$$\sin\theta_i = \frac{\lambda}{d_i},$$

then a concourse of complex and irregular grating produce diffraction pattern. The positive first order diffracted rays engender a virtual image on the pristine position. The −1 order diffracted rays engender real image that is called counterfeit image. The 0 order diffracted rays still spread according to the direction of reappearance light. We can see a stereoscopic image at the position of original meeting the direction of the +1 diffraction light.

3. The characteristic of hologram

(1) The hologram of the object being shot has the complete living three − dimensional stereo − sense. When people observe it at different angulation by moving eyes, just like facing the original object, they can see its different sides, another object that being shut out by the object being shot at some angulation can be seen at another angulation.

(2) For the photic information coming from every point of the object has been recorded in any small area of the holographic plate at different inclination of object lightwave, so with a block of shatter hologram, we can reappear the integrate stereo − sense of the object only and what we need is to take out any small fragment.

(3) The same holographic plate can be exposed or repeatedly recorded many times. Only change the position of the plate a little (such as rotating one angulation) or change the direction of the reference light beam, the same holographic plate can be repeated exposure and can reappear their own images not being interfered. If all the components of the optical circuit maintain fixedly strict, only to mak the object being shot get tiny displacement or deformation with the effect of the outside force, and make the holographic plate repeatedly expose, then when the object lightwave is reappeared, the object lightwave of the object being twice recorded will appear at the same time, and form the interference fringe which reflects the varied characteristic of the object conformation. This is the metric foundation of the holographic interfere.

(4) If the hologram is irradiated by the different wavelength laser, the reappearance image may be enlarged or reduced. According to the formula, the wavelength of the reappearance light is greater than the original reference light, the image is enlarged, conversely reduced.

(5) The object lightware reappeared by the hologram is a part of the reappearance beam of light, so the stronger the reappearance beam of light, the brighter the image of the reappearance object. It is pointed out that the adjustment may reach 10^3 times between the bright and the dark.

[Note]

To actualize the holographic record of the lightwave, the static state hologram must have three basic experimental requirements:

1. The well − coherenced light source

Thee helium neon laser is well − coherenced and the wavelength of the output laser beam is 632.8nm. If the width of the spectral line is $\Delta\lambda = 0.0020$nm, thus the coherent length is $L_m = \dfrac{\lambda^2}{\Delta\lambda} = 20$ (cm). Even if we have had the well coherenced light source, we must notice in the experiment that:

(1) Decrease the optical path difference of object light and the reference light to the best of our abilities. We must arrange the optical circuit appropriate in the experiment, while controlling the optical path difference in several centimeters.

(2) The intensity ratio of reference light and object light are chosen during 2 : 1 ~ 10 : 1, so we need to select the suitable splitter and decay plates. The light intensity of the reference light and object light can be measured by photocell mating and the galvanometer. In the non-oversize range of the light quantity, the photocurrent of the photocell directs ratio with the incident light flux. So, under this condition, when the incident lights are reference light and object light, the ratio of their photocurrent come into being in the photocell, that is the light intensity ratio of the reference light and object light.

2. The recording medium of the high resolution

The interference fringes recording on the holographic plate are often very dense. Example as the Fig. 18 – 2, if $\theta = 30°$, $\lambda = 632.8$nm, the separation distance $d = \frac{\lambda}{\sin\theta} = 1.3 \times 10^{-3}$ (mm) forming, scilicet there are near one thousand fringes per mm. The separation distance of the fringe will diminish with the augmenting angle. While the resolution ratio of the commonly holographic plate is only 100 strips per mm. So the hologram needs to carry on using the high resolution medium——holographic plate. The resolution ratio of this kind of holographic plate may be greater than one thousand strips per mm, but its sensitivity is very high, and the exposure time needed is longer than the common holographic plate (45s ~ 1min). The holographic plate used for He – Ne laser is the most sensitive to red light, so the operation should be carried under the light of sap green.

3. Favorable shock absorber

Because the interference is dense, there must be a very steady condition when recording the exposure, even if the slight vibrate or disturbance will make the optical path change at the order of magnitude of the wavelength, the fringe will be slurred. So the holographic laboratory is often placed far from the vibration source, all the components of the hologram optical circuit are collocated on the holographic shockproof desk, the object to be shot, apiece optic components and holographic plate must be rigidly fixed. At the same time, we must avoid major air current in the laboratory when shooting.

【Procedure】

1. Shooting the hologram of the still life

(1) Arrange the optical circuit as the Fig. 18 – 1 on the holographic desk, satisfy the requirements below:

① Make the object optical path and the reference optical circuit equality (how to calculate the optical path of the object light and the reference light?). For the sake of the convenience, do not put extender lens L_1 and L_2 in the object ptical path path temporarily, holographic plate can be replaced by other sheet glass.

② Put in extender lens L_1 and L_2 (make good use of the light energy of the laser, esp. in the object opticle path), and make sure that the object light beam and reference light beam would receive the uniform light from the object to be shot and the holographic plate. Avoid the extended object light beam illuminating directly onto the position of the holographic plate.

③ Make the intensity of light ratio of the reference light and object light lie in the appropriate range.

(2) Exposure

① Choose the exposure time according to the light intensity (given by the laboratory).

② Shut out the laser beam, install the holographic plate, the emulsoid side should face to the laser beam.

③ Rest a few minutes, then expose. We must take strict precautions against shock during exposure, all the optic components should be fixed strictly, there are no other stray light disturbing except for dark green light.

(3) Develop and the fix the holographic plate. We adopt the D – 19 developer to develop; and adopt F – 5 fixing bath to fix. The processing procedure is identical to ordinary holographic plate, but if can also proceed at the dark green light.

(4) We can prepare to observe the reappearance after washing and drying the holographic plate.

2. Observing the reappearance image of the hologram

(1) Select the reappearance – light to illuminate the hologram (make the sensitive emulsoid side face to the laser beam) coincident with the original reference light to observe the reappearance virtual image, to experience the stereo – character of the reappearance image (From which phenomenon can illustrate the reappearance image which you have observed is the stereo – image?), compare the size and the situation of the reappearance virtual image with the original object.

(2) Observe the reappearance virtual image through the ostiole, and alter the situation in which the hologram is covered by the ostiole to observe and write out the observing outcome.

(3) Observe the reappearance real image

Let the unextended laser shine the glass cardinal plane (the back of the emulsoid side) of the hologram directly, select the appropriate angulation, then receive the reappearance real image with the help of the flat frosted glass. Alter the point of incidence of the laser beam, observe the characteristic of the parallax. Alter the situation of the screen to observe the size of the real image and the diversification of the definition, only the situation where the image with the best display is the situation of the real image.

(4) Exchange the reappearance light with the sodium lamp or mercy vapour lamp, observe the variation of the reappearance of the object image and record.

(Ma Jiao, Su Ting – ting)

第四章 综合性与设计性实验

Chapter Four　Comprehensive and Designing Experiment

实验十九　指端光电容积法测量脉搏波

【实验目的】
（1）了解光电容积测量脉搏（PPG）的原理。
（2）了解光电传感器。
（3）测量人体指端脉搏波。

【实验器材】
指套式光电脉搏传感器、生物信号放大器、信号示波器、配套电源。

【实验原理】

1. 光电容积法的生理依据

常用的脉搏采集方法有压力传感器法、超声脉图法、光电容积法、电容传感器法、电声传感器法等。以上方法中，超声脉图法和光电容积法在目前临床应用中比较普遍。而电容、电声和压力传感器法多用于无创血压测量中的脉搏测量。实验中所用的光电容积法是利用光信号强弱的变化间接检测出脉搏信号。

由于每次心跳都有少量血液流入手指，使小动脉网扩张，然后经过毛细血管前括约肌进入毛细血管床，流入静脉后返回心脏。毛细血管前括约肌的阻力和毛细血管床的容量较大，使小动脉的搏动减弱。正常生理情况下，毛细血管和静脉不搏动，只有小动脉搏动。用一束光透照手指可检测这种搏动，心脏收缩时手指血容量最多，因而光吸收量也最大。心脏舒张期手指血流量最少，光吸收量也最小，所以，光吸收量的变化反应了这种变化。

2. 光电传感器的结构

当光源和光敏元件置于手指的同一侧（或两侧），光源发出的光照射在组织上，经反射（或透射）后被光敏元件接收，其示意图见图4-19-1。接收光强信号后，光敏元件将脉动的光强度信号转变为脉动的电信号。

透射式光电容积法将光敏接收器件和光源对称分布在指端上下两侧，这种方法可以较好地反映心率的时间关系，但相较于反射式而言不容易测出血液容积量的变化。本实验侧重于脉搏信号的测量，所以采用透射式光电传感器。

图 4-19-1　反射式和透光式光电传感器

3. 光电脉搏传感器的制作

光电脉搏传感器由于采用不同的光敏元件有着多种实现方法，其中光敏元件主要有光敏电阻、光敏二极管、光敏三极管和硅光电池。在传统的光电脉搏传感器设计中，通常采用的是独立光敏元件，利用半导体和光电效应改变输出的电流。通常光敏元件输出的电流极低，容易受到外界干扰，而且对后续的放大器的要求比较严格，需要放大器空载时的电流输出较小，避免放大器空载输出电流对脉搏信号测量的干扰，这样普通的放大器就不能直接应用在光敏元件的后端。目前许多新型的光敏元件将感光部件和放大器集成在同一个芯片内部，这种集成化的设计方式有效地克服了后端运算放大器空载电流输出对光敏部件输出电流的影响，而且芯片输出的电压信号可以通过外部的精密电阻进行调节，有利于芯片适应整体的电路设计，同时芯片的集成化设计也能够减小系统的功耗。

脉搏信号主要由动脉血的充盈引起，而血液中还原血红蛋白（Hb）和氧合血红蛋白（HbO_2）含量变化将造成透光率的变化。当氧合血红蛋白和还原血红蛋白对光的吸收量相等时，透射光的强度将主要由动脉血管的收缩和舒张决定，此时能够比较准确地反映出脉搏信号，如图 4-19-2 所示。因此，光源波长的选择也尤为重要。对于脉搏的测量，倾向于选择 805nm 附近的近红外光波段。

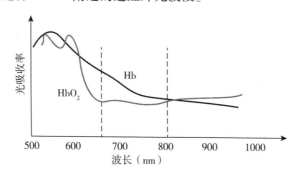

图 4-19-2　血液中血红蛋白的光波吸收曲线

4. 电路设计

在脉搏信号测量过程中，为了尽量减少光源供电波动对测量脉搏信号的影响，需要恒流电路来控制光源的稳定供电，使脉搏测量过程中发射光源发出的光强是恒定的。

光电传感器输出的电脉冲信号是非常微弱的信号，并且还伴有各种环境噪声干扰，所以要经过放大电路进行放大，同时还要滤除各种噪声干扰，因此，我们直接将光电脉搏传感器的输出信号接入生物信号放大器，进行信号的放大及滤波。

为了将脉搏波形完整显示出来，经过放大滤波去噪的脉搏波信号进入信号示波器中，整个采集装置如图4-19-3所示。示波器显示的横轴为时间轴，纵轴为电压值，波图便直观反映了脉搏信号随时间的变化规律，如图4-19-4所示。

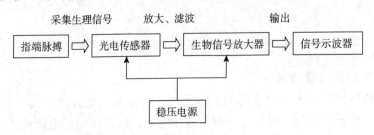

图4-19-3　光电容积法脉搏采集系统框图

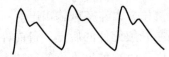

图4-19-4　指端脉波示意图

【实验步骤】

（1）按照各仪器的接口连接各仪器设备，检查接线是否有误。

（2）设备通电，稳压电源稳定后再进行测量。

（3）首先进行初步测试，调整示波器的X、Y轴，使波形能够充分显示在屏幕上。

（4）仪器预热并调试好后，进行正式测试，将手指深入光电传感器中，保持固定，完整显示8~10个波形后，锁定示波器波形，结束测量。

【数据记录及处理】

（1）从示波器上的脉搏波大致估算出心率（BPM）。

（2）由图4-19-5所示的局部脉波图，计算上升角 α_1（单位以度计），该角度反映流过指端微血管床血量的变化速度，顺应性好，则上升速度快，将数据记录在表4-19-1。

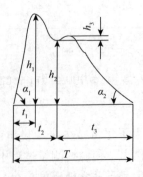

图4-19-5　脉波图局部放大示意图

表 4-19-1　α_1 角的测量

指标	主波幅值 h_1	快速射血期时值 t_1	脉动周期时值 T	上升角 $\alpha_1 = h_1 / t_1$
数值				

【注意事项】

（1）为降低噪声干扰，建议被测者端坐，不讲话，平和呼吸。

（2）光照射指尖部，并注意避免被周围环境光所干扰。

【思考题】

（1）光电容积法若改变光源波长，是否可以表征其他生理指标？国内外有哪些进展？

（2）思考一下在指端采用透射式和反射式结构的各自特点；对于其他部位，如何使用光电容积法测量生理指标？

（孙　言）

Experiment 19　Fingertip Pulse Wave Measurement Using Photoelectric Plethysmography

【Objects】

(1) To understand the principle of photoelectric plethysmography (PPG) for pulse measurement.

(2) To understand photoelectric sensors.

(3) To measure the pulse wave of the fingertip.

【Apparatus】

Finger-type photoelectric pulse sensor, biosignal amplifier, signal oscilloscope, supporting power source.

【Theory】

1. Physiological basis of photoelectric plethysmography

The commonly used pulse acquisition methods are: pressure sensor method, ultrasonic pulse method, PPG, capacitive sensor method, electroacoustic sensor method and so on. Among the above methods, the ultrasonic pulse method and the PPG are relatively common in current clinical applications. Capacitive, electroacoustic, and pressure sensor methods are often used for pulse measurement in non-invasive blood pressure measurements. The PPG uses the change of the intensity of the optical signal to indirectly detect the pulse signal.

Since each heartbeat has a small amount of blood flowing into the finger, the arteriolar network is dilated, and then blood passes through the precapillary sphincter into the capillary bed, into the vein and back to the heart. The resistance of the precapillary sphincter and the volume of the capillary bed are large, which weakens the pulsation of the arterioles. Under normal physiological conditions, capillaries and veins do not pulsate, and only small arteries pulsate. This pulsation can be detected by a light transilluminating finger. The finger blood volume is the most when the heart contracts, so the light absorption is also the largest. During diastole, finger blood flow is minimal and light absorption is minimal. Therefore, the change in the amount of light absorption reflects this change of blood.

2. Structure of photoelectric sensor

When the light source and the photosensitive element are placed on the same side (or both sides) of the finger, the light emitted by the light source is irradiated on the tissue, and is reflected (or transmitted) to the photosensitive element. The schematic diagram is as shown in Fig. 4 – 19 – 1. The light intensity signal is converted into a pulsating electrical signal.

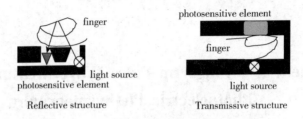

Fig. 4 – 19 – 1 Reflective and transmissive photoelectric sensors

The transmissive fingertip PPG symmetrically distributes the photosensitive device and the light source on the both sides. This method can better reflect the time relationship of heart rate, but it is not easy to measure the change of blood volume compared with the reflective structure. This experiment focuses on the measurement of pulse signals, so a transmissive photoelectric sensor is used.

3. Production of photoelectric sensor

Photoelectric pulse sensors have various implementation methods due to the use of different photosensitive elements, wherein the photosensitive elements mainly include photosensitive resistors, photodiodes, phototransistors and silicon photo cells. In conventional photoelectric pulse sensor designs, separate photosensitive elements are typically employed that utilize semiconductor and photoelectric effects to vary the output current. Usually, the output current of the photosensitive component is extremely low, and it is susceptible to external interference, moreover, the requirements for the subsequent amplifier are strict. It is necessary to reduce the output current when the amplifier is idle, in order to avoid the interference of the amplifier's no – load output current to the measurement of pulse signal, so that the ordinary amplifier cannot be directly applied to the back end of the photosensitive element. At present, many new types of photosensitive elements integrate the photosensitive components and the amplifiers inside the same chip. This integrated design effectively overcomes the influence of the no – load current output of the back-end operational amplifier on the output current of the photosensitive component. Moreover, the voltage signal output by the chip can be adjusted by external precise resistance, which is conducive to the chip adapting to the overall circuit design. The integrated design of the chip can also reduce the power consumption of the system at the same time.

The pulse signal is mainly caused by the filling of arterial blood, while the changes in the content of reducedhemoglobin (Hb) and oxyhemoglobin (HbO_2) in the blood will also cause changes in light transmittance. When the absorption of light by HbO_2 and Hb is equal, the intensity of transmitted light will be mainly determined by the contraction and relaxation of the arterial vessels, and the pulse signal can be more accurately reflected at this time, as shown as Fig. 4-19-2. Therefore, the choice of light source wavelength is particularly important. For the pulse measurement, it is preferred to select the near-infrared band near 805 nm.

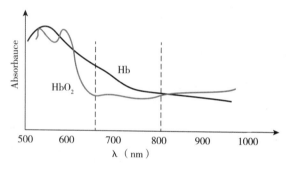

Fig. 4-19-2　Absorption curves of hemoglobin in blood

4. Circuit design

In the pulse signal measurement process, in order to minimize the influence of the power supply fluctuation on the measurement of the pulse signal, a constant-current circuit is needed to control the stable power supply of the light source, so that the intensity of the light emitted by the light source is constant during the measurement process.

The electric pulse signal output by the photoelectric sensor is very weak, and also accompanied by various environmental noises, so it needs to be amplified by the amplifying circuit, and at the same time, various noise signals are filtered out. Therefore, we directly put the output signal of the sensor into the biosignal amplifier for signal amplification and filtering.

In order to display the pulse waveform completely, the amplified and denoised pulse wave signal by amplification and filtering is entered into the signal oscilloscope. The block diagram of the whole acquisition device is shown in Fig. 4-19-3. The horizontal axis of the oscilloscope is the time axis, and the longitudinal axis is the voltage value. The waveform visually reflects the variation of the pulse signal with time, as shown in Fig. 4-19-4.

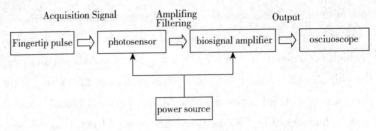

Fig. 4 – 19 – 3 Pulse acquisition system block diagram

Fig. 4 – 19 – 4 Sphygmogram of fingertip

【Procedure】

(1) According to the interface of each instrument, connect each instrument or equipment and check the connection.

(2) After the device is powered on and the power supply is stabilized, then the measurement is carried out.

(3) First perform a preliminary test to adjust the X – Y axis of the oscilloscope so that the waveform can be fully displayed on the screen.

(4) Perform a formal test, after the instrument is well warmed up and regulated. Put one finger deep into the photoelectric sensor, keep it fixed. After displaying 8 – 10 waveforms completely, lock the oscilloscope waveform and end the measurement.

【Data and Calculation】

(1) Estimate heart rate (using BPM) from the sphygmogram on the oscilloscope.

(2) From the local pulse diagram of Fig. 4 – 19 – 5, calculate the rising angle α_1 (in degrees), which reflects the change speed of blood flow through the fingertip capillary bed. If the compliance is good, the rising speed is fast. Fill the values in Tab. 4 – 19 – 1.

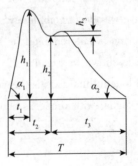

Fig. 4 – 19 – 5 Local diagram of sphygmogram

Tab. 4 – 19 – 1 α_1 degree measurement

parameter	main wave amplitude h_1	fast ejection duration t_1	pulsation cycle time value T	rising angle $\alpha_1 = h_1 / t_1$
value				

【Note】

(1) In order to reduce noise interference, it is recommended that thesubject sit upright, do not speak, and breathe peacefully.

(2) Illuminate the fingertips and take care to avoid being disturbed by ambient light.

【Questions】

(1) If the PPG method changes the wavelength of the light source, can it characterize other physiological indicators, and what progress has been made at home and abroad?

(2) Think about the characteristics of the transmissive and reflective structures at the fingertips. For other parts, how do you measure the physiological indicators using the PPG method?

(Sun Yan)

实验二十 血液流变学参数测定

血液流变学是研究血液、血管、血液组分的流变性质及其规律的新兴学科，是生物物理学的重要分支。血液流变学是物理学、化学、生物学以及医学和药学等学科相互交叉渗透融合的一门新兴的边缘学科。血液流变学参数测定在临床医学和药学应用中具有十分重要的意义。临床医学中常常将血液流变学参数异常与否作为诊断心脑血管疾病的参考依据；在药学应用中依据血液流变学参数指标来指导药物的研制和开发，如药物组方成分的筛选，药物疗效的判断并确定药物用量、服药方法和疗程等。血液流变学参数较多，如血液黏度，红细胞压积，红细胞变形性和聚集性，血小板聚集性以及血沉等。相关的测量检验仪器主要有：血液黏度计、红细胞变形测定仪、血小板聚集仪、血沉分析仪等。本实验重点测定血液黏度及红细胞沉降速率这两个参数。

一、血液黏度计测量血液黏度

【实验目的】

(1) 掌握旋转式黏度计和毛细管黏度计的工作原理、基本结构及其使用方法。

(2) 熟悉黏度计调校测定原理。

(3) 了解旋转式黏度计和毛细管黏度计性能指标、日常维护方法及常见故障排除。

【实验器材】

血液黏度计。

【实验原理】

血液黏度是临床血液流变学最重要的参数之一，血液黏度的大小直接影响血液循环中阻力的大小，必然影响血液在血管中的流动状态及流动规律，因此对血液黏度的测定具有十分重要的临床意义。

按工作原理，黏度计又分为毛细管黏度计和旋转式黏度计。前者按结构分为奥氏黏度计和乌氏黏度计；后者按结构可分为筒－筒式、锥－板式、锥－锥式以及棱球式黏度计等。按自动化程度黏度计可分为半自动黏度计和全自动黏度计。

1. 毛细管黏度计

（1）检测原理：按泊肃叶（Poiseuille）定律设计，即一定体积的牛顿液体，在恒定的压力驱动下，流过一定管径的毛细管所需的时间与黏度成正比。

（2）毛细管黏度计基本结构：包括毛细管、储液池、控温装置、计时装置等。

（3）毛细管黏度计的特点：价格低廉，操作简便，速度快，易于普及，测定牛顿流体黏度结果可靠，是血浆、血清样本测定的参考方法，但不能直接检测某剪切率下的表观黏度，不利于研究红细胞（RBC）、白细胞（WBC）的变形性和血液的黏弹性，不能反映全血等非牛顿流体的黏度特性。

2. 旋转式黏度计

（1）工作原理：以牛顿的黏滞定律为理论依据。主要包括以外圆筒转动或以内圆筒转动的筒－筒式旋转黏度计（又称 Couette 黏度计）和以圆锥体转动或以圆形平板转动的锥板式（Weissenberg）黏度计。

（2）锥板式黏度计工作原理：锥板式黏度计示意图见图4-20-1。它是同轴锥板构型，平板与锥体间充满被测样本，调速电机与圆形平板同速旋转，锥体与平板及马达间均无直接联系。当圆形平板以某一恒定角速度旋转时，转动的力矩通过被测样本传递到锥体；样本越黏稠，传入的力矩越大。当此力矩作用于锥体时，立即被力矩传感装置所俘获，并将其转换为电信号，其信号大小与样本黏度成正比。

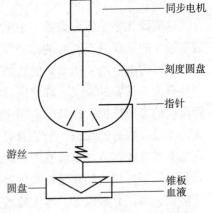

图 4-20-1　锥板式黏度计示意图

（3）锥板式黏度计的基本结构：由样本传感器，转速控制与调节系统，力矩测量系统，恒温系统等组成。

【测量仪器性能指标及实验步骤】

1. 旋转式黏度计的特点

能提供所需不同角速度下的剪切率；被测液体中各流层的剪切率一致，使液体在剪切率一致的条件下做单纯的定向流动；可以定量了解全血、血浆的流变特性，RBC

与 WBC 的聚集性、变形性等。价格较贵，操作要求更精细，但使用较为简单，是目前血液流变学研究和应用较为理想的仪器。

2. 仪器性能评价

准确度：以国家计量标准油为准，在剪切率（$1\sim200s^{-1}$）范围内分别用低黏度油（约 $2mPa\cdot s$）和高黏度油（约 $20mPa\cdot s$）测定其黏度，要求实际测定值与真值的相对偏差 <3%。

分辨率：是指黏度计所能识别出的血液表观黏度最小变化量。取比容在 0.40～0.45 全血测试。

高剪切率 $200s^{-1}$ 状态下，能反映出比容相差 0.02 时的血液表观黏度的变化。

在低剪切率 $5s^{-1}$ 以下状态，能反映出比容相差 0.01 时的血液表观黏度的变化。

重复性：取比容在 0.40～0.45 血样，测量 11 次，取后 10 次测定值计算变异系数（CV）值；在高剪切率时，血液表观黏度 CV<3%；在低剪切率时，血液表观黏度 CV<5%。

灵敏度与量程：测力传感器应具有 10mPa 灵敏度才能测定 $1s^{-1}$ 的血液黏度。

对于恒定剪切应力的黏度计，这一控制范围包括 100～1000mPa。

性能指标：①黏度测试范围；②剪切率变化范围；③黏度值重复性 CV<3%；④准确度 ±3% 等。

测试参数：①血浆黏度；②全血黏度；③血沉；④红细胞压积等。

【仪器介绍】

JN23－LB－2A 型自动锥板式黏度计（血液流变仪）。

1. 性能及特点

仪器采用锥板式原理。稳态测量，进口机芯耐磨损，免维护设计；机械自动保护系统，性能独特，环保型液晶显示窗，显示清晰。实测全血黏度和血浆黏度，重复性好，测试精度高，数据精确。采用空气自动校正仪器，随时可校正仪器，方法简洁准确。标准的 RS232 接口可联机或单机操作，方便快捷。仪器自动检测，数据自动传输，操作简便，标本用量少。自动清洗，采用双蠕动泵管路，自动排堵，性能先进，确保检测数据准确。标准的血液流变学分析软件，适用于各级临床检验部门。

2. 技术参数

测量范围：$1\sim30mPa\cdot s$；重复性偏差：±（3%～5%）。

切变率范围：$1\sim200s^{-1}$；样品用量：0.8ml；标准的 RS 232 接口；温度控制 37℃ ±0.2℃。

二、自动血沉分析仪测量红细胞沉降速率

【实验目的】

（1）掌握自动血沉分析仪的构成及测量原理。

（2）熟悉自动血沉分析仪的性能指标、使用方法、维护和保养以及质控规定。

（3）了解红细胞沉降曲线方程及红细胞沉降速率的传统测定方法；了解血沉分析

方法的进展。

【实验原理】

血沉分析仪。

【实验原理】

红细胞沉降率（erythrocyte sedimentation rate，ESR）是指红细胞在一定条件下沉降的速度，简称血沉。红细胞沉降率的传统测定方法是魏氏法（Westergren）。

20世纪80年代诞生自动血沉分析仪，20世纪90年代初又开发出了18°倾斜管方式的快速自动血沉仪，目前动态血沉分析仪也已问世。

工作原理：红细胞的沉降过程是一个包含力学、流变学及细胞间相互作用的复杂过程。

影响红细胞沉降的因素很多，主要包括红细胞的形态和大小；红细胞的变形性、聚集性；红细胞间的相互作用，红细胞压积；血浆介质和上升流动及沉降管的倾斜度等。

红细胞沉降曲线方程：

$$H = \frac{H_\infty}{1 + (t_{50}/t)^\beta}$$

通过对红细胞沉降时间过程的记录以及采用非线性最小二乘拟合方法，可得红细胞沉降曲线（$H-T$曲线）（图4-20-2）。

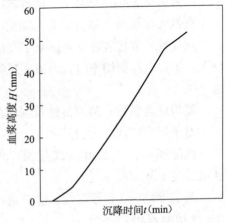

图4-20-2　红细胞沉降曲线（$H-T$曲线）

【测量仪器性能指标及实验步骤】

1. 仪器的读数原理

（1）将血沉管垂直固定在自动血沉仪的孔板上，使光电二极管沿机械导轨滑动，对血沉管进行扫描。如果红外线不能到达接收器，说明红外线被高密度的红细胞阻挡，一旦红外线能穿过血沉管到达接收器，接收器的信号就引导计算机开始计算到达移动终端时所需的距离。

（2）记录血沉管中的血液在时间零计时的高度，此后每隔一定时间扫描一次，记录每次扫描时红细胞和血浆接触的位置，并以计算机自动计算转换成魏氏法测定值的报告结果。

2. 仪器的性能特点

测量时间：18～60min；测量精度：≤1mm；定时精度：≤1min。

样品用量：1ml左右；同时测量样品数：10～40个；电源：AC220V±20V，50Hz。

标本采集：真空管或普通管。

3. 仪器使用和维护

（1）仪器的安装：应安装在清洁、通风处（室内温度应在15～32℃，相对湿度应≤85%），避免潮湿；应安装在稳定的水平实验台上（最好是水泥台）；禁止安装在高温、阳光直接照射处；应远离高频、电磁波干扰源。

（2）仪器的维护保养：使用过程中，要避免强光的照射，否则会引起检测器疲劳，计算机采不到数据；使用前要按程序清洗仪器，同时要定期彻底清洗并进行定期校检；仪器要保持水平状态。

红细胞沉降率 ESR 是某些疾病常用的参考指标，从 1988 年起，血沉与血浆黏度等作为监测急性期反应的手段之一，纳入血液流变学范畴。血沉分析仪实现了红细胞沉降的动态结果分析，对监测血沉全过程和研究红细胞沉降的机制等提供了新的数据。

红细胞沉降过程表现为悬浮、聚集、快速、缓慢沉降四个阶段。

血沉仪操作虽然简单，但其影响因素甚多，仪器的校正、操作正确与否等都会影响自动化分析测定结果的准确性。全面质量控制是提高检验质量的重要措施，操作步骤的标准化及质量控制品是一个急需解决的问题。

【仪器介绍】

1. 2068A 型血沉动态分析仪性能及特点

（1）测量方法简单便捷，随机插入样品，随时进行检测。

（2）自动输出结果，显示和打印样本号（序号）、魏氏法检测结果及全过程血沉曲线、检测时间和温度。

（3）仪器具有选择环境温度补偿功能：当选择该功能时，可对室温 18～30℃ 的检测结果根据血沉校正表修正到 18℃ 时的数值。

（4）仪器可储存不少于 150 个样本的检测结果，包括样本号（序号）、血沉值和动态曲线。

（5）有断电保护功能：突然断电时，自动保存已完成检测的检测结果。

（6）仪器配制内置热敏打印机，具有自动、手动打印选择功能；全中文操作界面设计，薄膜键盘操作，易于操作，通讯接口为 RS232 串行接口。

2. 技术参数

（1）仪器组成：本仪器由检测主机（含红外检测装置，微机控制系统，热敏打印机）和适配电源组成。

（2）检测范围：0～140mm/h（魏氏法结果）。

（3）检测相关性：相对于魏氏法结果的线性相关系数 $r \geq 0.98$。

（4）检测重复性：$CV < 3\%$。

（5）机箱温度测量准确度：在 15～30℃ 范围内，偏倚不超过 ±2.5℃。

（孙　言）

Experiment 20　Determination of Hemorheology Parameter

Hemorrheology is not only an important branch of biophysics but also the newly emerging subject which consists of the studies of blood, blood vessels, rheological properties and the laws of blood components. Hemorrheology is an emerging edge subject combined with physics, chemistry, biology, medical science, pharmacy and so on. The measurement of hemorrheology parameters has great significance in the clinical and pharmaceutical application. In clinical medicine, the diagnosis of cardiovascular disease are usually based on the hemorrheology parameters abnormalities. For pharmaceutical applications, we can research about drug investigation according to hemorrheology parameters, such as the screening of drug fraction, the judgment of drug efficacy, to ensure the dosage, methods of taking medicine and treatment. There are many hemorrheology parameters, such as blood viscosity, hematocrit, red blood cell deformability and aggregation, platelet aggregation, and erythrocyte sedimentation rate, and so on. The relevant test instruments are listed as follows: blood viscometer, red cell deformability analyzer, platelet aggregation analyzer, erythrocyte sedimentation rate analyzers. The experiment focuses on measuring two parameters: blood viscosity and erythrocyte sedimentation rate.

1. Measuring blood viscosity by blood viscometer

【Objects】

(1) Master operating principle, the basic structure and methods of using rotary viscometer and capillary viscometer.

(2) Be familiar with principle of viscometer calibration measurement.

(3) Know about the performance, maintenance and common trouble solving methods of rotary viscometer and capillary viscometer.

【Apparatus】

Blood viscometer.

【Theory】

Blood viscosity is one of the most clinically important parameters of hemorrheology, which will directly affects the resistance of the blood circulation, furthermore affects the flow state and the regularity of blood in the blood vessels. So it is very important clinical significance for us to measure the blood viscosity.

According to the principle, the viscometer can be divided into capillary viscometer and rotary viscometer. The former can be divided into Ostwald viscometer and Ubbelohde viscometer according to the structure: the latter can be divided into the cylinder – Cylinder, cone – plate, cone – cone and edge ball viscometer and so on. According to the degree of automation, viscometer can be divided into semi – automatic and fully automatic viscometer.

(1) Capillary viscometer

Test principle: according to the design of Poiseuille's Law, at a constant pressure, a volume of Newtonian liquid flow through a capillary tube which has a required diameter, the viscosity is proportional to the flowing time.

The basic structure of capillary viscometer: capillary, liquid storage tank, temperature control devices, timing devices and so on.

The features of capillary viscometer: low price, simplicity of operation, high speed, simplicity of spread, measure the results of Newtonian fluid viscosity in a reliable way, which is the reference method of samples measurement of plasma and serum, can not directly detect an apparent viscosity with some shear rate, which is not suitable to investigate the RBC (Red Blood Cell), WBC (White Blood Cell) deformation and blood viscoelastic and can not reflect the viscosity properties of non-Newtonian fluids of whole blood.

(2) Rotary viscometer

①Operating principle: Based on Newton's laws, there is mainly cylinder rotating or rotating cylinder barrel-drum rotation viscometer (also known as Couette viscometer) and cone rotating circular plate viscometer or the circle plate (also known as Weissenberg viscometer) viscometer.

②Operating principle of cone plate: It is a cone-plate configuration which filled with test samples between plate and cone, speed regulating motor and circular plate run at the same speed, there is no direct contact among the cone, plate and the motor. When the circular plate rotate with a constant angular velocity, rotational torque is transmitted from the measured sample to the cone; the more viscous samples are, the bigger the torque will be. When the torque acted on the cone is immediately captured by the torque sensing means, and converts it into an electric signal, the signal magnitude is proportional to the viscosity of the sample.

③The basic structure of cone plate viscometer: the sample sensor, speed control and regulation systems, torque measuring systems, thermostats and other components.

【Procedure】

(1) Features of rotary viscometer: Be able to provide the shear rates under different required angular velocity, the flow rate of each layer in tested liquid is consistent, so that the liquid do simple directional flow under the same shear rate, quantitatively understanding of whole blood, plasma rheological properties, RBC and WBC of aggregation, deformation and so on. More expensive, required more sophisticated operations, simple to use. It is the ideal instrument to do some research and application of hemorrheology.

(2) Instrument performance evaluation

Accuracy: Subject to national measurement standards, its viscosity was measured which shear rates within the ($1 \sim 200s^{-1}$) with a range of low viscosity oil (approximately $2mPa \cdot s$) and high viscosity oils (approximately $20mPa \cdot s$) respectively, requiring the relative deviation of the actual measured value and the true value less then 3%.

Resolution: It refers to the smallest variation of the apparent viscosity of blood

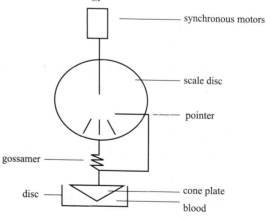

Fig. 4-20-1 A schematic view if a cone-plate viscometer

which the viscometer can identify. We takes a 0.40 ~ 0.45 specific volume to do whole blood test.

Under high shear rate of $200s^{-1}$, it can reflect the changement of the blood apparent viscosity when specific volume differs by 0.02.

When the shear rate is lower than $5s^{-1}$, it can reflect the changement of the blood apparent viscosity when specific volume differs by 0.01.

Repeatability: Get blood sample which specific volume is between 0.40 and 0.45, measure 11 times, calculate CV values by taking the lase 10 times measured value.

When the shear rate is higher, the apparent viscosity of the blood CV is less than 3%. When the shear rate is lower, the apparent viscosity of the blood CV is less than 5%.

Sensitivity range: 10mPa force sensor sensitivity can be measured with a blood viscosity $1s^{-1}$'s. For a constant shear stress viscometer, the control range is between 100mPa and 1000mPa.

Performance indicators: ① viscosity test range, ② shear rate range of variation, ③ viscosity repeatability CV < 3%, ④ accuracy ±3% and so on.

Test parameters: ① plasma viscosity, ② whole blood viscosity, ③ erythrocyte sedimentation rate, ④ hematocrit.

【Introduction to Apparatus】

(1) Performance and features of JN23 – LB – 2A automatic cone plate viscometer (rheology instrument)

Instruments designed by cone – plate principle, steady – state measurements, imported movement abrasion, maintenance – free. Mechanical automatic protection system, a unique performance, environmentally friendly LCD window, clear display. Measured whole blood viscosity, plasma viscosity, good repeatability, high precision, accurate data. Using air automatic correction instrument calibration, calibrating the instrument anytime, the method is simple and accurate. It is convenient and quick to use standard RS232 interface which is available to operating on – line or stand – alone. The instrument can detect automatically, transfer data simple and automatically, use less specimens. Self – cleaning, double peristaltic pump tubing, automatic Paidu, advanced performance, ensure accurate detection of the data. Hemorrheology standard analysis software for all levels of clinical testing department.

(2) The technical parameters

Measuring range: 1 ~ 30mPa · s. Repeatability deviation: ±(3% ~ 5%).

Shear rate range: $1 \sim 200s^{-1}$, sample volume: 0.8ml, standard RS – 232 interface, temperature control 37℃ ± 0.2℃.

2. Measurement of erythrocyte sedimentation rate by automatic ESR analyzer

【Objects】

(1) Master composition and measure the principle of automatic ESR analyzers.

(2) Familiar with performance indicators, use, care and maintenance, and quality con-

trol requirements of automated ESR analyzer.

(3) Know erythrocyte sedimentation curve equation and the traditional measurement method of erythrocyte sedimentation rate. Understand the progress of blood sedimentation analysis method.

【Apparatus】

Erythrocyte sedimentation rate analyzer.

【Theory】

Erythrocyte sedimentation rate (abbreviated as ESR) refers to erythrocyte sedimentation speed under certain condition which is short for "blood sedimentation". The traditional method of measuring erythrocyte sedimentation rate of is Westergren method.

1980s birth Automatic ESR analyzer. The early 1990s and developed a 18° birth fast automatic ESR analyzer by the tilted tube method. There is now dynamic ESR analyzer.

Operating principle: erythrocyte sedimentation is a complex process consists of mechanics, rheology and interaction between cells.

There are many factors to affect the erythrocyte sedimentation, including the shape and size of the red blood cell, deformability of red blood cells, the aggregation of red blood cell, the red blood cell interactions, erythrocyte between hematocrit, plasma medium and rising flow, inclination of sedimentation tube and so on.

Curve equation of erythrocyte sedimentation:

$$H = \frac{H_\infty}{1 + (t_{50}/t)^\beta}$$

Through the records of practice process of erythrocyte sedimentation to get erythrocyte sedimentation curve ($H-T$ curve) by non-linear least squares fitting methods.

【Procedure】

(1) Principle of instrument readings

① Fix blood sedimentation tube vertically to orifice plate of automatic blood sedimentation, make the photodiode sliding along the mechanical guide rail, scanning the sedimentation tube. If the infrared ray can not reach the receiver, indicating a high density infrared erythrocytes blocked infrared ray, once infrared ray can through blood sedimentation tube reaching the receiver, the computer is required to calculate the distance to reach the mobile terminal.

② Record ESR blood tube height in the blood at time zero, then scan at evear regular intervals, record the situation which red blood cells contact with plasma, and then the computer automatically calculates to convert the conclusion to reported assay results of Westergren.

(2) Characteristics of instrument performance

Measuring time: 18~60 minutes, accuracy: ≤ 1 mm, timing accuracy: ≤1min.

Sample usage: about 1 ml, simultaneously measure the number of samples: 10 to 40, supply: AC 220V ±20V, 50Hz.

Specimen collection: vacuum or ordinary pipe.

(3) Operation and maintenance of instrument

① Installation of Instruments: Equipment should be installed in a clean, well-ventilated place (room temperature should be at 15~32℃, relative humidity should be ≤ 85%), avoid moisture. It should be installed at a stable level test bench (preferably a cement platform). Prohibit the installation in high temperature, direct sunlight department, should stay away from high-frequency, electromagnetic interference sources.

② Maintenance of the instrument: In the process of using we should avoid strong light, otherwise it will cause the fatigue of detector, less computer data mining.

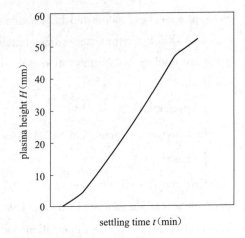

Fig. 4-20-2 Erythrocyte sedimentation curve ($H-T$ curve)

According to the program to clean the equipment before using, while do regular cleaning and school inspection, and keep the instrument maintaining the level of the state.

Erythrocyte sedimentation rate ESR is a common reference for some certain diseases, since 1988 erythrocyte sedimentation rate and plasma viscosity start to be the means of monitoring the acute phase response, they are included in the scope of blood rheology. ESR analyzer achieves a dynamic analysis of the results of erythrocyte sedimentation, providing new data of the whole process of monitoring the ESR, erythrocyte sedimentation mechanism studies and so on.

ESR analyzer to achieve a dynamic analysis of the results of erythrocyte sedimentation, the whole process of monitoring the ESR, erythrocyte sedimentation mechanism studies provide new data. Erythrocyte sedimentation process performance assume four stages: suspension, aggregation, fast, slow subsidence.

Although ESR analyzer operation is simple, but due to many factors such as instrument calibration, proper instrument operation or not and so on that will affect the accuracy of the measurement results of automated analysis, a comprehensive quality control inspection is an important measure to improve the quality, standard operating procedures and quality control product is an urgent problem.

【Introduction to Apparatus】

(1) Dynamic performance and features of 2068A-type ESR analyzer

① The measurement method is simple and convenient, randomly inserted samples ready for testing.

② The automatic output, display and print sample number (serial number), and the detection result of the whole process of Westergren ESR curve method, the detection time and temperature.

③ The instrument has a select ambient temperature compensation function. When you select this function, the test will correct the results for the temperature of 18～30℃ to 18℃ according to ESR correction table.

④ The instrument can store the test results of more than 150 samples, including sample number (serial number), ESR value and the dynamic curve.

⑤ Power protection, when electricity is cut suddenly, it will save test results automatically which have been completed testing.

⑥ Built–in thermal printer equipment with automatic, manual print selection function. Chinese user interface design, membrane keyboard, easy to operate, the communication interface is RS 232 serial interface.

(2) The technical parameters

① The apparatus consists of: the instrument detected by the host (including infrared detection devices, computer control systems, thermal printers) and the adapter power supply.

② Detection range: 0～140 mm/h (Westergren method results).

③ The correlation detection: with respect to the Westergren method results correlation coefficients $r \geq 0.98$.

④ Test repeatability: $CV < 3\%$.

⑤ Chassis temperature measurement accuracy: bias does not exceed ±2.5℃ in the range of 15～30℃.

(Sun Yan, Qi Xia)

实验二十一　利用压力传感器测定人体血压

【实验目的】

(1) 了解血压测量的方法。

(2) 了解压力传感器的基本原理。

(3) 测量人体血压。

【实验器材】

仪器内袖带充气系统、袖带、压力传感器、柯氏音传感器、标准压力表、气体压力传感器、注射器、橡皮管。

【实验原理】

血压是指血管内流动的血液对单位面积血管壁的侧压力，一般所说的血压是指体循环的主动脉血压。由于大动脉中血压的降落很小，则上臂肱动脉处所测得的血压值基本上可以代表主动脉血压，因此，通常测量血压值是以肱动脉血压为标准。临床上习惯的写法是"收缩压/舒张压 kPa (mmHg)"。

血压测量方法有间接测压法和直接测压法。间接测压通常仅能测得收缩压和舒张压，而直接测压法可以测得血压波形曲线。由于测量方法和部位不同，测量值不能直接比较。

（一）间接测压法

间接测压法是一种非创伤性血压测量技术，用该法测量的血压称为无创血压。间接测压法是利用袖带充气到一定压力时，完全压迫动脉血管并阻断动脉血流，然后随着袖带压力减小，动脉血管将开通。根据这个原理，可以通过测量血管恰好开通或关闭时的袖带内压，测量收缩压和舒张压。测量的方法有听诊法、搏动法、超声多普勒法等。

1. 电子柯氏音法

听诊法是用听诊器听柯氏音的方法进行测压，如图 4-21-1 所示。电子柯氏音法是用柯氏音传感器代替医生的听诊器。测血压时，给袖带加压直到使动脉壁闭合，然后再逐步降低气袖压力，当气袖压降到某个值时，血流冲过阻断，血管中开始有血液流动，柯氏音传感器检测到脉搏（第一声柯氏音），此时的气袖压是收缩压，气袖压继续下降，直到柯氏音消失（最后一声柯氏音），此时的气袖压是舒张压。

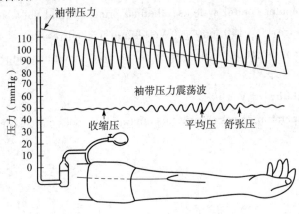

图 4-21-1 柯氏音测压法

2. 压力传感器基本原理

生物医学传感器是感知生物体内各种生理、生化、病理信息，按照一定规律将其不失真地转换为电信号的装置。压力传感器是生物医学传感器里的一种，它可以把压强转换成电学量。在生物医学工程中，压力传感器可用于非侵入式监测血压。

集成压力传感器是以硅为主要材料，把用来感受压力的硅应变膜、应变电阻和集成应变信号的桥式电路和放大输出电路等电路集成于一个芯片上的器件。

本实验所用气体压力传感器 MPX2050，是一种压阻元件的桥，MPX2050 加上 +5V 的工作电压，气体压强范围为 0~40kPa，则它随着气体压强的变化能输出 0~15mV 的电压。

实验仪器应在实验前验证仪器特性。若在压力传感器的输出端接一只数字电压表，在压力（压强）输入端，通过三通管接一只压力表，改变输入气体压强 P，可从数字电压表读出与之相应的输出电压值 U，作图 U-P，可得气体压强与输出电压的线性关系，于是可用标准压力表定标组装的数字式气体压强计，即把压力传感器和数字电压表组合起来，构成一只数字血压计。由于气体压力表为精密微压表，测量范围应在全范围的五分之四，即 32kPa，微压表 0~4kPa 为精度不确定范围，则实际测量范围应为 4~32kPa。

（二）直接测压法

直接测压法是一种有创伤性血压测量技术，用该法测得的血压称为有创血压。

1. 测量原理

有创血压的测量是通过将测压导管插入被测体的血管内进行的,将导管经皮由动脉穿刺,置入被测部位的血管内并固定;导管内的压力传感器将压力信号转换成电信号,在监视器上显示动脉压力变化的动态波形。

2. 测压过程

临床上常用的液体耦合法,是把血管内的血压经过充满液体的导管,传递到体外的压力传感器。测压装置如图 4-21-2 所示,测压导管的一端插入体内的血管中,另一端与压力传感器相连。导管内注满生理氯化钠溶液(生理盐水),由于流体具有压力传递作用,血管内的压力将通过导管内的液体被传到外部的压力传感器上,再由传感器把压力信号转换成电信号送入血压监护仪,获得血管内压力变化的动态波形,从而得到收缩压、舒张压和平均动脉压。

有创血压的测量多用于重症手术和重症监护病房(ICU),虽然操作复杂,患者有一定的痛苦,但能对血压进行实时监测,准确度高,具有很重要的临床价值。

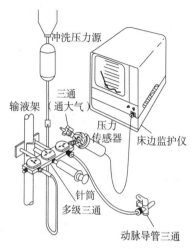

图 4-21-2　直接测压法测压装置

【实验步骤】

由于有创血压的测量有一定的危险,且实验室条件有限,本实验只对间接测压法进行操作,测量无创血压。

首先检查仪器标准配置,应包括:电源线、连接管路套件等。打开仪器开关,预热 5min,待仪器稳定后开始实验,实验仪器面板如图 4-21-3 所示。

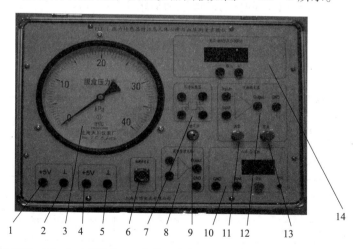

1. 实验传感器供电电源 +5V;2. 电源接地;3. 指针压力表;4. 实验电源 +5V;5. 电源接地;6. 脉搏传感器接口;7. 压力传感器,①接电源 +5V;②输出(+);③电源接地;④输出(-);8. 脉搏波形整理电路;9. 指针压力表和传感器进气口;10. 脉搏显示模块;11. 定标放大器零点调节旋钮;12. 定标放大器电路模块;13. 定标放大器增益调节旋钮;14. 数字电压表显示直测单位为 mV,经定标放大器后显示单位为 kPa

图 4-21-3　实验仪器面板

1. 气体压力传感器特性测量

（1）在气体压力传感器 MXP2050 输入端①接上直流电压源 $\boxed{+5V}$、③直流工作电压源 $\boxed{\perp}$，输出端②接数字电压表 $\boxed{+}$、④接数字电压表 $\boxed{-}$。

（2）用橡皮管将注射器的针孔与压力表连通，注意连接前把活塞拉至 80ml 位置，然后缓慢推进活塞以改变管路内气体压强。

（3）记录压力表的指示压力（4～32kPa 间测 8 点），以及与此相应的气体压力传感器的输出电压。

（4）画出气体压力传感器的输出电压 U 与压强 P 的关系曲线。

2. 数字式压力表（血压计）的组装和定标

（1）用接线将气体压力传感器 MPX2050 的输出端 $\boxed{2}$ 与定标放大器的 $\boxed{Input+}$ 输入端连接，传感器 MPX2050 的输出端 $\boxed{4}$ 与定标放大器的 $\boxed{Input-}$ 输入端连接；再将定标放大器的 Output 与数字电压表输入（＋）相连接，$\boxed{GND}$ 与数字电压表输入（－）相连接。

（2）反复调整气体压强为 4kPa 与 32kPa 时放大器的零点与放大倍数，使放大器输出电压在气体压强为 4kPa 时为 40mV，在气体压强为 32kPa 时为 320mV。

（3）将放大器的零点与放大倍数调整好后，此时电压表显示的电压实为压强值，单位为 kPa，组装好的数字式压力表可用于人体血压或气体压强的测量。

3. 血压的测量

（1）将袖带绑定在上臂，并把柯氏音传感器放在袖带内肱动脉处。

（2）袖带连接管通过三通接头与仪器进气口连通，用压气球向袖带压气至 20kPa 时，打开排气口缓慢排气，当听到第一次柯氏音时，记下压力表的读数，即为收缩压，最后一次听到的柯氏音所对应的压力表读数为舒张压。

【数据记录及处理】

数据记录于表 4-21-1 和表 4-21-2。

表 4-21-1　MPX2050 气体压力传感器输出特性

气体压强 P（kPa）	4	8	12	16	20	24	28	32
输出电压 U（mV）								

表 4-21-2　血压测量值

	收缩压（kPa）	舒张压（kPa）
1		
2		
3		

【思考题】

（1）什么是收缩压和舒张压？为什么用肱动脉处测得的血压表示主动脉血压？

（2）怎样用水银血压计和电子血压计测量人体血压？

(3) 气体压力传感器由哪几部分组成？

【知识链接】

传感器概述：在信息化高度发展的今天，人们对信息资源的需要日益增长，促进了信息产业的发展。信息采集系统的前端器件——传感器，其作用越来越重要。

1. 传感器的定义和组成

（1）定义：根据国家标准规定，传感器的定义为：能感受规定的被测量并按照一定规律转换成可用输出信号的器件或装置。

（2）组成：传感器通常由敏感元件、转换元件及相应的转换电路组成，如图4-21-4所示。

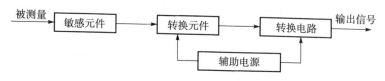

图4-21-4 传感器组成框图

敏感元件：能直接感受或响应被测量的部分并以确定关系输出某一物理量。如弹性敏感元件将力转换为位移或应变输出。

转换元件：将敏感元件输出的非电物理量（如位移、光强等）转换成电路参数（如电阻等）或电量。

转换电路：将电路参数转换成适于测量的电量，如电压、频率等。

从传感器的定义中表达出三层含义：它是由敏感元件和转换元件构成的一种检测装置；能按一定规律将被测量转换成电信号输出；传感器的输出与输入之间存在确定的关系。

2. 传感器的分类

传感器种类繁多，由于与许多学科有关，并有各种各样的设计原理，所以还没有统一的分类方法。目前按照我国制定的传感器分类体系中，可分为三大门类，下含12个小类，分类如下。

（1）物理量传感器：包括力学量传感器、热学量传感器、光学量传感器、磁学量传感器、电学量传感器、声学量传感器和射线传感器。

（2）化学量传感器：包括气体传感器、离子传感器和温度传感器。

（3）生物量传感器：包括生化量传感器和生物量传感器。

传感器还有其他的分类方法。按输出信号性质分类：可分为模拟和数字传感器；按结构分类：可分为结构型、物性型和复合型传感器；按功能型分类：可分为单功能、多功能和智能传感器；按转换原理分类：可分为机电、光电、热电、磁电和电化学传感器；按能源分类：可分为有源和无源传感器。

由于一种被测量可用不同类型的传感器检测，而同一原理的传感器也可以测量不同的量，并且随着技术的发展，新型传感器不断出现，难以将传感器精确分类，因此分类方法各不相同。

3. 传感器的作用和发展趋势

（1）传感器的作用：在信息化高度发展的社会，先要解决的是获取准确可靠的信息，而传感器是获取信息的重要途径和手段。无论是基础科学研究还是工业生产都需要获取大量信息来监测和控制整个科研或生产过程，这些都需要相应的传感器，如果没有传感器，现代的基础科学研究和工业生产就失去了基础。由此可见，传感器技术在发展经济和推动社会进步方面的重要作用是十分明显的。

（2）传感器的发展趋势：传感器技术是迅猛发展的高新技术，它与通信技术、计算机技术构成信息产业的三大支柱之一，社会的发展对传感器提出越来越高的要求。现代传感器技术的发展趋势：一是开发基础研究，重点研究传感器的新材料和新工艺；二是实现传感器的微型化、阵列化、集成化和智能化。

（邓岩浩）

Experiment 21　Human Body Blood Pressure Measurement Using a Pressure Sensor

【Objects】

(1) To understand blood pressure measurement method.
(2) To understand basic principle of the pressure sensor.
(3) To measure human blood pressure.

【Apparatus】

Inflatable system of cuff, cuff, pressure sensor, Korotkoff sound sensor, standard pressure gauge, gas pressure sensor, syringe, and rubber tube.

【Theory】

Blood pressure is the lateral force of the intravascular blood flow per unit area of the vessel wall, the blood pressure generally called is the aortic blood of the greater circulation. Since the blood pressure attenuation in the aorta is small, blood pressure values measured at brachial artery on upper the arms basically represents aortic blood pressure. Therefore, the blood pressure measured at the brachial artery is usually the standard one. It is written on clinical practice as "systolic pressure/diastolic pressure in kPa (mmHg)".

Blood pressure measurement includes indirect manometry and direct manometry. Indirect manometry usually measures systolic and diastolic blood pressure, and direct manometry can measure blood pressure waveform curve. Due to the different measurement methods and sites, the measured values are not comparable directly.

1. Indirect manometry

Indirect manometry is a non-invasive blood pressure measurement technology. The blood pressure measured using the method is called non-invasive blood pressure (NIBP). Indirect manometry uses a cuff to inflate a certain pressure to completely oppression arteries and blocks blood flow in the arteries. The arteries then open as the cuff pressure is reduced. According to this principle, the systolic and diastolic pressures are measured by measuring the pressures within the cuff when the blood vessels just open or close. The measurements include auscultation method, pulse method and ultrasonic Doppler method.

(1) Electronic Korotkoff sound method: Auscultation method uses a stethoscope to listen to Korotkoff sound to perform manometry, shown in Fig. 4 – 21 – 1. Electronic Korotkoff sound method uses a Korotkoff sensor instead of a doctor's stethoscope. When measuring blood pressure, inflate the cuff until the arterial wall is closed, and then gradually reduce the cuff pressure. When the cuff pressure drops to a threshold value, the blood crosses the blocked vessels and begin to flow. The Korotkoff sound sensor detects a pulse (the first Korotkoff sound heard), at this point, the cuff pressure is systolic blood pressure. As the cuff pressure continues to drop until the Korotkoff sounds disappeared (the last Korotkoff sound heard), at this point, the cuff pressure is diastolic blood pressure.

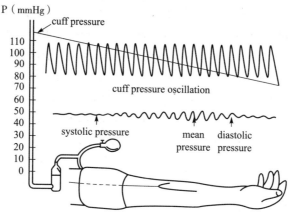

Fig. 4 – 21 – 1   Korotkoff sound manometry

(2) Pressure sensor fundamentals: Biomedical sensors are devices which sense a variety of physiological, biochemical and pathological information, and convert them into electrical signals without distortion in accordance with certain rules. The pressure sensor is a biomedical sensor, which can convert pressure into electrical signal. In biomedical engineering, the pressure sensor can be used for non – invasive monitoring of blood pressure.

Integrated pressure sensors is an integrated circuit chip based on silicon as the main material, which integrates the silicon strain film used to sense pressure, strain resistors, bridge circuit and an amplifier.

The gas pressure sensor MPX2050 used in the experiment is a piezoresistive element bridge. When MPX2050 operates with a +5 V power source, it can output a voltage of 0 ~ 15mV as gas pressure changes in the range of 0 ~ 40kPa.

Laboratory instruments must be calibrated for their characteristics before the experiment. Enter the gas pressure P, the corresponding output voltage U can be read out from a digital voltmeter if the pressure sensor output is connected to a digital voltmeter, and the pressure input is connected to a pressure gauge by a T-joint. The linear relationship between output voltage and gas pressure can be obtained by plotting $U-P$ curve. Therefore, a digital blood pressure meter can be assembled using a digital gas pressure meter, that is calibrated by a standard pressure gauge, which is the pressure sensor and a digital voltmeter are combined. Since the gas pressure meter is a precision micro-pressure gauge, its measuring range should be in four-fifths of the full range, that is, 32kPa. In addition, the range of 0~4kPa of the micro-pressure gauge is uncertainty for the accuracy, the actual measurement range should be 4~32kPa.

2. Direct manometry

Direct manometry is a traumatic blood pressure measurement technology. Blood pressure measured using this method is called invasive blood pressure (IBP).

(1) Measurement principle: Invasive blood pressure is measured by inserting a test tube into the blood vessel. The tube punctures the artery through the skin, and is fixed into the blood vessel of the test site. The tube pressure sensor converts the pressure signal into an electrical signal, and the dynamic waveform of the arterial pressure changes is displayed on a monitor.

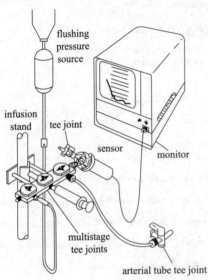

Fig. 4-21-2 Direct manometry devices

(2) Measurement process: The fluid coupling method used clinically is to pass the intravascular blood pressure into an external pressure sensor through a fluid-filled tube. Measurement devices are shown in Fig. 4-21-2. One terminal of the tube is inserted into the vessel in body, the other terminal is connected to the pressure sensor. The tube filled with saline, as the fluid has the effect of pressure transmission. The intravascular pressure will be transmitted on the external pressure sensor through the fluid in the tube, and then the sensor converts the pressure signal into an electrical signal fed to the blood pressure monitor to obtain the dynamic waveform in intravascular pressure changes, thereby obtaining the systolic pressure, diastolic pressure and mean arterial pressure.

Invasive blood pressure measurement is used for intensive surgery and Intensive Care Unit (ICU). Although the operation is complex and painful for patients, the blood pressure can be monitored in real time with high accuracy. Therefore, invasive blood pressure measurement has a very important clinical value.

【Procedure】

Because invasive blood pressure measurement has some risk and laboratory conditions limits, only indirect manometry is operated to measure noninvasive blood pressure in this experiment.

1. Measure gas pressure sensor characteristics

(1) Connect input ① of the gas pressure sensor MXP2050 to DC voltage powersupply $\boxed{+5V}$, input③ to DC voltage power supply $\boxed{\perp}$, output② to digital voltmeter $\boxed{+}$, output④ to digital voltmeter $\boxed{-}$.

(2) Connect syringe pinhole and pressure gauge with a rubber tube, note to pull the plunger to the 80ml position before connecting. Then slowly push the plunger to change the gas pressure inside the tube.

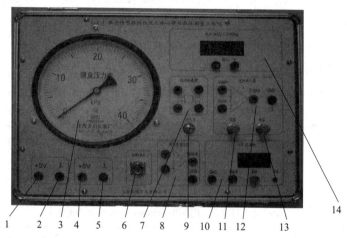

1. +5 V power supply for experimental sensor; 2. Supply ground; 3. Pointer pressure gauge; 4. +5 experimental power; 5. supply ground; 6. pulse sensor interface; 7. pressure sensor, ①connected +5 V power supply, ②output (+), ③power ground, ④output (-); 8. pulse wave shaping circuit; 9. inlet connector of the pointer pressure gauge and the sensor; 10. pulse display module; 11. scaling amplifie zero adjustment knob; 12. scaling amplifier circuit module; 13. scaling amplifier gain adjustment knob; 14. digital voltmeter display direct measurement in mV, displayed after scaling amplifier in kPa

Fig. 4 - 21 - 3　Experimental instrument panel

(3) Record pressure gauge indication (measure 8 points between 4 ~ 32kPa), and the corresponding output of the gas pressure sensor voltage.

(4) Draw a curve illustrating gas pressure sensor output voltage U vs. the pressure P.

2. Digital pressure gauge(sphygmomanometer) assembly and calibration

(1) Connect the output $\boxed{2}$ of the gas pressure sensor MPX2050 to the scaling amplifier $\boxed{\text{Input}+}$, the output $\boxed{4}$ of the sensor MPX2050 to the scaling amplifier $\boxed{\text{Input}-}$ using wires; then connect the scaling amplifier output to the digital voltmeter input (+), the GND to the digital voltmeter input (-).

(2) Repeatedly adjust the amplifier zero and gain until the amplifier output voltage is 40mV when the gas pressure is 4kPa, and the amplifier output voltage is 320mV when the gas pressure is 32kPa.

(3) After adjusting the amplifier zero and gain, the voltage voltmeter should display the real value of the pressure in units of kPa. The assembled digital pressure gauge can be used for human blood pressure or gas pressure measurements.

3. Blood pressure measurement

(1) Bind the cuff on the upper arm, and place the Korotkoff sound sensor at the brachial artery under the cuff.

(2) Connect the cuff tube to the inlet connector of the instrument through a tee joint. When the cuff pressure is inflated to 20kPa by squeezing the inflation bulb, open the air release button to deflate slowly. As you hear the first Korotkoff sound, record the reading as systolic blood pressure. The pressure gauge reading corresponding to the last Korotkoff sound is the diastolic pressure.

【Data and Calculations】

Date and calculations are shown in Tab. 4-21-1 and Tab. 4-21-2.

Tab. 4-21-1 MPX2050 gas pressure sensor output characteristic

gas pressure P (kPa)	4	8	12	16	20	24	28	32
output voltage U (mV)								

Tab. 4-21-2 Blood pressure measurement

	systolic pressure (kPa)	diastolic pressure (kPa)
1		
2		
3		

【Questions】

(1) What are the systolic pressure and diastolic pressure? Why does the blood pressure measured at brachial artery indicate aortic pressure?

(2) How to measure human blood pressure with a mercury sphygmomanometer and electronic sphygmomanometer?

(3) How many parts does the gas pressure sensor consist of?

【Knowledge Links】

Sensor overview: Today's information is developing rapidly, and the people's need for information resources is continually growing, so that it promotes the development of information industry. The role of the sensors is increasingly important as front-end devices of data acquisition system.

1. Sensor definition and composition

(1) Definition: According to national standard, the sensor is defined as: a part or a device which senses a measured variable and converts it into a usable output signal in accordance with certain rules.

(2) Composition: The sensor typically consists of a sensitive component, a conversion component and the corresponding conversion circuit shown in Fig. 4 -21 -4.

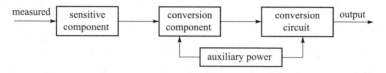

Fig. 4 -21 -4　Block diagram of sensor

Sensitive component: this component is able to sense or respond to a measured variable directly, and to output a physical variable in accordance with a certain relationship. For example, an elastic sensitive component converts the force to a positional shift or a strain output.

Conversion component: the component converts non – electrical physical variable (such as positional shift, light intensity, etc.) of the sensitive component output into a circuit parameter (such as resistor, etc.), or an electrical variable.

Conversion circuit: the conversion circuit converts the circuit parameter into an electrical variable suitable to be measured, such as voltage, frequency, etc.

Three meanings expressed from the sensor definition: it is a detecting device constituted by a sensitive component and the conversion component, is able to convert a measured variable into an electrical signal output in accordance with a rule, there is a certain relationship between the output and input of the sensor.

2. Sensor classification

There are many types of sensors related to many disciplines, and have various design principles, therefore there is no uniform classification. According to China's current classification system, the sensors can be divided into three categories, shown as follows, including 12 small classifications under the categories.

(1) Physical sensors include mechanical sensors, thermal sensors, optical sensors, magnetic sensors, electrical sensors, acoustic sensors and radiation sensors.

(2) Chemical sensors include gas sensors, ion sensors and temperature sensors.

(3) Biological sensors include biochemical sensors and biological sensors.

There are other sensor classifications. By signal outputs: it can be divided into analog and digital sensors. By structure: it can be divided into structured, physical and composite sensors. By functions: it can be divided into single – function, multi – functional and smart sensors; by conversion principles: it can be divided into electromechanical, photoelectric, thermoelectric, magnetoelectric and electrochemical sensors. By energy: it can be divided into to active and passive sensors.

Since a measured variable can be detected by different types of sensors, a sensor with the same principle can also measure different variables, and as the development of technology, new sensors continue to emerge, it is difficult to accurately classify the sensors. Therefore, the classifications are different.

3. The effect and development trend of the sensors

(1) Sensor's effect: In a society where information is highly developed, the first thing to be solved is to obtain accurate and reliable information, and the sensors are an important way and means to obtain information. Either basic scientific research or industrial production need to gain a lot of information to monitor and control the entire researches or production processes and require appropriate sensors. If there are no sensors, the modern basic scientific research and industrial production will lose its foundation. Thus, the importance of the sensor technology to develop our economy and to promote social progress is signicant.

(2) Development trend of the sensors: Sensor technology, as a rapid developing high-tech, is one of the three pillars of the information industry constituted with communication technology and computer technology. Social development increases the demands of sensors. Modern sensor technology trends include: ①to develop basic research focusing on new materials and new processes, and ②to achieve miniaturized, array, integrated, and intelligent sensors.

(Deng Yan – hao)

实验二十二 用箱式电位差计测量温差电动势

【实验目的】

(1) 了解箱式电位差计的原理和结构，掌握用箱式电位差计测电动势的方法。

(2) 掌握直流复射式检流计的使用。

(3) 通过测量热电偶的温差电动势，作出热电偶的温差电动势与温度差间的关系曲线，运用图解法求出热电偶温差系数并对热电偶进行定标。

【实验器材】

UJ25 型直流电位差计、待测电池、热电偶、AC15 型直流复射式检流计、标准电池、工作电源（两节1.5V 干电池）、温度计、电热杯、保温杯、导线等。

【实验原理】

（一）箱式电位差计的工作原理

箱式电位差计是利用补偿法原理制成的精密而使用方便的仪器，可以用来精确测量微小电势差（电压或电动势），也可间接测量其他很多物理量。它虽有多种型号，但一般都包括三个部分，其工作原理图如图 4-22-1 所示。

1. 工作电流回路

主要由 E、R_n、R_1、R、K_0 等组成。其中 E 为工作电源，调节 R_n 可以改变该电路中的工作电流 I，以使 I 达到设计电路时已定好的恒定标准值。

2. 校准回路

主要由 E_S、R_S、G、K_1、K_2（S）等组成。该电路的作用是用标准电池电动势 E_S 把工作电流回路中的电流 I 校准为已设计好的恒定标准值，以使在测量回路中能直接读出待测数值。例如，若测量时温度为20℃，查得此时标准电池电动势 $E_S = 1.01830\text{V}$，则选取标准电阻的值 $R_S = 101.830\Omega$，然后接通 K_0、

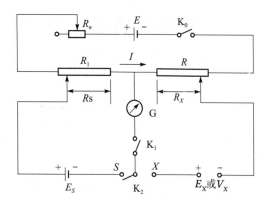

E，工作电源；R_n，工作电流调节电阻；E_S，标准电池；G，检流；K_0、K_1，单极开关；R，可变电阻；R_1，调定工作电流用可变电；E_X（或 E_X），待测电动势或电压；K_2，单极转换开头

图 4-22-1 箱式电位差计的工作原理图

K_1，将 K_2 合向 S 侧，调节可变电阻 R_n 改变工作电流 I 的大小，直到检流计 G 指针不偏转为止，这时称工作电流 I 已校准为标准值，其大小为

$$I = \frac{E_S}{R_S} = \frac{1.01830}{101.830} = 0.0100000(\text{A})$$

此后，在测量过程中 I 保持这一定值，R_n 不能再改变。

3. 测量回路

主要由 E_X、R_X、G、K_1、K_2（X）等组成。当 K_2 合向 X 侧时，因工作电流 I 已校准为标准值，所以，调节 R_X 使 G 指向零，即 IR_X 与被测电动势 E_X（或电压 V_X）补偿时，则有

$$E_X(\text{或} V_X) = IR_X = 0.0100000 R_X \qquad (4-22-1)$$

由式（4-22-1）可知，一定的 E_X（或 V_X）对应一定的 R_X 值。于是可以从精密电阻箱 R 的转盘上直接读出待测的电动势或电压数值。

（二）热电偶

两种不同金属（如铜和康铜）组成一个闭合回路。当两个接触点处于不同温度时，接触点间将产生电动势，回路中会出现电流，此现象称为温差电现象，产生的电动势称为温差电动势。这种由两种不同金属焊接并将接触点放在不同温度下的回路称为温差电偶。温差电偶的温差电动势大小由热端和冷端的温差决定，其极性热端为正极，冷端为负极，如图4-22-2所示。其关系为

$$E_t = \alpha(t_1 - t_0) + \frac{1}{2}\beta(t_1 - t_0)^2 + \cdots \qquad (4-22-2)$$

式中，E_t 为温差电动势；t_1 为热端温度；t_0 为冷端温度；α 和 β 是由构成热电偶的金属材料决定的常数。当冷热端温差不大时，$\alpha \gg \beta$，上式可简化为

$$E_t = \alpha(t_1 - t_0)$$

故温差电动势 E_t 与冷热端温差 $t_1 - t_0$ 成线性关系。比率系数 α 称为温差热电系数，

在国际单位制中,温差热电系数的单位为伏每开,记为 $V \cdot K^{-1}$。

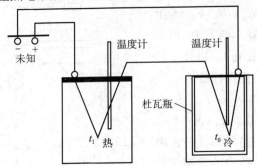

图 4-22-2 热电偶原理图

热电偶可制成温度计,即先将 t_0 固定(例如放在冰水混合物中),用实验方法确定热电偶的 $E-t$ 关系,称为定标。定标后的热电偶与电位差计配合可用于测量温度。与水银温度计相比,温差电偶温度计具有测量温度范围大(200～2000℃)、灵敏度和准确度高,便于实验遥测和 A/D 变换等一系列优点。

【实验步骤】

(1) 在电位差计使用前首先将"N、断、X_1、断、X_2"转换开关放在左边"断"的位置,并将其下方的三个电计按钮("粗""细""短路")全部松开。然后将工作电源(电压为 3V)、标准电池、待测电池(未知 1)、热电偶(未知 2)按正负极接在电位差计相应的端钮上(极性切不可接错),检流计在电计旋钮上无极性要求。

(2) 计算标准电池由于受温度影响而在某一温度下的标准数值。计算公式如下:

$$E_S(t) = E_S(20) - 0.0000406(t - 20) - 0.00000095(t - 20)^2$$

式中,$E_S(t)$ 在 t℃时标准电池的电动势;$E_S(20)$ 为 20℃时标准电池的电动势。计算结果存留到 0.00001V。

(3) 检流计零点调节:检流计接通电源前,应先检查电源插头是否插在"220V"插口(在后面)里,电源开关置于"220V"一侧。接通电源后,在标尺上应该有光标(光点影像)出现。如果找不到,可将检流计的"分流器"置于直接处,若有光标摇晃不停时,可按电位差计上的"短路"按钮,调节"零点调节器",将光点调至标度盘中部。光标出现后,先将"标盘活动调零器"(标尺盘上的一个金属小柱体)移动至左右均有活动幅度的位置,然后调节"零点调节器"(粗调),使光标线移近标尺的零刻度线上(注意光标走向与"零点调节器"的旋回方向相反)。如果相差较小时再移动标盘上的小柱体(零点细调),使光标线与标尺零线重合。零点调好后,将检流计分流器直接旋至"×0.01"档,即后面的调节先从最低灵敏度档开始。

(4) 调整工作电流:将"N、断、X_1、断、X_2"转换开关旋至"N"(标准)位置,将步骤(2)中计算好的 $E_S(t)$ 值的第五、六位数在标准电池补偿钮上旋出。按下电位差计左下方"粗"按钮,调节工作电流的"粗"旋钮及"中"旋钮,使检流计指零,然后将检流计分流器从最低灵敏度档("×0.01"档)逐步调至高灵敏度档,并按下左下方"细"按钮,依次调节工作电流的"细""微"旋钮,直到检流计在高灵敏度档时，按"细"按钮指零。此时的工作电流即可认为是：0.00001A（即本仪器的工作电流）。上述调节过程中，若发现检流计受到冲击，摆动太大时，应迅速按下"短路"按钮。

（5）测试热电偶温差电动势。将转换开关旋到"未知1"（"未知2"），开始测试。先在杜瓦瓶中放入一定量冷水，并将盖子盖好。读出插入瓶中温度计的示值作为冷端温度 t_0。将开水倒入电热杯，盖好盖子，插入其中温度计的示值作为热端温度 t_1。按下电位差计上的"粗""细"按钮，调节测量十进盘使检流计电流为零，此时立即记下 t_1，再记下测量十进盘示值之和，即为热电偶在该瞬时温差下的温差电动势 E_t。将数值记录在表格 4-22-1 中。

上述方法是测量 1.911110V 以下的电动势或电压的方法，若测量值比其大时可串联标准分压箱，本实验不涉及。

（6）重复步骤（5），测出 8 组以上数据。以 Δ（$= t_1 - t_0$）为横坐标，E 为纵坐标，作出热电偶 $E - \Delta t$ 定标曲线，并用图解法求出热电偶温差系数 α。

（7）测试待测电池的电动势，方法同上。

实验结束后，应将检流计置"短路"档以保护检流计。电位差计转换开关旋至"断"位置，并将其下方的三个电计按钮（"粗""细""短路"）全部松开。

【数据记录及处理】

表 4-22-1 用 UJ25 型电位差计测热电偶温差电动势

	1	2	3	4	5	6	7	8
热端温度 t_1（℃）								
冷端温度 t_0（℃）								
$t_1 - t_0$（℃）								
温差电动势 E_t（mV）								

$$\alpha = \underline{\hspace{4cm}} \text{mV} \cdot \text{K}^{-1}$$

【注意事项】

（1）箱式电位差计必须先粗校（测），再细校（测），保护检流计。
（2）电位差计校准好之后，限流电阻 R_n 不可再调动。
（3）每测一组数据后，都应再次校准电位差计，实验时应注意提醒和检查。

【思考题】

（1）电位差计是利用什么原理进行测量的？
（2）使用电位差计测量电势差之前要进行哪些操作？

【仪器介绍】

1. 箱式电位差计

箱式电位差计类型很多，下面简单介绍本实验所用 UJ25 型直流电位差计，图 4-22-3 为这种电位差计面板图。在电位差计面板的上方有 13 个端钮，供接"检流计""标准电池""待测电势差""工作电池""屏蔽"之用。左下方有 3 个按钮，分别标有"粗""细""短路",供在对检流计进行粗调、细调及短路时接通检流计之用。它们相当于图 4-22-1 中开关 K_1 的作用。

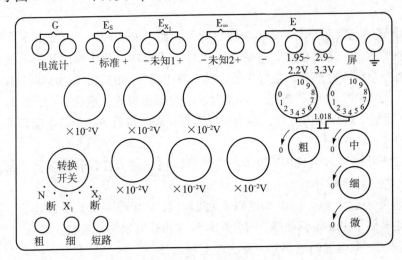

图 4-22-3 箱式电位差计面板图

在此三按钮上方有一旋钮,标有"N、断、X_1、断、X_2"字样,供选择用,分别旋到"N""X_1""X_2",即接通标准电池、未知 1、未知 2。它相当于图 4-22-1 中开关 K_2 的作用。

右上方标有数字 0~10 的两个旋钮,是供补充标准电池第五、第六位数值之用,把它们直接旋至 t ℃时标准电池电动势 $E_S(t)$ 的值后,就相当于图 4-22-1 中选取了标准电阻 R_S 值。

右下方标有"粗""中""细""微"四个旋钮,供调节工作电流用之,它们相当于图 4-22-1 中的工作电流调节电阻 R_n。

面板中间六个大旋钮,其下方都有一个小窗孔,被测电动势或电压的数值由此示出。它们相当于图 4-22-1 中测量回路的可变电阻 R,而当该电路达到补偿时所调至的电阻 R_X 与待测电动势 E_X(或电压 V_X)相对应,于是从转盘的小窗孔中能直接读出待测电动势或电压数值。

2. 直流复射式检流计

AC15 型直流复射式检流计的面板结构如图 4-22-4 所示,其使用方法和注意事项分述如下。

(1) 待测电流由面板左下角标有"+"和"-"的两个接线柱接入(有的是三个接线柱,可接"-"和"1"两个接线柱),一般可以不考虑正负。检流计电源插口在仪器背面,有 AC220V 和 AC6V 两种,在接通电源前,要特别注意电源的选择开关应和实际电源相符(本实验用 AC220V)。

(2) 实验时,先接通电源,看到光标后将分流器旋钮从"短路"档转到"×0.01"档,看光标是否指"0",若光标不指"0",应使用零点调节器和标盘活动调零器把光标调到"0"点。若找不到光标,先检查仪器的小灯泡是否发光,若小灯泡是亮的,轻拍检流计,观察光标偏在哪边,若偏在左边,逆时针旋转零点调节器;若偏在右边，则顺时针旋转零点调节器，使光标露出并调零。

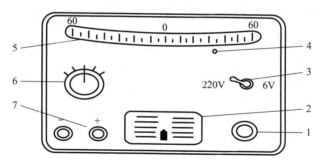

1. 零点调节器；2. 灯泡盖板；3. 电源标志及开关；4. 活动调零器；5. 标盘；6. 分流器；7. 接线柱

图 4 – 22 – 4　直流复射式检流计面板图

（3）测量时，检流计的"分流器"应从最低灵敏度档（×0.01档）开始，或者把"分流器"旋钮直接转到指定的档位"直接"档上，对检流计进行调节。当实验结束时必须将分流器置于"短路"档，以防止线圈或悬丝受到机械振动而损坏。

（李玉娟　张红艳）

Experiment 22　Measurement of Thermal Electromotive Force with the Box – type Potentiometer

【Objects】

（1）To understand the principle and structure of the box type potentiometer, and grasp the method of measuring EMF with the box type potentiometer.

（2）To be familiar with the use of galvanometer.

（3）By measuring the thermal electromotive force between the thermocouple, draw the relationship curves of the thermocouple between the thermal electromotive force and the temperature difference. By using the graphical method, we can get the thermocouple temperature coefficient, and the thermocouple is calibrated.

【Apparatus】

UJ25 type direct – current potentiometer, cell to measure, thermocouple, AC15 type galvanometer, working power source (two dry cells of 1.5V), spot or digital galvanometer, the standard battery, DC power supply, thermometer, electric cup, vacuum cup, wire etc.

【Theory】

1. The principle of the box – type potentiometer

The box type potentiometer is a precision and convenient instrument that made with the principle of compensation method. It can be used to accurately measure minute electric potential difference, or indirect measure some other physical quantity. Though it has many types, it commonly concludes three parts, its operational principle showed in Fig. 4-22-1.

(1) The circuit of working current: Consist of E、R_n、R_1、R、K_0 etc. E is working power source, adjust R to make working current I attain the standard value that has been set when the circuit designed.

(2) Adjusting circuit: Consist of E_S、R_S、G、K_1、K_2 (S) etc. The action of this circuit is to adjust I to the standard value with the electromotive force E_S of the normal element, then the measured value can be read directly from the measuring circuit.

(3) Measuring circuit: Consist of E_X、R_X、G、K_1、K_2 (X) etc. Close K_2 to X, because the working current has been adjusted to standard value, when you adjust R_X to make G point to zero, namely IR_X compensate for measured electromotive force E_X (or voltage V_X), we know that

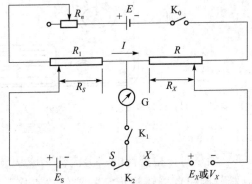

E – power supply; R_n – current regulation resistance;
E_S standard cell; G – galvanometer; K_0, K_1 – unipolar switch;
R – variable resistor; R_1 – setting current variable resistor;
E_X (or V_X) – measured electric potential or voltage;
K_2 – single pole switch

Fig. 4-22-1 The principle diagram of box potentiometer

$$E_X(\text{or } V_X) = IR_X = 0.0100000 R_X \qquad (4-22-1)$$

From (4-22-1) it shows that certain E_X (or V_X) corresponds to a certain value of R_X. So you can read directly the electromotive force or voltage value to be tested from the turntable of the precision resistance box.

2. Thermocouple

Two different metals (such as copper and constantan) form a closed loop. When the two contact points are at different temperatures, an electromotive force will be generated between the contact points, and the circuit current occurs. The phenomenon is known as thermoelectric phenomena, and the electromotive force generated is called thermal electromotive force. This consists of two different metals welding and the contact points on the circuit at different temperatures are called as a thermocouple. Thermocouple electromotive force size is determined by the temperature difference between the hot and cold ends of the temperature. The hot end is positive polarity, the cold end is negative, as shown in Fig. 4-22-2, and their relationship is as follows:

$$E_t = \alpha (t_1 - t_0) + \frac{1}{2}\beta (t_1 - t_0)^2 + \cdots \qquad (4-22-2)$$

Where E_t is the thermal electromotive force, t_1 the temperature of the hot end, t_0 the temperature of the cold end, α and β are constituted by a metallic material thermocouple decision. When the temperature difference between the hot and cold side is not obvious, $\alpha \gg \beta$, the above equation can be simplified as:

$$E_t = \alpha(t_1 - t_0)$$

Therefore, thermal electromotive force E_t and the temperature difference between the hot and cold side $t_1 - t_0$ have formed a linear relationship. The ratio of the coefficient α is called pyroelectric coefficient. In the international system of units, the unit of temperature difference pyroelectric coefficient is Volt per Kelvin, denoted $V \cdot K^{-1}$.

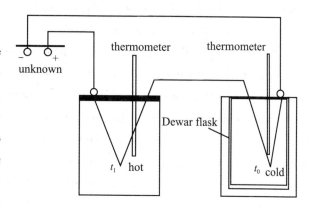

Fig. 4 – 22 – 2　Measuring thermal electromotive force

Thermocouple thermometers can be made. To this end, firstly fix t_0 (for example, in ice – water mixture), secondly determine the relationship between the thermocouple $E - t$ in the experimental method, called the scaling. After scaling, thermocouple can be used to measure temperature with potentiometer. Compared with the mercury thermometer, thermocouple thermometer have a wide range of measuring temperature (– 200 to 2000℃), high sensitivity and accuracy, telemetry and easy test A/D conversion and other advantages.

【Procedure】

(1) Before using the potentiometer, put the "N、X_1、Break、X_2、Break" commutation switch to the place of "Break", and loose the three button ("coarse" "fine" "short circuit"). Then connect the working power source (3V), the normal element and the cell to measure to the corresponding button (do not connect the wrong polarity). The galvanometer do not have the requirement of polarity on potentiometer.

(2) Calculate the standard figures of the normal element in a certain temperature. The equation is:

$$E_S(t) = E_S(20) - 0.0000406(t - 20) - 0.00000095(t - 20)^2$$

$E_S(t)$ is the electromotive force of the normal element in t℃, $E_S(20)$ is that in 20℃. The result retains to 0.00001V.

(3) Adjust the zero set of the galvanometer: Inspect whether the plug of the power source socket in the socket of "220V" (on the back), and whether the power switch turn to the side of "220V", before switch on the power source. When we switch on the power source, there will be a cursor on the ruler. Begin to adjust the galvanometer from the minimum sensitivity.

(4) Adjust the working current: Put the "N、X_1、X_2、Break" commutation switch to the place of "N" (Normal), screw the button of compensation of the normal element to the value of the fifth and sixth numerical digit of $E_S(t)$ that calculated in the step two. Adjust the galvanometer from the minimum sensitivity to the higher, until it points to zero when we press the button of "fine". Then the current is 0.00001 A (namely the working current of the

instrument). If the finger of the galvanometer swings too strongly in the course above, we will press the button of "short circuit" quickly.

(5) Measure the thermal electromotive force between thermocouple. Put the switch knob to "Unknown 1"("Unknown 2") to make the determination. Put a large of cold water in the Dewar flask and cover the lid. Read the value of thermometer inserted into the bottle and record it as a cold end temperature t_0. Pour the boiling water into the electric cup, cover the lid, and insert a thermometer, which shows the value of the hot end temperature t_1. Turn the switch knob of the potentiometer to "coarse", "fine" button, adjust the six small windows making galvanometer current zero, at this time write down t_1 immediately, then write down the sum of the six small windows as the value of the thermocouple thermoelectric force E_t in the instantaneous temperature difference. The values is recorded in the Tab. 4-22-1.

The method above is to measure the electromotive force or voltage less than 1.911110V, if the value measured is larger than this, we will install the normal voltage box in series. This experiment does not involve that.

(6) Repeat step 5 and measure data over eight groups. Make the calibration curve of thermocouple $E - \Delta t$ taking $\Delta (= t_1 - t_0)$ as the abscissa, E as the ordinate. Then calculate the thermocouple temperature coefficient by the graphic method.

(7) Determine the thermoelectric force of the battery to be measured.

After the experiment, we put the galvanometer to "short circuit" to protect it. Put the switch of the potentiometer to "break", and loose the three button ("coarse" "fine" "short circuit"), break the power source.

【Data and Calculations】

Tab. 4-22-1 Measure the thermal electromotive force between thermocouple with UJ25 type potentiometer

	1	2	3	4	5	6	7	8
hot end temperature t_1 (℃)								
cold end temperature t_0 (℃)								
$t_1 - t_0$ (℃)								
thermoelectric force E_t (mV)								

$\alpha = _____$ mV·K^{-1}

【Note】

(1) Box-type potentiometer must be firstly rough calibrated (measured), then fine calibrated (measured), to protect the galvanometer.

(2) After calibrating the potentiometer, the limiting resistor R_n is no longer changed.

(3) After measuring each set of data, you should re-calibrate potentiometer again. In the experiment you should remind and check.

【Questions】

(1) What principle does the potentiometer measuring use?

(2) What operations will be done before measuring position voltage by using potentiometer?

【Introduction to Instruments】

1. Box-type potentiometer

Fig. 4-22-3 is the front panel diagram for the UJ25 type direct——current potentiometer. On the top part of the front panel, there are 13 connection ends, and they are for galvanometer, standard battery, subjected electric potential difference, working battery, and shield respectively. On the bottom left, there are three buttons, and they are "coarse", "fine", and "short circuit". When connecting to galvanometer, these buttons are for rough adjustment, finer adjustment, or short circuit respectively. They are equivalent to switch K_1 in Fig. 4-22-1.

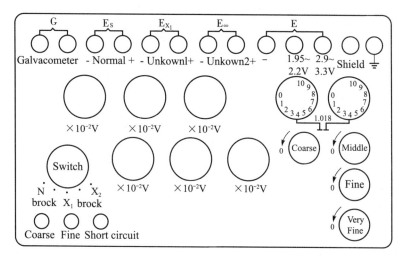

Fig. 4-22-3 Front panel diagram of box type potentionmeter

Right above these three buttons, there is a knob with selectable position marked of N、Break、X_1、Break、X_2. Turning to N, X_1, or X_2 means connecting to standard battery, unknown 1, unknown 2 respectively. It is equivalent to switch K_2 in Fig. 4-22-1.

There are two knobs on the up right corner with number 0-10 marks. They are used to complement the fifth and sixth digits for standard battery. When turning them into standard battery electromotive force $E_S(t)$ at $t℃$, it is equivalent to select standard resistance R_S in Fig. 4-22-1.

To the bottom right, there are four turning knobs marked with "coarse", "medium", "fine", and "ultrafine". They are used to adjust working current, and they are equivalent to working current adjustment resistance R_n in Fig. 4-22-1.

In the middle of front panel, there are six big turning knobs, and there is a small window below each knob. The measured electromotive force or voltage value is displayed here. They are equal to the adjustable resistance R in the measurement circuit of Fig. 4-22-1. When the circuit reaches balance, the resistance R_X is correlated to the measured electromotive force E_X (or voltage V_X). The value of the measured electromotive force E_X (or voltage V_X) can be directly read from the small window of each dial.

2. Direct current duplicate shoot galvanometer (AC15type)

The panel structure of AC15 type DC duplicate shoot galvanometer is shown in Fig. 4-22-4. The using method and attentions are as follows respectively:

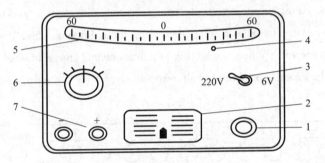

1. zero adjuster; 2. lamp cover; 3. power signs and switch; 4. Activity zero control; 5. standard disk; 6. the current divider; 7. terminals

Fig. 4-22-4 The panel DC duplicate shoot galvanometer

(1) The current to be measured is accessed from the two terminals, the lower left corner of the panel labeled " + " and " - " (some have three terminals, which can be accessed from " - " and "1"). In general, it needn't consider polarity. The power jack is on the back of galvanometer, where are two types include AC220V and AC6V. Before the power is plugged in, you should particularly pay attention to that the selected switch matches the actual power switch (in this experiment, with AC 220V).

(2) During the experiment, first to connect the power, when seeing the cursor, change "the current divider" from the "short" to " × 0.01" to watch whether the cursor refers to "0". If the cursor does not refer to "0", you should use "zero adjustment" and "activity zero control" to transfer the cursor to "0". If you can't find the cursor, check whether the small bulb lights in the instrument. If the small bulb is light, pat galvanometer lightly to see cursor tending to which side. If it's tending to the left, rotate the zero adjuster counter clockwise. If it's tending to the right, then rotate the zero adjuster clockwise. That is to make the cursor expose and zero.

(3) When measuring, you should start from the lowest sensitivity (×0.01) of the galvanometer or switch "the current divider" directly to the designated position "direct" so as to

adjust galvanometer. At the end of the experiment, "the current divider" must be placed in the "short" to prevent the hanging wire or coil damaged by mechanical vibration.

(Li Yu – juan, Zhang Hong – yan)

实验二十三　人耳听觉听阈的测量

【实验目的】
（1）掌握听觉听阈的测量方法。
（2）学会测定人耳的听阈曲线。

【实验器材】
听觉实验仪、立体声耳机、半对数坐标纸等。

【实验原理】

1. 声强、声级、响度级和等响曲线（包含听阈曲线和痛阈曲线）

声波是机械纵波。频率在 20～20000Hz 之间的声波，能够引起人的听觉称为声波。描述声波能量的大小常用声强和声强级两个物理量。声强是单位时间内通过垂直于声波传播方向的单位面积的能量，大小正比于波的振幅的平方，用符号 I 来表示，其单位为 W/m^2。引起人的听觉的声波，不仅有一定的频率范围，还有一定的声强范围。能够引起人的听觉的声强范围大约为 $10^{-12}\sim1W/m^2$。声强太小，不能引起听觉；声强太大，将引起痛觉。

由于可闻声强的数量级相差悬殊，通常用声级来描述声波的强弱。规定声强 $I_0=10^{-12}W/m^2$ 作为测定声强的标准，I_0 通常取普通人能听到的最小强度，这个最小强度称为听阈。某一声强 I 的声级用 L 表示：

$$L = \lg \frac{I}{I_0}$$

声级 L 的单位名称为贝尔，符号 B。通常用分贝（dB）为单位，1B＝10dB。这样上式可表示为

$$L = 10 \times \lg \frac{I}{I_0}(dB)$$

声音响度是人对声音强度的主观感觉，它与声级有一定的关系，声级越大，人感觉越响。

一般来说响度随着声强的增大而增加，但两者不是简单的线性关系，因为还与频率有关，不同频率的声波在人耳中引起相等的响度时它们的声强（或声强级）并不相等。在医学物理学中，用响度级这一物理量来描述人耳对声音强弱的主观感觉。它是选取频率为 1000Hz 的纯音为基准声音，并规定它的响度级在数值上等于其声强级的数值（但是单位不相同），然后将被测的某一频率声音与此基准声音比较，若该被测声音听起来与基准音的某一声强级一样响，则这基准音的响度级

就是该声音的响度级。

人耳对不同频率的声音敏感程度不同。不同频率的声波在人耳中引起相等的响度时它们的声强（或声强级）并不相等。以频率的常用对数为横坐标，声强级为纵坐标，绘出不同频率的声音与1000Hz的标准声音等响时的声强级与频率的关系曲线，得到的曲线称为等响曲线。图4-23-1表示正常人耳的等响曲线。通过对很多人的研究做出的曲线，每条曲线代表的声音具有相同响度。每条曲线上标记的数字代表着响度级（单位是方），它在数值上等于1000Hz时的声级（单位为dB）。比如，标记40的曲线代表普通人听到1000Hz 40dB声级声音的等响线。从这条40dB的曲线我们可以看出，100Hz的声音必须在62dB下才能和1000Hz 40dB的声音具有同样响度。图4-23-1中最下面的一条曲线（标记为0方）作为频率的函数，代表着只有听力非常好才能听到的最轻柔的声音。能引起听觉的最小声强称为听阈，对于不同频率的声波听阈不同，听阈与频率的关系曲线称为听阈曲线。注意，耳朵对频率在2000~4000Hz的声音最为敏感，这在演讲和音乐中很常见。同样需要注意的是，1000Hz的声音在0dB时可以被听见，然而，100Hz的声音至少在40dB才能被听见。

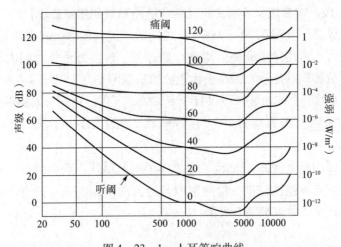

图4-23-1 人耳等响曲线

随着声强的增大，人耳感到声音的响度也提高了，当声强超过某一最大值时，声音在人耳中会引起痛觉，这个最大声强称为痛阈。对于不同频率的声波，痛阈也不同，痛阈与频率的关系曲线称为痛阈曲线。

由图4-23-1可知，听阈曲线即为响度级为0方的等响曲线，痛阈曲线则为响度级为120方的等响曲线。在临床上常用听力计测定患者对各种频率声音的听阈值，与正常人的听阈进行比较，借以诊断患者的听力是否正常。

人耳听觉听阈测量实验仪由信号发生器、功率放大电路、频率计、数字声强指示表（dB表）等组成。

调节衰减旋钮（含粗调和微调）可改变功率、从耳机中得到不同分贝声音，衰减越多、声强级越小。用此仪器可测量人耳（左或右）对于不同频率、不同声强声音的听觉情况。本测量仪测得的声强（dB）指示是相对值，当测量者在1000Hz时，调节声强，使声强（dB）指示为0（dB），然后调节校正旋钮，使自己刚刚能听到，此时声强为0dB。该测量实验仪的声强指示范围为-5~55dB，只能满足实验室听阈测量。仪器面板如图4-23-2所示。

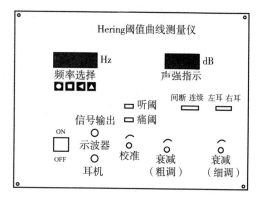

图4-23-2 实验仪器面板示意图

2. 仪器设置键的使用说明

（1）复位键：复位信号频率，仪器设定复位（初始）频率为1000Hz。

（2）确认键：任何设置后必须按下确认键，设定的频率才能有效输出。仪器对设置频率值进行限制，如设置频率值高于20000Hz则输出有效频率只能为20000Hz，如设置频率低于20Hz则输出有效频率只能为20Hz。

（3）选位键：频率数字显示有5位，分别为个、十、百、千、万。选位键能按次序分别选中其中一位，被选中的一位数码管会闪烁，这时只能对闪烁的被选中的位进行修改操作，修改完成后，按下确认键闪烁就会停止，输出有效频率。

（4）"加1"键：对被选中位的数字进行修改，按下"加1"键，就会对选中的位的数字进行加1，每按1次数字加1，依次改变数字为0~9。

【实验步骤】

（1）熟悉听觉实验仪面板上的各键功能，接通电源，打开电源开关，指示灯亮，预热5min。

（2）在面板上将耳机插入，把仪器各选择开关按到选定位置。

（3）被测者戴上耳机，背向主试人（医生）和仪器（或自行测试）。

（4）测量

① 按说明要求选择测量频率（仪器初始为1000Hz）。

② 调节"衰减"旋钮（衰减粗调和微调2个旋钮），使声强指示为0dB。调节"校准"旋钮，使被测者刚好听到1000Hz的声音（整个听阈测量实验内"校准"旋钮不能再调节）。

③ 选定一个测量频率，用渐增法测定：将衰减旋钮调至听不到声音开始，逐渐减小衰减量（可交替调节粗调和微调），当被测人刚听到声音时主试人（或自己）停止减小衰减量，此时的声强（或声强级）为被测人在此频率的听觉阈值，其衰减分贝数用 L_1 表示。

④ 同一个频率用渐减法测定：步骤基本同③，只是将衰减旋钮先调在听得到声音处，然后再开始逐渐增大衰减量，直到刚好听不到声音时为止，与步骤③一样，对相应同一频率的声音，可得到相同的听觉阈值，其衰减分贝数用 L_2 表示。

⑤ 令 $L_{测} = (L_1 + L_2)/2$（负值）——所测频率衰减分贝数的平均值（相对声强）。

⑥ 改变频率，重复①~⑤步骤，分别对64Hz、128Hz、256Hz等9个不同的频率进行测量，得到右耳或左耳9个点的听觉阈值，连起来便是听阈曲线。

（5）作听阈曲线：以频率的常用对数为横坐标（并分别注明测试点的频率值），声强级值为纵坐标，在计算纸上用上面所得数据定点，连起来便为听阈曲线。

【实验记录】

将实验数据填入听阈曲线测量记录数据表（表 4 – 23 – 1）。

表 4 – 23 – 1 听阈曲线测量记录数据表

频率（Hz）	64	128	256	512	1k	2k	4k	8k	16k
L_1（dB）									
L_2（dB）									
$L测 = (L_1 + L_2)/2$									

（张　翼）

Experiment 23　Determination of the Human Ear Auditory Threshold Curve

【Objects】

(1) To master auditory threshold measurement methods.

(2) Learn to determinate auditory threshold curve.

【Apparatus】

Auditory experiment instrument, stereophone, semi-logarithmic paper etc.

【Theory】

1. sound intensity

Sonnd level, loudness level and equal loudness curve (including hearing threshold of curve and pain threshold of curve).

Sound wave is mechanical wave of frequencies from 20Hz to 20000Hz which is audible by human ears. We usually use sound intensity and sound intensity level to describe the magnitude of acoustic energy. Sound intensity (symbol I) is defined as the energy transported by a wave per unit time across unit area perpendicular to the energy flow, and is proportional to the square of wave amplitude. Intensity has unit of power per unit area, or watts/meter^2 (W/m^2). The sound wave which is audible by human ears not only has a certain frequency range but also intensity. Intensity range is from 10^{-12} W/m^2 to 1W/m^2. The smaller intensity can't cause hearing, the bigger intensity can cause pain. Since the range of the audible sound intensity is very large in order, it is always measured instead in terms of sound level. Set the intensity of the faintest audible sound $I_0 = 10^{-12}$ W/m^2 as a basic standard. The intensity I_0 is usually taken as the minimum intensity audible to an average person, which is called the threshold of hearing. The sound level L of intensity I is defined with common logarithm as

$$L = \lg \frac{I}{I_0}$$

The unit of the sound level is called **bel** with symbol **B**. However, decibel (**dB**) is always used in practice and then above equation can be rewritten as

$$L = 10 \times \lg \frac{I}{I_0} (\text{dB})$$

The loudness of sound is the subjective feeling of human beings, its unit is phon. It is related to the sound level somewhat proportionally, the larger the sound level, the louder the sound heard.

Generally, loudness increases with sound intensity, but they are not linear, because it's also related to frequency. When different frequency sound wave cause same loudness, their sound intensity (or sound intensity level) is different. In medical physics, we use loudness level to describe our subjective feeling to the strength of sound. It choose 1000Hz pure tone as reference sound, and set that its loudness level is numerically equal to its sound intensity level (different unit). Then, compare the test sound with reference sound. If test sound has same loudness with one sound intensity of pure tone, the loudness level of pure tone will equal to the test sound's loudness level.

The ear is not equally sensitive to all frequencies. To hear the same loudness for sounds of different frequencies requires different intensities. Take frequency's logarithm as horizontal axis, sound intensity as vertical axis, draw the relationship curve of sound intensity level and frequency when different frequency sound has same loudness with 1000Hz pure tone. It is equal loudness curve. Fig. 4 - 23 - 1 is normal human's equal loudness curve.

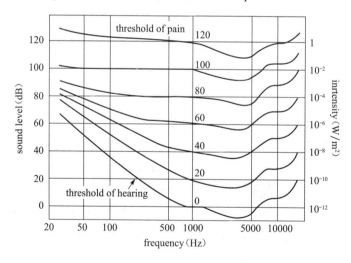

Fig. 4 - 23 - 1 Human's equal loudness curve

Studies averaged over large numbers of people have produced the curves shown in Fig. 4 - 23 - 1. On this graph, each curve represents sounds that seem to be equally loud. The number labeling each curve represents the loudness level (the unit is called phon), which is numerically equal to the sound level in dB at 1000Hz. For example, the curve labeled 40 represents sounds that are heard by an average person to have the same loudness as a 1000Hz sound with a sound level of 40dB. From this 40 - phon curve, we see that a 100Hz tone must be at a level of 62 dB to sound as loud as a 1000Hz tone of only 40 dB. The lowest curve in Fig. 4 - 23 - 1 (label 0 phon) represents the sound level, as a function of frequency, for the softest sound that is just audible by a very good ear (the "threshold of hearing"). The hearing threshold is minimum sound intensity that can cause hearing. For different frequency sound wave, hearing threshold is different. The relationship curve of frequency and hearing threshold is hearing threshold curve. Note that the ear is most sensitive to sounds of frequency between 2000 and 4000Hz, which are common in speech and music. Note that whereas a 1000Hz sound is audible at a level of 0 dB, a 100Hz sound must be at least 40 dB to be heard.

With increasing sound intensity, the loudness of the sound is also increased, when the sound intensity exceeds a certain maximum, the sound can cause pain in the ear, the maximum sound intensity called pain threshold. For sound waves of different frequencies, pain threshold is different, pain threshold and frequency curve is called the pain threshold curve. As shown in Figure 23 - 1, the hearing threshold curve is the equal loudness curve which loudness level is 0 phon, the pain threshold curve was equal loudness curve which loudness level is 120 phon. In clinical, audiometer are used to measure patients's hearing threshold at different frequency, enable to diagnose if the patients's hearing is normal.

Audition threshold determination instrument consists of signal generator, power amplifier, frequency counter, digital sound intensity instruction meter, etc.

Attenuation adjustment knob (rough and fine tuning) can change power to get different decibel voice. The more it attenuates, the smaller sound intensity level would be. With this instrument can measure the human ear (left or right) for different frequencies, different sound intensity sound hearing. The sound intensity (dB) measured by instrument is relative value. When the surveyor at 1000Hz, adjust the sound intensity, make sound intensity (dB) is indicated as 0 (dB), then adjust the calibration knob, make it just can be heard, at this point the sound intensity is 0dB. The experimental apparatus's measure scope is 5 ~ 55dB. Only can meet the requirement for laboratory to measure hearing threshold.

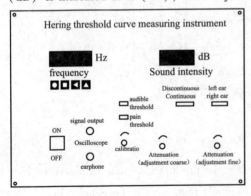

Fig. 4 - 23 - 2 Experimental instrument panels schematic diagram

2. Instrument panel as shown in Fig. 4 - 23 - 2

(1) Reset button: Reset signal frequency, Instrument settings reset (initial) frequency is 1000Hz.

(2) Enter button: You must press the Enter key after any setting, then setting frequency can be output effectively. Instrument settings to limit the frequency value, if setting

the frequency is higher than the 20000Hz, the output effective frequency will be only 20000Hz. If setting the output frequency is lower than 20Hz, the output effective frequency will be only 20Hz.

(3) Select key: Frequency figures show five numbers respectively, one, ten, hundred, thousand, ten thousand. Select keys can be chosen sequentially with one, LED under the one selected will blink. At this time only for the blink of selected bits edit operation. After editing, press the enter key will stop flashing, output the effective frequency.

(4) Plus 1 button: Editing selected digits, press the plus one button, will plus one to selected digit, press one time will plus one, change the number of 0~9 in order.

[Procedure]

(1) Familiar with the features of every keys on the instrument panel, power on, turn on the power switch, indicator light, warm five minutes.

(2) Insert earphone into the panel, put each selector switch to the selected position.

(3) Testees wear the earphone, facing away from the experimenter (doctors) and instrument (self – test)

(4) Determination

① Follow the instructions required to select the measurement frequency (instrument is initially set to 1000Hz).

② Adjust "attenuation" knob (coarse and fine two attenuation knob) enables the sound intensity is indicated as 0dB. Adjust the "calibration" knob to make sure the testees hear 1000Hz sound (During the entire auditory threshold measurement experiment "calibrate" knob can no longer be adjusted).

③ Select a measurement frequency, use incremental method to determine: start with attenuation knob at non – sound, and gradually reduce the amount of attenuation (alternately adjustment coarse and fine) until testees just hear the sound experimenter (or himself), then to stop reducing the amount of attenuation. At this point the sound intensity (or sound intensity level) is equal to the hearing threshold measured at this frequency, the attenuation decibels is represented by L_1.

④ The same frequency use decreasing method: steps basically same with step ③, reverse operation, start with attenuation knob at hearing sound, and gradually increase the amount of attenuation (alternately adjustment coarse and fine), until testees can't hear the sound then to stop increasing the amount of attenuation. Corresponding to the same frequency sounds, get the same hearing threshold, the attenuation in decibels is represented by L_2.

⑤ Let L measured = $(L_1 + L_2)/2$ (negative) average of attenuation decibels of sound (relative to sound intensity).

⑥ Change the frequency, repeat the steps from 1 to 5, 9 different frequency (64Hz, 128 Hz, 256 Hz⋯) are measured, get 9 points of the right or left ear hearing threshold, then we will get the curve of hearing threshold.

(5) Draw audition curve: Take logarithmic frequency as the horizontal axis (respectively indicate the frequency of the test point value), the sound intensity level is the vertical axis, use obtained data to draw points on calculation paper, line them together, then get the hearing threshold curve.

【Data and Calculation】

Date of threshold curve measurements is shown as Tab. 4-23-1.

Tab. 4-23-1 threshold curve measurements recorded data table

frequency (Hz)	64	128	256	512	1k	2k	4k	8k	16k
L_1 (dB)									
L_2 (dB)									
$L_{measured} = (L_1 + L_2)/2$									

(Zhang Yi)

实验二十四 热敏电阻温度计的制作

【实验目的】

(1) 了解热敏电阻的特性,掌握用热敏电阻测量温度的基本原理和方法。
(2) 熟悉非平衡电桥的输出特性。
(3) 设计测量温度范围为 25~75℃ 的温度计。

【实验器材】

惠斯通电桥、直流稳压电源、电阻箱、滑动变阻器、微安表、热敏电阻、温控仪、烧杯等。

【实验原理】

1. 热敏电阻的温度特性

热敏电阻是利用半导体电阻值随温度呈显著变化的特性制成的一种热敏元件。热敏电阻具有负的电阻温度系数,即电阻值随温度的升高而降低,产生这种现象的原因是由于半导体中的自由电子数目随温度的升高而按指数剧烈地增加,因而导电能力迅速增强,虽然原子振动也会加剧并阻碍电子的运动,但这种作用对导电性能的影响远小于电子被释放而改变导电性能的作用,所以温度上升会使电阻下降。由于热敏电阻的阻值具有随温度变化而变化的性质,因此可将其作为一个感温元件,以阻值的变化来体现环境温度的变化。

2. 惠斯通电桥原理

惠斯通电桥的原理电路图见图 4-24-1。

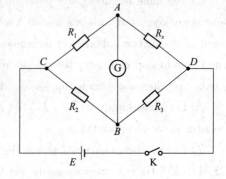

图 4-24-1 惠斯通电桥的原理电路图

R_1、R_2、R_3 和 R_x 为四个"桥臂"上的电阻。适当调节电阻值(例如改变 R_3 的大小),可使"桥"A、B 两点的电位相等,此时通过检流计 G 的电流 $I_g = 0$,这称为电桥平衡,可得

$$R_x = \frac{R_1}{R_2} R_3 \qquad (4-24-1)$$

式(4-24-1)即为惠斯通电桥的平衡条件,也是用来测量电阻的原理公式。欲求 R_x,调节电桥平衡后,只要知道 R_1、R_2 和 R_3 的阻值,即可由(4-24-1)式求得其阻值。

如果 R_1、R_2 和 R_3 保持不变,当 R_x 发生变化时,A、B 两点的电位不再相等,这时电桥处于非平衡状态,此时 A、B 间有电流通过。根据基尔霍夫定律可以求得通过检流计处的电流表达式:

$$I_g = \frac{(R_1 R_3 - R_2 R_x) U_{CD}}{R_g (R_1 + R_x)(R_2 + R_3) + \Delta} \qquad (4-24-2)$$

式中,$\Delta = R_1 R_2 R_3 + R_1 R_2 R_x + R_1 R_3 R_x + R_2 R_3 R_x$;$R_g$ 为用作检流计的微安表头的内阻。可以看出,在其他参数不变的条件下,I_g 为 R_x 的单值函数。如果以热敏电阻 R_T 代替 R_x,由于 R_T 的阻值受温度 T 的影响,因此就会形成 I_g 和 T 之间的单值对应关系,这正是制作热敏电阻温度计的基础。

【实验步骤】

1. 用平衡电桥测热敏电阻的 $R_T - T$ 曲线

(1) 热敏电阻温度计的设计电路图如图 4-24-2 所示。将直流稳压电源 E 的输出电压调为 4.5V,调节滑动变阻器 R 使 U_{CD} 在 1.5~2V 之间(用万用表测量)。

(2) 将电键 S 接到热敏电阻 R_T 端,图 4-24-2 中检流计用量程为 50μA 的微安表,R_3 为电阻箱,取 $R_1 = R_2$,即倍率为 1,且"桥臂"上四个电阻的阻值要基本接近。

(3) 把热敏电阻放在烧杯中,用温控仪改变并测量水温。

(4) 接通电桥电源,调节 R_3,使电桥平衡,此时 $R_T = R_3$。逐步提高水温,测出不同温度下的热敏电阻阻值(从 25℃ 开始至 75℃,每隔 5℃ 测量一次),并将测量的 R_T 数据记录于表 4-24-1 中。

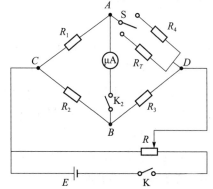

图 4-24-2 热敏电阻温度计的设计电路图

(5) 逐步降低水温,重复(4)操作,并取升温过程和降温过程中在该温度下的电阻的平均值作为热敏电阻在该温度时的阻值。由表 4-24-1 中的数据就可绘得热敏电阻的电阻-温度特性曲线。

表 4–24–1　电阻–温度特性及温度计定标测量数据表

T (℃)	R_T (Ω) 升温	R_T (Ω) 降温	$\overline{R_T}$ (Ω)	I_g (μA)
25				
30				
35				
40				
45				
50				
55				
60				
65				
70				
75				

2. 温度计的调试

（1）用一个四位旋钮式的电阻箱代替 R_T 接入 S、D 两点，R_1 和 R_2 保持不变。

（2）根据前面热敏电阻的温度特性曲线的测量，将电阻箱调至热敏电阻温度计下限温度（25℃）时的阻值（$R_{25℃}$），调节 R_3，使电桥平衡，此时 R_3 调节完毕，应有 $R_3 = R_{25℃}$。

（3）再将电阻箱调至热敏电阻温度计上限温度（75℃）时的阻值（$R_{75℃}$），调节滑动变阻器 R，使微安表满偏，此时 U_{CD} 的调节完毕。

（4）最后将电键 S 接到电阻箱 R_4 端，调节 R_4，使微安表满偏，此时 R_4 调节完毕，应有 $R_4 = R_{75℃}$。R_4 是用于校正满刻度电流用的。

3. 温度计的定标

（1）将电键 S 重新接回 R_T 端的电阻箱，调节电阻箱，使其阻值分别为热敏电阻在 30℃、35℃、…70℃一系列温度下的阻值，并在表 4–24–1 中记录微安表的相应示数。

（2）利用表 4–24–1 中的数据绘制 T–I_g 曲线，此时热敏电阻温度计的定标完毕。

（3）撤掉电阻箱，接入热敏电阻 R_T，此时热敏电阻温度计的制作完成。

4. 测量自己手的温度

（1）先将电键 S 接到电阻箱 R_4 端，然后接通电源观察微安表是否满偏，如果不满偏则调节滑动变阻器 R 使其满偏，以校准温度计。

（2）将电键 S 再接到热敏电阻 R_T 端，用手心握住 R_T，待温度稳定后从微安表中读出电流值，根据绘制的 T–I_g 曲线即可得到手的温度。

【思考题】

如何才能提高改装热敏电阻温度计的精确度？

（李玉娟）

Experiment 24 The Assembly Make of Thermistor Thermometer

【Objects】

(1) Have acquaintance with the features of thermistor thermometer. Master the principle and method of temperature measurement by thermistor thermometer.

(2) Be familiar with the output features of non equilibrium bridge.

(3) Design a thermometer with the temperature ranging from 25 ~75℃.

【Apparatus】

Wheatstone bridge, cocurrent regulated power supply, resistance box, slide rheostat, microampere appearance, thermistors, temperature controller, beaker and et al.

【Theory】

1. The temperature properties of thermistor

Thermistor is a kind of thermosensitive element based on the features of semiconductor resistance value's temperature properties. Thermistor owns the property of negative temperature coefficient of resistance, which means resistance value becomes stronger as temperature arises. The reason is that there will be more free electron in semiconductor which increases conductivity, not withstanding the obstruction of atom motion. The impact of atom motion has much less influence on conductivity than that imposed by free electron, that is to say resistance value declines as temperature arises. As thermistor owns the property of negative temperature coefficient of resistance, it is also used as a temperature – sensing device, which tests the changing of surrounding temperature.

2. Principle of wheatstone bridge

Fig. 4 – 24 – 1 shows the principle of wheatstone bridge. R_1, R_2, R_3 and R_x are resistances on four bridge arms. With the adjustment of resistance value (the value of R_3), equal to the potential of point A and B can be reached. The electric current on galvanometer is $I_g = 0$, which is called bridge balance. As a conclusion:

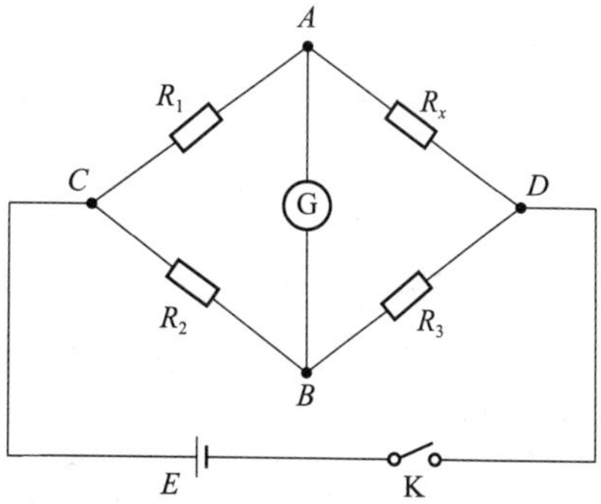

$$R_x = \frac{R_1}{R_2} R_3 \quad (4-24-1)$$

Fig. 4 – 24 – 1 Circuit diagram for bridge principle

Formula (4 – 24 – 1) shows the equilibrium condition of wheatstone bridge. Also, it serves as the principle of formula to measure resistance. R_x can be concluded from formula (4 – 24 – 1) with the adjustment of bridge balance and desiring the value of resistance of R_1, R_2 and R_3.

If R_1, R_2 and R_3 remain at the same value, the bridge is in non – equilibrium condition when the value of R_x changes and the electric potential of A、B no longer stays equal. Based on Kirchhoff law, the electric current on galvanometer can be concluded as:

$$I_g = \frac{(R_1 R_3 - R_2 R_x) U_{CD}}{R_g (R_1 + R_x)(R_2 + R_3) + \Delta} \qquad (4-24-2)$$

In the formula (4-24-2), $\Delta = R_1 R_2 R_3 + R_1 R_2 R_x + R_1 R_3 R_x + R_2 R_3 R_x$, R_g serves as the internal resistance of the galvanometer microampere meter. If the other parameters remain, I_g is the one-valued function for R_x. In replacement of R_x by R_T, as the value of R_T changes with Temperature T, I_g and T forms the relation as unit value. This is also the basic of producing the thermistor thermometer.

【Procedure】

1. Measuring the thermistor $R_T - T$ curve with bridge balance

(1) The circuit diagram of thermistor thermometer is shown as Fig. 4-24-2. Set the output voltage of DC stabilized power supply E to 4.5V, adjust the slide rheostat and set U_{CD} between 1.5V and 2V (measured by multimeter).

(2) Set keyset S to thermistor R_T, the galvanomete in Fig. 4-24-2 is a micro ammete with the measuring rate of 50μA. R_3 is a resistance box. If $R_1 = R_2$, the magnification was 1 and the values of four resistances on the bridge arm should be nearly the same.

(3) Set thermistor into a beaker. Measure and change the water temperature by temperature controller.

(4) Connect the bridge power. Adjust R_3 to balance the bridge. At this time, $R_T = R_3$. Increase the temperature of water gradually, measure the thermistore resistance value in different temperature (From 25℃ to 75℃, measure the temperature every 5℃). Log in data recording of R_T in Tab. 4-24-1.

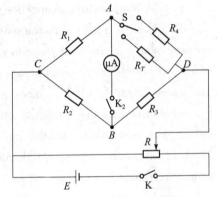

Fig. 4-24-2 Circuit diagram of the rmistor temperature

(5) Reduce the water temperature gradually. Repeat the above operation and set the average resistance value during the procedure of raising and reducing water temperature as the thermistor value under that certain temperature.

Tab. 4-24-1 Resistance to temperature characteristic and the thermometer calibration measurement table

T (℃)	R_T (Ω)		$\overline{R_T}$ (Ω)	I_g (μA)
	warming	cooling		
25				
30				
35				
40				

T (℃)	R_T (Ω)		$\overline{R_T}$ (Ω)	I_g (μA)
	warming	cooling		
45				
50				
55				
60				
65				
70				
75				

2. Debugging the thermometer

(1) Connect a four knob type resistance box to spot S and D instead of R_T, R_1 and R_2 remained the same.

(2) Based on the temperature feature curve of the previous thermistor's, set resistance box to the minimum temperature (25℃) value ($R_{25℃}$). Adjust R_3 to keep bridge balance. After the adjustment of R_3, $R_3 = R_{25℃}$ can be concluded.

(3) Then set resistance box to the maximum temperature (75℃) value ($R_{75℃}$). Adjust the slide rheostat R until the microammeter turns to its full slide. Then the adjustment for U_{CD} is completed.

(4) Lastly, set key S to resistance box R_4. Adjust R_4 to make the microammeter to its full slide. Then the adjustment for R_4 is completed. $R_4 = R_{75℃}$ can be concluded. R_4 is for the correction of the full scale current.

3. Calibration the thermometer

(1) Replace key S to the resistance box on R_T side. Adjust resistance box to set the thermistor values to the resistance ones under the series of 30℃, 35℃, ⋯70℃. Log in the microammeter data in Tab. 4-24-1.

(2) According to the data in Tab. 4-24-1, draw the $T - I_g$ curve. The thermistor thermometer calibration is completed.

(3) Remove the resistance box and join in the thermistor R_T. Then the thermistor thermometer is produced completely.

4. Measure the temperature of your hands

(1) Set key S to resistance box R_4 and connect the power to see whether the microammeter goes to its full side. If not, adjust slide rheostat R to make the microammeter goes to its full side in order. Then the thermometer is corrected completely.

(2) Set key S to thermistor R_T again. Hold R_T in your hand and read the current value from microammeter. According to the $T - I_g$ curve, hand temperature can be known.

【Questions】

How to improve the accuracy of thermistor thermometer?

(Qi Xia)

第五章 计算机仿真实验

Chapter Five　The Computer Simulation Experiment

实验二十五　虚拟仪器基础

【实验目的】
(1) 了解虚拟仪器的概念。
(2) 了解图形化编程语言 LabVIEW，熟悉 LabVIEW 开发环境并学习基本的编程。

【实验器材】
计算机（含操作系统）、LabVIEW 软件、示波器。

【实验原理】
虚拟仪器（virtual instrument，VI）是基于通用计算机硬件平台通过专用测试软件实现的一种计算机仪器系统。虚拟仪器最突出的一个外部特征是没有传统仪器的实物面板，所有的控制旋钮和指示器都"安放"在计算机的显示屏上，是一个虚拟的面板或称之为"软面板"。虚拟仪器利用了通用计算机强大的计算和处理能力，配合不同的传感器和接口卡实现信号的输入，同时利用键盘、鼠标和显示器来完成操作控制和信号显示等功能。用户自己可以设计定义虚拟仪器的面板以及测试功能。使用时，只需要使用鼠标和键盘就可以完成测试测量操作。

LabVIEW 是一种图形化的编程语言，又称为"G"语言。使用这种语言编程时，基本上不写程序代码，取而代之的是流程图，使编程简单直观。LabVIEW 自带了丰富的函数库，可以实现数据采集、设备控制、数据分析和显示等功能。

虚拟仪器程序是指使用 LabVIEW 开发平台编制的程序，简称 VI。VI 包括三个部分：程序前面板（Panel）、框图程序（Diagram）和图标/连接器。

程序前面板用于设置输入数值和观察输出量，用于模拟真实仪表的前面板。在程序前面板上，输入量被称为控制（Controls），输出量被称为显示（Indicators）。控制和显示是以各种图标形式出现在前面板上，如旋钮、开关、按钮、图表、图形等，这使得前面板直观易懂。

每一个程序前面板都对应着一段框图程序（即流程图程序）。框图程序用 LabVIEW 图形编程语言编写，可以把它理解成传统程序的源代码。框图程序由端口、节点、图框和连线构成。其中端口被用来同程序前面板的控制和显示传递数据，节点被用来实现函数和功能调用，图框被用来实现结构化程序控制命令，而连线代表程序执行过程中的数据流，定义了框图内的数据流动方向。

【实验步骤】

1. 初步熟悉 LabVIEW 开发环境的基本操作和简单编程方法

由 开始 → 程序 → National Instruments LabVIEW 7.0，启动 LabVIEW 程序。选择 NEW VI 进入 LabVIEW 环境。首先看到的是灰色的前面板。点击 Windows → Show Black Diagram，可以显示框图程序面板。在前面板，点击 Windows → Show Controls Palette/Show Tools Palette 可以显示工具模板和控制模块（图 5-25-1）。旋钮、开关等控制量在控制模板上，选择这些图标并将其放到前面板上，那么相应的端子和图标会出现在流程图上。通过这些控制量图标可以通过前面板控制程序中的数据，或者将程序运行结果显示出来。在框图程序面板窗口中，选择 Windows→Show Functions Palette 来显示功能模板，利用功能模板提供的循环、数学运算、比较以及公式节点等功能函数可以创建框图程序。每个模块的详细说明参见相关参考书以及软件。

2. 创建一个模拟温度测量程序，熟悉 LabVIEW 程序基本概念，观察程序运行结果并解释程序每部分功能

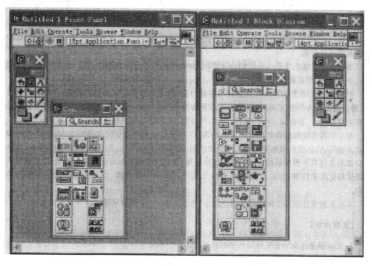

图 5-25-1　LabVIEW 开发环境图

假设传感器的输出电压和温度成正比。例如，当温度为 80°F 时，传感器输出电压为 0.8V，同时程序提供摄氏温度和华氏温度显示选择开关。为简单起见，我们用软件代替了 DAQ 数据采集卡，使用 LabVIEW 提供的 Demo Read Voltage 子程序来仿真电压测量，然后把所测得的电压值转换成摄氏或华氏温度读数。Demo Read Voltage VI 子程序模拟从数据采集卡的 0 通道读取电压，程序再将读数乘以 100.0 转换成华氏温度读数，或者再把华氏温度转换成摄氏温度。前面板图和框图程序图见图 5-25-2 和图 5-25-3。

（1）创建前面板：用 File 菜单的 New VI 选项打开一个新的前面板窗口。在控制模板中的 Numeric 子模板中选择 Thermometer，并将其放入前面板窗口。在高亮的文本框中输入"温度计"。在控制模板中的 Boolean 子模板中选择 Vertical Slide Swich（或其他开关），并将其放入前面板窗口。在文本框中输入"温度值单位"。使用标签工具 A，在开关的"条件真"（True）位置旁边输入自由标签"摄氏"，再在"条件假"（False）位置旁边输入自由标签"华氏"。在控制板中的 Numeric 子模板中选择 Numeric Indicator，并将其放入前面板窗口，在高亮的文本框中输入"温度值"（图 5-25-2）。

（2）创建框图程序：打开框图程序窗口，在功能模板中找到下列对象并将其放入框图程序窗口中。

图 5-25-2　前面板图

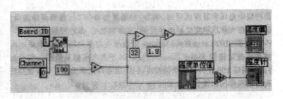

图 5-25-3　框图程序图

Demo Read Voltage VI 程序。功能模板显示设为 Advanced 模式时，在 Select a VI 子模板选择文件 D：\Program Files\National Instruments\LabVIEW 7.0\vi. lib\tutorial. llb。打开后选择其中的 Demo Read Voltage。该程序模拟从 DAQ 卡的 0 通道读取电压值。

Multiply（乘法）功能（Numeric 子模板）、Subtract（减法）功能（Numeric 子模板）、Divide（除法）功能（Numeric 子模板）。

Select（选择）功能（Comparison 子模板）。根据温标选择开关的值输出华氏温度（当选择开关为 False）或者摄氏温度（选择开关为 True）数值。

将所需要的对象放入框图程序窗口中之后，把图标移至如图 5-25-3 所示的位置，再用连线工具连接起来，并在需要的位置创建数值常数（用连线工具，右击你希望连接一个数值常数的对象连线端子，选择 Create Constant 功能。如要修改常数值，用标签工具双点数值，再写入新的数值）。整个程序创建完毕。

（3）运行程序：选择前面板窗口，运行 VI 程序。点击连续运行（Run）按钮，使程序运行于连续运行模式。观察程序运行结果并解释程序每部分的功能。停止程序运行。用文件菜单的 SAVE 功能保存上述文件。

（4）关闭程序。

3. 创建一个频率幅度可调的方波、正弦波和三角波信号发生器，学习 LabVIEW 基本函数的用法

该信号发生器能够根据用户选择生成指定频率和幅度的方波、正弦波（占空比变可调）和三角波，并在前面板上显示波形，用户可以控制波形的显示。

在框图程序图增加一个 DAQ 卡输出函数，就可以将波形输出并通过示波器观察。掌握较快的同学可以选作这部分内容。

操作步骤如下：

（1）创建前面板：放置一个 Waveform Graph（Controls→Graph）用于显示波形，其标签改为"显示屏"。选中 Scale Legend 和 Graph Palette（对着该对象点击右键，Visible Items 中）。

放置一个开关。放置3个控制量，标签分别修改为频率、振幅、占空比，缺省值分别设为1.00、1.00和50.00（用标签工具输入数值后，右键选择设置精度和缺省值）。放置一个枚举量（Controls→Ring & Enum→Enum），枚举值设为"正弦波""方波""三角波"（用标签工具输入字符串后，右键选择 Add Item After），如图5-25-4所示。

（2）创建框图程序：首先放置一个 While 循环，将所有图标包含在内，将开关与 While 循环的循环条件（Loop Condition）相连，同时将 Loop Condition 设置为 Continue If True。

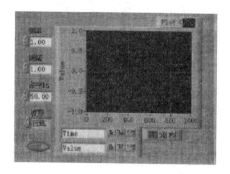

图5-25-4　信号发生器前面板图

再加入一个 Case 结构，Case 结构可以包含2个或更多的子程序（Case），由与 Case Selector（即边框上的问号）相连接的外部控制量决定程序执行哪一个子程序，控制量可以是布尔量、整数或者字符串等。将"波形"图标与 Case Selector 相连。

通过选择结构框上方的 Selector Label 可以选择相应的子程序。

在 Selector Label 选择为正弦波时，在功能模板中选择 Analyze → Waveform Generation → Sine Wave. vi，将生成的正弦函数加入到 Case 子程序中。将输出跨过 Case 结构连接到显示屏。

说明：在模拟状态下，信号频率以每秒周期数为单位（Hz），但在数字系统中常采用数字频率，单位是周期数/采样点，也成为标准频率，它们之间的转换关系如下：

模拟频率 = 数字频率 × 采样频率

数字频率 = 模拟频率/采样频率（采样频率就是每秒采样的点数）

Sine Wave. vi 以及下面的 Square Wave. vi 和 Triangle Wave. vi 等模块中采用的是数字频率，因此要将前面板输入的模拟频率转换为数字频率再输入到相应的输入端。

分别选择 Selector Label 为"方波"和"三角波"，并在 Case 结构中加入相应的方波生成函数（Analyze → Signal Processing → Signal Generation → Square Wave. vi）和三角波生成函数（Analyze → Signal Processing → Signal Generation → Triangle Wave. vi），并完成连线。注意："占空比"图标只连接到方波模块的 duty 端，如图5-25-5所示。

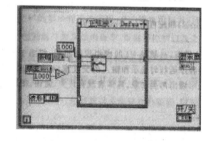

图5-25-5　信号发生器框图程序图

（3）运行程序：运行该程序，调节前面板控制量，观察波形的改变，并解释，最后命名并保存该程序。

（4）选做内容：速度快的同学可做选做内容，通过接口卡输出所生成信号并利用示波器观察，具体步骤此处省略。

【思考题】

（1）虚拟仪器系统与传统仪器有什么区别？请简要说明。

（2）虚拟仪器的出现对于测量仪器的发展有什么影响？对于仪器使用者来说，意味着什么？

（3）虚拟仪器的软件开发环境有哪些类型？什么是"G"语言？

（4）你对虚拟仪器系统在物理实验中的应用有何设想？能否结合具体实验给出一个简单的设计思路。

【知识链接】

LabVIEW 简介：使用 LabVIEW 开发平台编制的程序成为虚拟仪器程序，简称 VI。VI 包括三个部分：程序前面板、框图程序和图标/连接器。由于 LabVIEW 的功能强大、内容丰富，为使初学者能尽快掌握 LabVIEW 的基本内容，下面对它的一些基本功能和结构做简单介绍。

（一）基础知识

1. LabVIEW 程序构成

双击 LabVIEW 快捷图标，出现启动画面，单击其中的 NEW VI，打开一个新的 LabVIEW 程序，可以看到它由前面板和流程图组成。

（1）前面板窗口：前面板窗口是图形用户界面，也就是 VI 的虚拟仪器面板，相当于试剂仪器的控制面板，它将用户和程序连接起来，是程序运行时显示和输入的交换窗口。如图 5-25-4 所示的信号发生器的前面板，上有用户输入和显示输出两类对象，具体表现有开关、旋钮、图形以及其他控制和显示对象。

（2）流程图窗口：流程图窗口提供 VI 的图形化源程序，相当于实际仪器箱内的东西，在流程图中程序员用图形语言编写 LabVIEW 程序源代码，以控制和操纵定义在前面板上的输入和输出功能。如图 5-25-4 所示的信号发生器流程图，上面包括前面板上的控件的连线端子，还有一些前面板上没有，但编程必须有的东西，如函数、结构、连线等。

下面介绍工具条上常用按钮的功能。

运行（Run），如果 VI 有编译错误，此按钮将变成 。

连续运行（Run Continuously）。

异常中止执行（Abort Execution）。

加亮执行（Highlight Execution），只有流程图中有。进入加亮执行时，变成 ，此时，流程图中的数据变亮，同时显示 VI 执行中的一些中间数据。

提示：使用了 ，会使 LabVIEW 占用大量 CPU，使计算机其他操作变慢，建议一般不要使用。

2. 控制选项板功能介绍

只能在前面板窗口中使用，通过前面板窗口 Windows → Show Controls Palette 打开，也可以在前面板窗口中空白处单击右键打开。该选项板用来给前面板设置各种所需的输出显示对象和输入控制对象。模块图标右上角有"▶"表示有子选项板，单击模块图标可以访问子选项板。

在控制选项板上我们可以看到几个常用的模块，我们对它们做一些介绍，其他模块可以在 All Controls 下找到。

数值量（Numeric），数值的控制和显示。包括数字式、指针式显示表盘及各种输入框。

布尔量（Boolean），逻辑数值的控制和显示。包含各种布尔开关、按钮以及指示灯等。

数组和簇（Array&Cluster），数组和簇的控制和显示。

图形（Graph），显示数据结果的趋势图和曲线图。

3. 函数选项板功能介绍

只能在流程图窗口中使用，通过流程图窗口 Windows→Show Function Palette 打开，也可以在路程图面板中空白处单击右键打开。函数选项板提供创建流程图的工具，其与控制板基本相同，模块图标右上角有"▶"表示有子选项板，单击模块图标可以访问子选项板。

结构（Structure），包括程序控制结构命令，如循环控制以及全局变量和局部变量。

数组（Array），包括数组运算函数、数组转换函数以及常数数组等。

比较（Comparison），包括各种比较运算函数，如大于、小于、等于。

数据采集（Dtata Acquisition），包括数据采集硬件的驱动，以及信号调理所需的各种功能模块。

信号分析（Analysis），信号发生，时域及频域分析功能模块及数学工具。

提示1：LabVIEW 中，你可以随时获得帮助，用 Hlep→Show Context Help 打开帮助窗口（Context Help）快捷键为 Ctrl+H，当把鼠标放到任何感兴趣的模块对象上时，就会在帮助窗口中显示相应的帮助信息。

提示2：在任何一个控制或是函数模块上单击右键，都会出现弹出菜单可以方便地对模块进行编辑。

4. 工具选项板功能介绍

通过 Windows→Show Tools Palette 打开和关闭。它提供了各种用于创建、修改和调试 VI 程序的工具。当从选项板内选择了任一种工具后，鼠标就会变成该工具相应的形状。下面介绍常用工具按钮的功能（其他按钮功能）。

操作（Operate Value），用于操作前面板的控制和显示，使用它向数字或字符串控制中键入值时，工具会变成标签工具。

选择（Position/Size/Select），用于选择、移动或改变对象的大小。当它用于改变对象的连框大小时，会变成相应形状。

标签（Edit Text），用于输入标签文本或者创建自由标签。当创建自由标签时它会变成相应形状。

连线（Connect Wire），用于在流程图程序上连接对象。如果联机帮助的窗口被打开时，把该工具放在任一条连线上，就会显示相应的数据类型。

5. 数据线、数据流和数据类型

（1）数据流工作方式：在流程图上，模块之间的连线就是数据线。数据通过数据线在模块之间传递。LabVIEW 不像一般语言按照语句的顺序一行一行地执行，它是依靠在数据线上传递的数据来控制程序的，只有当模块要求的输入数据完全到达这个模块时才能执行，然后向其所有的输出端口输出数据，这些数据在沿数据线流向其他模块。这就是 LabVIEW 的数据流工作方式。前面介绍过，可以通过加亮 执行观察它的数据流是如何工作的。

（2）数据线：连线时，LabVIEW 会提示该接口的名称，只能在同一数据类型的端口之间连线，不同类型数据连线的形状和颜色也不同。错误的连线会表示成黑色连线，这时你应将连线工具移动到它上面，LabVIEW 会给出两端的数据类型信息，你可以检查并改正。添加新的模块和连线可能会对已设置好的部分有影响，而且原来的错误连线也可能对新添加的连线有影响。表 5-25-1 为常用连线类型。

表 5-25-1 常用连线类型

类型	标量	一维数组	二维数组	颜色
整形				蓝色
浮点型				橙色
布尔型				绿色
字符型				红色

（3）数据类型：LabVIEW 的基本数据类型有 5 种，即：Numeric（数值），Boolean（逻辑），String（字符串），Enum（枚举），还有一种叫 Ring（环形枚举），和 Enum 类似，可以循环枚举。

Numeric 类型的数据按精度又分若干种类型，与标准 C++ 的数据类型基本是一致的。其代表符号直观的表现其类型（表 5-25-2）。可以在数值对象上点击右键，通过弹出菜单中 Representation 修改。

表 5-25-2 LabVIEW 的数值数据类型

符号	I8	I16	I32	U8	U16	U32
类型	8 位整型（短整型）	16 位整型（单字型）	32 位整型（长整型）	8 位无符号整型	16 位无符号整型	32 位无符号整型
符号	SGL	DBL	EXT	CSG	CDB	CXT
类型	单精度浮点型	双精度浮点型	128 位扩展浮点型	单精度复数	双精度复数	128 位扩展复数型

提示：在流程图上，你会看到不同颜色的模块，不同颜色代表不同的数据类型，其定义与连线一致。

6. 控制量与显示量

在 LabVIEW 中，一个数字量、布尔量、字符量等都有控制量和显示量的区别。

控制量：用于控制程序，它相当于仪器上的控制按钮，如开关、旋钮等。

显示量：用于显示程序运行的结果，它相当于仪器上的显示部件，如显示屏、指示灯等。

在流程图窗口中，我们可以看到控制部件的外框比较粗，显示部件的外框比较细。要实现控制量与显示量之间的转换，只需在流程图上的模块上点击右键，选择 Change to Indicator （或者 Change to Control ）就可以了，如图 5-25-6 所示。

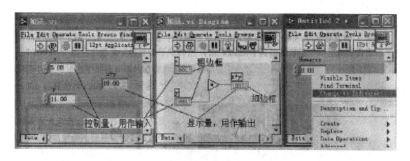

图 5-25-6 控制量与显示量

（二）几个要用到的模板

1. 组和簇（array & cluster）

（1）组：组是同类型元素的集合。一个组可以是一维或者多维，每维最多可有 $2^{31}-1$ 个元素。可以通过组索引访问其中的每个元素。和 C 语言一样，索引的范围是 $0 \sim n-1$，第一个元素的索引号是 0，第二个是 1，以此类推。组的元素可以是数据、字符串等，但所有元素的数据类型必须一致。

组的建立，创建一个组有两件事要做：

① 建一个组的"壳"（shell），将 Controls → Array & Cluster → Array 模块放在前面板中；

② 在这个壳中置入组元素（数、字符串开关等）。

我们将 Controls → Numeric → Digital Control 模块放在组框中，就形成了一个数组，其他类型的组的建立类似，只要在步骤②中组框中放入你想要建的组的元素就可以了。要改变组显示元素的多少，把鼠标换成选择模式，移到组框的角上，鼠标变成 ，这时，你可以任意拖动改变组。要改变组的维数，和改变数组显示元素多少的方法一样，在鼠标选择模式下移到组索引框的角上，鼠标将变成 ，拖动改变组的维数；也可以在组索引框上点击右键，选择 Add Dimension。

（2）簇：簇是另一种数据类型，它的元素可以是不同类型的数据。它类似于 C 语言中的 Struct。使用簇结构可以把分布在流程图中各个位置的数据元素结合起来，这样可以减少连线的拥挤程度，用于错误处理。

簇的建立：与组的建立相同，同样是先建立簇的"壳"（ Controls → Array & Cluster → Cluster ），然后在壳中放元素，可以放不同类型的数据。需要注意的是：向簇中放置对象，必须都是控制对象或者都是显示对象，不能混在一起；簇中的数据和其放置顺序有关而和其放置的位置无关。第一个放进去的对象就是元素 0，以此类推。删除其中的一个后，其顺序将自动调整。

2. 图表和图形

可以将数据以图表或图形方式显示出来。

（1）图表：通常 Y 轴为数据轴，X 轴为表示时间的值或数据点序号。

下面我们来看一个随机数的例子（图 5-25-7）。

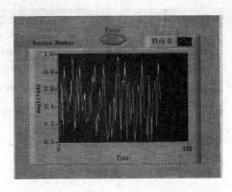
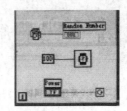

图 5-25-7　随机数的前面板和流程图

每循环一次，随机数模块产生一个数，图表会对这些数自动编号，在图形框中显示出来，y 表示数值，x 表示数的序号。它保持以前的数据，并追加新的数据，使图表保持更新。

可以在图形框上点击右键改变图标的性质，如在 Advance → Update Mode 中改变图表的更新模式；在 Visible Items 中选择显示标签、图例、图形模板等。同学可以自己试试其功能。

（2）图形：图形与图表相似，它们的不同是：图形不能显示新追加的数据，而图表可以。

以下是一个正弦余弦波的例子（图 5-25-8）。

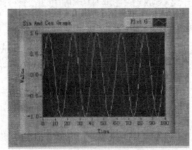
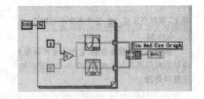

图 5-25-8　正弦余弦波前面板和流程图

通过例子我们可以看出，图形显示的是将要显示的数据形成数组，然后一次显示出来。不能像图表那样用追加新的数据的方法来显示数据。

3. 结构

（1）循环结构：LabVIEW 给大家提供了两种循环结构：一个是 For 循环，另一个是 While 循环。两个循环与大部分计算机语言中的循环结构非常相似。二者都是在 Functions → Structures 下。

For 循环结构：如图 5-25-9 所示 "N" 是指循环次数，"i" 指的是循环计数，i 从 0 执行到 N-1。可以从 Functions → Structures 中选中 For 结构对象，在流程图面板上拖拉就可以将要进行循环的对象包含进去。当然也可以直接对要（或不要）进行循环的对象进行拖拉使其进入（或离开）循环结构。如果要删除循环结构直接选中按 Delete 键即可。

While 循环结构：如图 5-25-9 所示，右边带箭头的小图标是用来接循环条件的，当调节为"真"时，开始继续执行下一次循环，为"假"时停止执行循环条件。这里循环条件的检验是在循

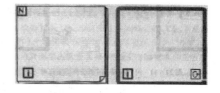

图 5-25-9　For 循环和 While 循环

环体执行完成时，因此循环至少执行一次。"i"的作用也是指示循环次数。

循环结构的数据传递：如果需要有数据在循环结构的内外传输，可以在循环体内的端子与循环体外的端子连线，这时循环体边框上出现小的黑方块（有些书上称为隧道）。在循环执行期间隧道不能进行数据传输，只能在循环结束时将最后一次循环的结果输出。所以要对每一次循环的结果进行观察，我们就应该将显示端子放在循环体内部。

移位寄存器（Shift Register）：对于 For 和 While 循环，移位寄存器是很重要的，它可以把每次循环的结果传递给以后的循环。寄存器口通过在循环结构的左右边框的右键菜单上的 Add Shift Register 生成。它由一对方向相反、分别位于循环结构左右边框上的两个小三角形构成。右端存储每次循环完成的输出结果，移位寄存器将这个结果转移到左边的小三角上，作为下次循环的输入数据。

（2）Case 结构：与一般机器语言中的条件语句非常相似，LabVIEW 提供了 Case 结构。如图 5-25-10（a）所示，左边带有问号的小方框是用来连接 Case 结构选择端的值的，如图（b）和（c）所示。如果连接的是代数值，则上边框指示的是数字，如果是逻辑型，则显示的是 True 或 False 两种条件。Case 结构每次只能显示一个子图，单击上面中间的箭头显示条（或点击箭头）可以选择不同的子图。对于数值型结构，只要在边框上点击右键，通过 Add Case 就可以增加条件子图。输入输出与循环结构一样，不同的是需要在输入数值的同时输入判断条件。

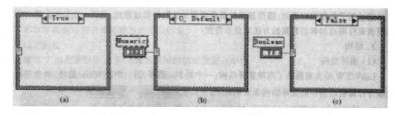

图 5-25-10　Case 结构

例：求平方根（图 5-25-11）。

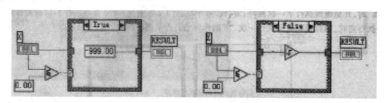

图 5-25-11　用 Case 结构求平方根的流程图

(3) 顺序结构：顺序结构就是按照顺序依次执行的结构，如图 5 - 25 - 12 所示，通过 Add Frame After 、 Add Frame Before 项可以生成新的标号（对应新的一个子图）图，图右边的图形所示。不同的子图可以按照一

图 5 - 25 - 12 顺序结构

定顺序执行，也是每次只能显示一个子图，可以通过上边框或箭头来选择。输入输出也是用的隧道，但是隧道每次只能处理（输入或输出）一个数据源，如果顺序结构每次执行每个子图时都要向隧道输入输出数据，那么就必须是多源的数据结构。因此顺序结构的数据只有在整个结构全部执行完以后才能输出。

(4) 公式节点：公式节点是一种可以直接在流程图中输入数学公式并进行计算的结构，它是一个大小可变的方框。在使用公式节点时要通过它的边框的输出输入节点传递数据，在边框上点击右键就会有 Add Input 和 Add Output ，分别选择即可显示节点，在输入输出节点上输入变量名（注意要与公式内使用的变量一致）。

公式节点所用的操作符和功能在公式节点的帮助窗口中可以查询。

例：求平方根（语句后面要加"；"号，而且还要注意同一变量的大小写），如图 5 - 25 - 13 所示。

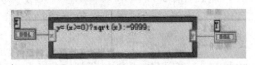

图 5 - 25 - 13 用公式节点求平方根的流程图

4. 波形输入输出

LabVIEW 配上 NI 公司的 DAQ 卡以及 DAQ 卡的驱动程序，可以完成数据的采集和输出等多种功能。LabVIEW 的 DAQ 程序包括模拟输入、模拟输出、计算器操作、数字输入、数字输出等，我们可以在 Functions 面板下的 Data Acquisition 中看到。在图中 6 个模块下，每个模块的子模块又分成 3 种，顶层的是 Easy Vis，中间的是 Intermidiate VIs，下面的是 Advanced Vis。

NI 公司 DAQ 卡提供 3 个输出和 16 个输入，测量电压 -10 ~ 10V。

模拟波形输入：简单的波形输入的模块有单通道波形输入和多通道波形输入，它们都在 Functions → Data Acquisiton → Analog Input 下。

单通道波形输入：虚拟示波器就用到了这个模块，采集外部信号的流程图如图 5 - 25 - 14 所示。

设备号（Device）：在 NI DAQ 设置工具中设定。该参数告诉 LabVIEW 你使用什么卡，它可以使 DAQ VI 自身独立于卡的类型，也就是说，如果你稍后使用了另一种卡，并赋予它同样的设备号，你的 VI 程序可正常工作而无须修改。

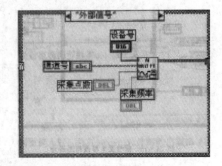

图 5 - 25 - 14 示波器流程图部分

通道号（Channle）：指定数据样本的物理源（表 5 - 25 - 3）。例如，一个卡有 16

个模拟输入通道，你就可以同时采集 16 组数据点。

采集点数（Number of samples）：采集数据的总的总数，对于模拟型号，在采集时转换成数字信号。

采集频率（Sample rate）：每秒采集数据的点数。

表 5 – 25 – 3　通道及其对应的字符串

通道	字符串
通道 1 通道 0 ~ 5 通道 1，8 以及 10 ~ 13	1 0∶5 1，8，10∶13

多通道波形输入：可以同时进行多个通道的输入，端口设置和单通道波形相似，只是通道控制字符串不同。

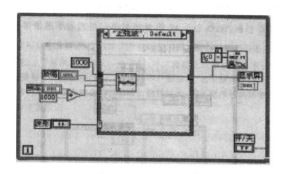

图 5 – 25 – 15　信号发生器流程图部分

模拟波形输出：和模拟波形输入一样，简单的波形输出模块有单通道波形输出和多通道波形输出，它们都在 Fuctions → Data Acquisiton → Analog Output 下，我们通过信号发生器的例子来了解这个模块，其流程图如图 5 – 25 – 15 所示，其各个端口的意义与模拟波形输入相似，具体可以看 LabVIEW 的帮助。

（赵　喆　孙　言）

Experiment 25　The Basic of Virtual Instrument

【Objects】

(1) To understand the conception of Virtual Instruments.

(2) To understand the Graphical Programming Language—LabVIEW and study the simple programming with LabVIEW.

【Apparatus】

Computer (including OS), LabVIEW environment, oscilloscope.

【Theory】

The Virtual Instrument (VI for short) is proposed with the use of computers in test and measurement. VI is a computer instrument system that is based on general computer hardware platform and implemented by a special test software. Unlike the traditional instruments, the most outstanding external characteristic of VI is that there is no material panel. All the buttons and indicators are "planted" on the computer's screen, and it is a virtual panel or in other words "software panel". Using the powerful calculation and processing ability of computers, and with different sensors and interface card, VI implements the signal input. Meanwhile, using the keyboard, mouse and monitor, VI implements the operation control and signal display etc. Users themselves can also customize the panel of VI as well as the test function. And they can complete the test and measuring task only by keyboard and mouse.

LabVIEW is a graphical programming language, which is also known as G language. When using this language, we don't have to write the program code, we can use the flow chart instead. It makes the programming simple and intuitive. The LabVIEW integrated with abundant function library, can fulfill the functions such as data acquisition, equipment control, data analysis and display etc.

The virtual program is the program developed by LabVIEW development platform, we call it VI for short. VI consists of three parts: Panel, Diagram and Icon/Connector.

The panel is used to set values and observe the output, so as to simulate the real instrument's front panel. In the panel of the program, the input is called as "Controls", the output as "Indicators". Controls and Indicators in the panel are of different icons, for example, knob, switch, button, chart, graph etc. This makes the panel more intuitive and easier to understand.

Every panel program corresponds to a block diagram program (flow chart program). Block diagram program is written by graphical programming language of LabVIEW, you can consider it as the source code of common program. Block diagram program is composed of ports, nodes, drawing frames and the connect line. Port is used to transfer data with the Controls and Indicators of panel, node is used to implement the function and call other features, drawing frame is used to implement the control commands of structured program, and the connect line represents the data flow in the process of program execution as well as defines the flow direction of the program.

【Procedure】

1. To be familiar with the basic operation of the LabVIEW development environment and the basic programming method

Click Start → Programs → National Instruments LabVIEW 7.0 to start the LabVIEW. Select the "NEW VI" to enter the LabVIEW develop environment. We can see the gray front panel firstly. Click Windows → Show Basic Diagram to show block diagram front panel. In the front panel, click Windows → Show Controls Palette/Show Tools Palette to show tools template and controls template (Fig. 5-25-1). Control items like knob and switch are in the controls template, we can select these item's icons and put them into the panel, then the corresponding ports and the icons will appear on the flow diagram. These icons's information can be shown by the data of the front panel or the program's executive result. In the panel window of the block diagram, Click $\boxed{\text{Windows}}$ → $\boxed{\text{Show Functions Palette}}$ to show function template. We can use the template to implement many functions such as loop, mathematical operations, cooperation, etc. And some of the functions can create block diagram. The Detail instruction of every module can be found in relevant reference books or software manuals.

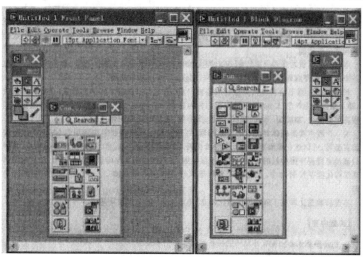

Fig. 5-25-1 Labview develop environment

2. To create a simulate program of measuring temperature, and to be familiar with the basic concept of LabVIEW. We need to observe the executive result of the program and to explain the functions of every part

Assume that the sensor output is proportional to the voltage and temperature. For example, when the temperature is 80 °F, the sensor output voltage is 0.8V. Meanwhile, the program provides the switch to select Celsius or Fahrenheit for display. For the sake of simplicity, we have substituted software for the DAQ (date acquisition card), and used the Demo Read Voltage of LabVIEW to simulate the voltage measurement, then converted the voltage value to Celsius or Fahrenheit degree. Demo Read Voltage VI simulates the process of getting the voltage value from the 0 channel of DAQ, then the value will be multiplied by 100.0 to show as Fahrenheit number. Of course, if we have selected Celsius, the program will convert the Fahrenheit number to Celsius number. The front panel and the block diagram are in Fig. 5-25-2 and Fig. 5-25-3.

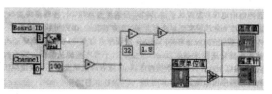

Fig. 5-25-2 The front panel   Fig. 5-25-3 Diagram program

The experimental procedures are as follows:

(1) To create the front panel: Open a new front panel window by using the New VI options in menu File. Select the thermometer in the sub template Numeric of the control template, and put it into the front panel window. Input "thermometer in the highlighted text box". Select the Vertical Slide Switch (or other switches) in the sub template Boolean of the control template, and put it into the front panel window. Input "the temperature value unit" in the text box. By using the label A, input the free label "C" next to the position "condition True" of the switch, and then input the free label "fahrenheit" next to the position "condition false". Select Numeric Indicator in the Numeric template of the control template, and put it into the front panel window. Input "the temperature value" in the highlighted text box (as shown in Fig. 5-25-2).

(2) To create the diagram program: Open the diagram program window. Find the following objects in function template and put it into the diagram program window:

Demo Read Voltage VI program. When the function template is set to the Advanced mode, select the file D: \ Program Files \ National Instruments \ LabVIEW 7.0 \ vi. lib \ tutorial. llb in Select a VI sub template. After opening, select the Demo Read Voltage. The simulate program read voltage values from the 0 channel of card DAQ.

Multiply function (Numeric sub template)、Subtract function (Numeric sub template), Divide function (Numeric sub template).

Select function (Comparison sub template). Output the Fahrenheit value (when the optional switch is False) or the Celsius value (when the optional switch is True) according to the optional switch for thermometric scale.

After putting the objects needed into the diagram program window, firstly move the icon to the position as shown in Fig. 5-25-3, secondly connect together with the connection tool, and then create a numerical constant in the required position (with the connection tool, right-click the object connection terminal about a numerical constant which you want to connect, and select Create Constant function. If you want to modify the constant value, double click numeric with the label tool, and write a new numerical value). The whole program has been created.

(3) Run the program: Select the front panel window, run the VI program. Click the run button (Run) to make the program run in continuous mode. Observe the program running results and explain the functions of each part of the program. Stop running. Save the file with SAVE function in the file menu.

(4) Close the program.

3. To create a signal generator include the rectangular wave, the sine wave and triangle wave with an adjustable frequency and amplitude. Learn the basic functions usage of LabVIEW

The signal generator can generate selected frequency and amplitude of the rectangular wave, the sine wave (duty ratio adjustable) and the triangular wave according to user's choice, and the waveform will be displayed on the front panel, which the user can control.

Add a DAQ card output function in the diagram program, the waveform can be output and observed through the oscilloscope. The students who master faster can choose the content of this part.

The experimental steps are as follows:

(1) To create the front panel: Place a Waveform Graph (Controls → Graph) to display the waveform, whose label is revised for "screen". Select the Scale Legend and Graph Palette (right - click the object and choose Visible Items).

Place a switch. Place three control quantities, their labels are revised respectively for frequency, amplitude and the duty ratio, whose default values are set to 1.00、1.00 and 50.00, 1 and 50(input numerical value with label tool, right - click and choose to set precision and default values). Place an enumeration(Controls → Ring & Enum → Enum), and the enumeration value is set to "sine wave", "square wave" and "triangle" (after inputting a string with the label tool, right - click and choose Add Item After), as shown in Fig. 5 - 25 - 4.

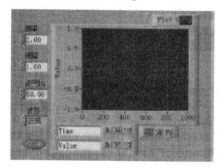

Fig. 5 - 25 - 4 The front panel of the signal generator

(2) To create diagram program: Place a While loop at first, with all icons are included, then connect the switch and the While loop conditions (Loop Condition) together, and meanwhile set the Loop Condition to Continue If True.

Add a Case structure. The case structure can contain two or more subroutines (Case), and the external control quantities connected to the Case Selector (i. e., the question mark on the frame) determine which subroutine will be executed. The control quantities can be Boolean, integer or string and so on. Connect the "waveform" icon and Case Selector together.

Through selecting the Selector Label at the top of the structure frame you can choose the corresponding subprogram.

When selecting the sine wave in Selector Label, choose Analyze → Waveform Generation → Sine Wave. vi in a function template. Add the sine function generated to the Case subroutine. Connect the output to the display screen across the Case structure.

Note: in the simulation condition, the signal frequency is in cycles per second as unit (Hz), but in the digital system it often chooses the digital frequency. The unit is the cycle number/sampling point, also called standard frequency. Their transform relationship is as follows:

Simulation frequency = digital frequency × sampling frequency

Digital frequency = Simulation frequency/sampling frequency (sampling frequency is sampling points per second)

In Sine Wave. vi module, the Square Wave. vi and Triangle Wave. vi below, it chooses the digital frequency, therefore, you must covert the simulation frequency input from the front panel to a digital frequency and then input them to the corresponding input terminal.

Respectively select the Selector Label as the "square wave" and "triangle", add the corresponding square wave generating function in the Case structure (Analyze → Signal, Processing → Signal Generation → Square Wave. vi) and the triangular wave generating function (Analyze → Signal, Processing → Signal Generation → Triangle Wave. vi), and finish the connection. Attention: the "the duty ratio" icon is connected only to the duty terminal of the square wave module, as shown in Fig. 5-25-5.

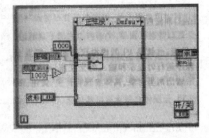

Fig. 5-25-5 The block diagram of the signal generator

(3) To run the program: Run the program, adjusting the control quantities on the front panel, observe the wave changes and explain, last name and save the program.

(4) Choose content: The students who work faster can choose this part. Output the generated signal through the interface card and observe the wave using the oscilloscope. The specific steps are omitted here.

【Questions】

(1) What distinctions are there between the virtual instrument system and the traditional instruments, please explain briefly.

(2) What is the effect of the development of virtual instrument for measuring instruments? To the instrument users, what does it mean?

(3) How many types do the software development environment of virtual instrument have? What is the "G" language?

(4) What do you imagine the application of virtual instrument system in physics experiment? Whether do you can give a simple design combining the specific experiment?

【Introduction to LabVIEW】

The program using the LabVIEW development platform has become a virtual instrument program, referred to as VI. VI consists of three parts: panel, diagram and icon/connector. Because LabVIEW is powerful and rich in content, so that beginners can master the basic contents of the LabVIEW as soon as possible, some of the basic functions and structure are introduced as follows: the details can make reference to the literatures.

1. Basic knowledge

(1) LabVIEW program: Double click the LabVIEW icon, the startup screen will appear. Click the NEW VI, and it will open a new LabVIEW program, which is formed by panel and diagram as seen.

① The panel window: The panel window is a graphical user interface, which is the virtual instrument panel of VI, which is equivalent to the control panel of actual instrument. It connects the user and the program, which is the exchange window between indicators and the input when the program is run. As shown in Fig. 5-25-4, there are two categories, the user input and indicators output in the panel of the signal generator. It includes switch, knob, graphics and other controls and indicators.

② The flow-process window: The flow-process window provides the graphical source program of VI, which is in corresponding to the actual things in the instrument box. In flow-process, a programmer write the source code of LabVIEW while using the graphical language to control and manipulate the input and output functions which are defined on the panel. As shown in Fig. 5-25-4, the signal generator consists of the connection terminal on the panel, and there must be some things, such as functions, structure, connections etc, but not on the front panel.

Button functions commonly used:

⇨: Run. If VI is in compilation error, this button will become ⇨.

⟳: Run Continuously.

◉: Abort Execution

💡: Highlight Execution, only exists in flow-process. When entering highlight execution, it will become 💡. At this time, the data in flow-process gets bright, and it also shows some intermediate data of VI in implementation.

Note: When using ⟳ to make the LabVIEW occupy a large part of CPU, which will make computer operate other programs slowly. It is suggested not to use ⟳.

(2) Controls palette and functions: It can be used only in the panel window. To view the palette, select Windows → Show Controls Palette. You also can display the Controls palette by right-clicking an open area on the front panel. Use the Controls palette to place controls and indicators on the front panel. Module icon located on the top right corner has sub options, and click the icon to access the sub options.

In the controls palette, we can see some common modules. We make some introductions to them, and other modules can be found in All Controls.

▭: Numeric, numerical control and indicate, including digital、pointer indicators and various input place.

▭: Boolean, logical numerical control and indicate, including a variety of Boolean switch, button and an indicator lamp etc.

▭: Array&Cluster, controls and indicators of arrays and clusters.

▭: Graph, indicate the data results of trend charts and graphs.

(3) Functions palette and functions: Only used in the flow – progress of window, the functions palette is opened through the flow – progress window $\boxed{\text{Windows}} \rightarrow \boxed{\text{Show Function Palette}}$, can also be right – click opened in blank at the flow – progress panel. Functions palette provides tools to create flow – progress. It is basically same with the control panel. Where in the top right corner of the module icon there is said "▶" sub palette, click the module icon to access sub options.

▫: Structure, including commands of the program control structure, such as cycle control and global variables and local variables.

▫: Array, including the array operation functions, array conversion functions and constant array etc.

▫: Comparison, including various comparison and operating functions, such as more than, less than, and equal to.

▫: Dtata Acquisition, including the driver of data acquisition hardware, and various functions modules required for signals regulation.

▫: Analysis, the analysis function module for signal generation, time domain and frequency domain and mathematical tools.

Note 1: In LabVIEW, you can get help in time. With $\boxed{\text{Help}} \rightarrow \boxed{\text{Show Context Help}}$, to open the help window (Context Help), the shortcut is Ctrl + H. When putting the mouse on any interested module, it will display the corresponding help information in the help window.

Note 2: Right – click in either a control or function module, it will display the pop – up menu, with which you can easily edit the module.

(4) Tools palette and functions: Open and close the tools palette through $\boxed{\text{Windows}} \rightarrow \boxed{\text{Show Tools Palette}}$. It provides various tools to create, modify and debug VI program. When you choose any kind of the tools, the mouse arrow will turn into a shape of the corresponding tools.

Introductions to the common tool buttons (the others seen in help)

▫: Operate Value, to control and indicate the operation of the front panel, with which type a number or string as control value, the tool will become label tool.

▫: Position/Size/Select, to select, move or change the size of objects. When used to change the frame size of the object, it will turn into the corresponding shape.

▫: Edit Text, to input a text or create a free label. When creating a free label, it will turn into the corresponding shape.

▫: Connect Wire, to connect the object in the flow – progress program. When the help window is opened, put the tool on any line, it will show the corresponding data type.

(5) Data lines, data flow and data types

① The operation mode of the data flow: In the flow – progress, the interconnections among modules are data lines. Data is transfered between modules through data lines. Unlike the general language executing the operation in accordance with the order line by line, Lab-

VIEW is to control the program relying on the transferred data on data lines. Only when the input data required completely reach this module, it will execute, and then output data to all its output terminals. These data flow to other modules along the data lines. This is the data flow way of LabVIEW. As introduced in front, you can observe how the data flow work through brightening ![] to execution.

② The data lines: When connecting lines, LabVIEW will prompt the name of the interface. It is only permitted to connect between the data terminals of the same type. The shapes and colors of data lines are different in different types. Wrong connections will be expressed as a black dotted line, at this time you should move the connection tools on it, and the LabVIEW will give out the data type information on both ends, so you can check and correct. It may impact the setted parts when adding new module and connecting and then the original error connections may also have influence on the new added lines. Tab. 5 – 25 – 1 provides the common connection types.

Tab. 5 – 25 – 1 Common wire types

types	scalar	one – dimensional arrays	two – dimensional arrays	color
integer	———	———	———	blue
float	———	———	———	orange
boolean	———	———	———	green
string	———	———	———	red

③ Data types: There are five basic data types of LabVIEW: Numeric, Boolean, String, Enum, there is an ertra type called Ring which is very similar with Enum and can be cycled.

The data of Numeric type can be divided into several types according to the precision which is basically the same with the data types of standard C + +. The symbols directly express their types (as shown in Tab. 5 – 25 – 2). You can right – click on numerical value, and change through the pop – up menu Representation.

Tab. 5 – 25 – 2 Numerical data type of LabVIEW

Symbol	I8	I16	I32	U8	U16	U32
Type	8 bit integer (short integer)	16 bit integer (single word)	32 bit integer (long integer)	8 bit unsigned integer	16 bit unsigned integer	32 bit unsigned integer
Symbol	SGL	DBL	EXT	CSG	CDB	CXT
Type	Single precision float	double precision float	128 bit extension float	single precision complex	double precision complex	128 bit extension complex

Note: In the flow – progress, you can see different colored modules. Different colors represent different data types, whose definition is agreement with connection.

(6) Control and indicator: In LabVIEW, all the numeric, Boolean, string and so on have the difference between the control and the indicator.

Control: to control program, it is equivalent to the control button, such as switch, knob etc.

Indicator: to indicate the running results of program, it is equivalent to the display component of instruments, such as display, indicating lamp etc.

In the flow - progress window, we can see the frame of control parts is thicker, and that of display parts is finer. In order to achieve a conversion between control and indicator, it only needs right - click on the flow - progress of a module and select [Change to Indicator] (or [Change to Control]), as shown in Fig. 5 - 25 - 6.

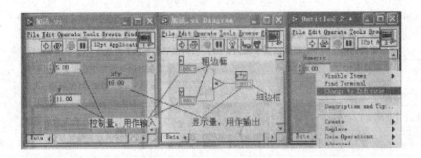

Fig. 5 - 25 - 6 Control and indicator

2. Several templates to be used

(1) Array & Cluster

① Array: Array is a collection of elements in the same type. An array can be one - dimensional or multidimensional, each dimension can have up to $2^{31} - 1$ elements. Each element can be accessed through the array index. Same to the language C, the index range is $0 \sim n - 1$: the index of the first element is 0, the second is 1, and so on. The elements of an array can be data, string and so on, but the data types of all the elements must be consistent.

Set up an array: there are two things to do to create an array:

Create a shell for an array: put the module [Controls]→[Array and Cluster]→[Array] on the front panel;

Place the elements in the shell (number, string switch etc.).

We put the module [Controls]→[Numeric]→[Digital Control] in an array box, forming an array, and which is similar to the other array types. As long as you put the elements into the array you want to build in the step. To change the display number of elements, you can change the mouse into the selection mode and move it to the corner of array frame, the mouse will became ▪, then you can drag arbitrarily to change an array. The method to change the dimension is the same with the one to change number of elements. In the model the mouse select, moving down to the corner of the index frame, the mouse will become ![icon], drag to change the dimension; You can also right - click on the array index box, and select Add Dimension.

② Cluster: Cluster is another data type, whose elements can be different types. It is similar to Struct in language C. Using cluster can make the data elements distributed in various locations of the flow - progress combinate together, so it can reduce the congestion degree to connect, for error handling.

Set up a Cluster: the same to an array, first create a "shell" for cluster (Controls → Array and Cluster → Cluster), and then place the elements in the shell, and the data can be different types. Attention: when placing the objects in the cluster, they must be all the control objects or the indicate objects, which cannot be mixed together. The data in the cluster have to do with the place order but not their locations. The first object putted is element 0, and so on. Deleting one of them, the order will be adjusted automatically.

(2) Chart and graph: The data can be indicated in a chart or a graph.

① Chart: Usually the axis Y represents data value, the axis X represents time or data number.

Let us look at an example of random number below (Fig. 5 -25 -7).

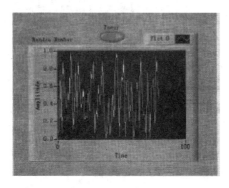 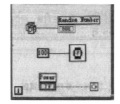

Fig. 5 -25 -7 The front panel of random number and diagram

In each cycle, the module of a random number generates a number, and the chart can number automatic ally and display them in the picture frame, y represents numeric, x represents the number. It maintains the previous data, and append new data to keep the chart updated.

You can right - click in the picture frame to change the characters of the chart, such as changing the update mode of the chart in Advance → Update Mode, selecting display labels、illustrations, graphics template and so on in Visible Items. You can try the functions.

② Graph: Graphs and charts are similar, but they are different: the graph can not display the new added data.

Let us see an example for sine or cosine wave (Fig. 5 – 25 – 8).

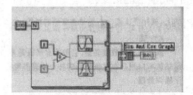

Fig. 5 – 25 – 8 The front panel of sine and cosine wave and diagram

Through this example we can see, what the graph displays is to form the data displayed to an array, and then displays one – off. It is unlike chart adding new data to display data.

(3) Structure

① Cyclic structure: LabVIEW provides two cyclic structures: one is For cycle, and the other is While cycle. The two cycles are very similar to the most computer language in the cycle structure. These two are in the Functions → Structures.

The "For" cyclic structure: as shown in Fig. 5 – 25 – 9, "N" refers to the cycle number, and "i" refers to the cycle count, which runs from 0 to n – 1. You can select the object of For structure from Functions → Structures, and drag the object in the flow chart on the panel, the circulating object can be included. Of course, it can also be directly (or not) dragged to circular object to enter (or leave) the circulation structure. If you want to delete the cycle structure, directly select the Delete key.

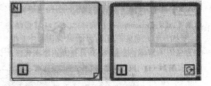

Fig. 5 – 25 – 9 For cycle and While cycle

The While cyclic structure: as shown in Fig. 5 – 25 – 9, the small icon on the right with an arrow is used to access the cycle condition. If the control is "true", it begins to proceed to the next cycle; if "false", stop the execution. This test for cycle conditions is when the execution of the loop is completed, the loop will execute at least once. The role of "i" is to indicate cycle number.

The data transfer of cyclic structure: if there is a need for data transmission in the internal and external loop structure, connect the terminals between the inner and outer of the loop cycle, when it appears as small black – square on the border of the cycle (some books called tunnel). Data transmission cannot be done in the tunnel during the loop execution, only output the last cycle results at the end of the cycle. So if each cycle results were observed, we should place the display terminals inside of the loop.

Shift Register: for "For" and "While" cycles, the shift register is very important, which can convey the results of each loop to the following cycles. The register mouth is generated through the left and right borders of the loop structure using the Add Shift Register of the right – click menu. It consists of a pair of opposite and small triangles which located on the left and right borders in circulation structure. The right end storages the output results when every cycle is completed and the shift register will transfer this result to the small triangle on the left, as the input data in the next cycle.

② Case Structure: Very similar to conditional statements in general machine languages, LabVIEW provides the Case structure. As shown in Fig. 5 – 25 – 10 (a), the left small box with a question mark is used to connect the choice terminal value in Case structure, as shown in figure (b) and (c). If the connection is the algebraic value, the frame indicates digital, if logical, it indicates the two conditions: True or False. Case structure can only display a sub – graph, click on the middle arrow display bar (or click on the arrow) can choose different sub – graphs. For the numerical type structure, just right click on the frame, through the Add Case can increase the conditional sub – graph. The input and output is similar to the loop structure, differently it is required to enter judgment condition at the same time to enter the input values.

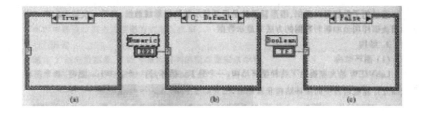

Fig. 5 – 25 – 10   Case structure

Example: find the square root (Fig. 5 – 25 – 11).

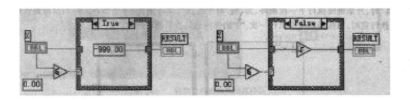

Fig. 5 – 25 – 11   Calculate the square root with case structure

③ Sequence structure: Sequence structure is to execute in accordance with the order, as shown in Fig. 5 – 25 – 12, through the Add Frame After and Add Frame Before it can generate a new label diagram (corresponding to a new sub graph), as shown on the right in figure. Different sub – graphs can be executed in sequence, and each can only display a sub graph, which can be chosen by the frame or arrow. The input and output also use the tunnel. But the tunnel only processes (input or output) a data source. If each execution to each sub graph in se-

Fig. 5 – 25 – 12   Sequence structure

quence structure must input and output data to the tunnel, then it must be multi-source data structure. So the data of the sequence structure is output only after all the whole structure is executed.

④ Formula node: The formula node is a structure, which can be directly in the flow-progress input mathematical equation and calculation. It is a frame of variable size. When using the formula node, to transfer data through the input and output nodes of its border, right click on the border will appear $\boxed{\text{Add Input}}$ and $\boxed{\text{Add Output}}$, respectively choose to display the node, and input variable name in the input and output nodes (note to be in accordance with the variable in formula.).

The operators and functions in the formula node can be queried in the help window of the formula node.

Example: for square root (add ";" behind statement, and pay attention to the case of the same variable), as shown in Fig. 5-25-13.

Fig. 5-25-13 The flow chart of calculating square root with the Formula Node

(4) Input and output waveform: LabVIEW with DAQ card and DAQ card driver of NI company can complete various functions of data collection and output. The DAQ program of LabVIEW includes simulation input, simulation output, counter operation, digital input, digital output and so on, which we can see under $\boxed{\text{Data Acquisition}}$ in the $\boxed{\text{Functions}}$ panel. Under the six modules in figure, the sub module of each module is divided into three kinds, in which the top is Easy VIs, the middle is Intermediate VIs and the bottom is Advanced VIs.

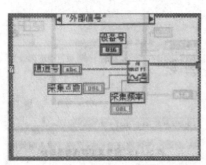

Fig. 5-25-14 The flow chart of oscilloscope

DAQ card of NI Company provides 3 outputs and 16 inputs, measuring voltage from -10 to 10V.

Simulation waveform input: simple waveform input module includes single channel wave input and multi-channel input waveform. They are all in the following $\boxed{\text{FunctionsData}}$ $\boxed{\text{Acquisition}}$ → $\boxed{\text{Analog Input.}}$

Single channel wave input : virtual oscilloscope used this module, the flow chart for acquisition of external signals is as shown in Fig. 5-25-14.

Device number (Device): set with the setting tools in the NI DAQ. This parameter tells LabVIEW what card you use, which can make the DAQ VI itself independent of the card type, that is to say, if you use another card later, and give the same device number, your VI program can work properly without modification.

Channel number (Channel): specify the physical source of data samples (Tab. 5-25-3). For example, a card has 16 simulation input channels, then you can collect 16 groups of data points at the same time.

Tab. 5-25-3　channel and the corresponding string

channel	string
channel 1	1
channel 0 ~ 5	0:5
channel 1, 8 and 10 ~ 131	1, 8, 10:13

Number of samples: the total amount of data, for simulation model, it can be converted to digital signals in acquisition.

Sample rate: acquisition points per second.

Multi-channel waveform input : it can undertake multiple channel inputs at the same time, and the port Settings are similar to the single channel waveforms, just different in control strings.

Simulation waveform output: similar with the simulation waveform input, the simple waveform output module includes single channel output waveform and multi-channel waveform output, they are all in the Fuctions → Data Acquisiton → Analog Output. Let us know the module through the example of the signal generator, and the flow chart is shown in Fig. 5-25-15. The meanings of the every ports are similar to the simulation waveform input, specificly it can be seen with the help of LabVIEW.

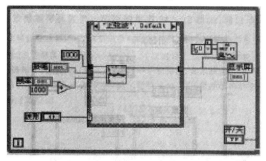

Fig. 5-25-15　The flow chart of signal generator

(Zhang Hong-yan, Su Ting-ting)

实验二十六　测试二极管并确定其极性

【实验目的】

(1) 使用 NI ELVIS 测试并确定二极管的极性。

(2) 做出二极管的特性曲线。

【实验器材】

NI ELVIS 3.0 教育平台软件、发光二极管、计算机（含操作系统）。

【实验原理】

半导体二极管是一种极性元件，通常其一端有带状标记注明是负极，另一端称为正极。尽管根据二极管的封装不同，有许多种方法标注极性，但是有一点始终不变，即如果在正极加上正电压使二极管正向导通，电流能够流通。

【实验步骤】

1. 完成以下步骤，使用 NI ELVIS 测试二极管并判断其极性

（1）启动 NI ELVIS 仪器启动界面，选择 DMM。

（2）点击按钮【▶▌】，开始测试。

（3）将一个发光二极管的两端连接到工作站 DMM（电流）的 HI 和 LO 引脚。

当二极管阻止电流通过时，显示器显示的值与没有接入二极管时显示的值相同（即电路开路）。当二极管导通时，发光二极管发光并且显示器所显示的电压低于开路电压。试着在两个方向使电流通过红色发光二极管。如果看到发光二极管发光，二极管中连接到 LO 即黑色香蕉接头的一端是负极。

（4）继续使用这个简单的测试方法测试其他二极管，确定它们的极性。对于整流二极管，在其正向导通时，显示器显示的电压小于3.5V，并且显示 GOOD，如图5-26-1所示。在其反向截止时，显示器显示开路电压（~3.5V），并且显示 OPEN，如图5-26-2所示。

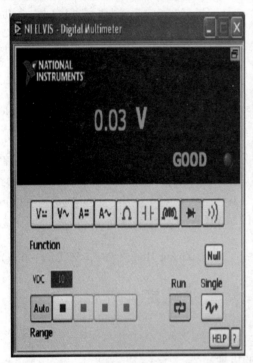 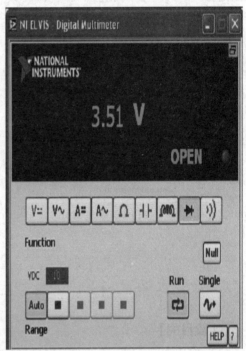

图5-26-1 硅整流二极管的正向导通显示　　图5-26-2 硅整流二极管的反向截止

（5）运行这个测试观察结果：测试的工作原理是显示器显示二极管中通过约1mA微小电流所需要的电压。在正向导通区域，电压值很小，并且与制造二极管的材料相关。在反向截止方向，没有电流通过，测试器显示开路电压，约3.5V。

2. 二极管的特性曲线

二极管的特性曲线是通过元件的电流关于二极管两端电压的函数图形，它能很好地展示二极管的电子特性。完成以下步骤做出二极管的特性曲线。

（1）将硅二极管接在 DMM（电流）接头的两端。确定负极连接到黑色香蕉头输入。

（2）启动 NI ELVIS 仪器启动界面，选择两端电流–电压分析器，从而打开一个新的软件前面板,可以显示被测元件的 ($I-V$) 曲线。软件前面板在二极管两端加入测试电压,从一个开始电压值以一定步长逐渐增加到终了电压值,其中的参数都可以选择设定。

(3) 对于硅二极管,设置以下参数

开始电压　　　　　　　-2.0V
终了电压　　　　　　　+2.0V
递增步长　　　　　　　0.1V

(4) 设置两个方向的最大允许电流,确保二极管中通过的电流不会达到可能引起损坏的大小。

(5) 点击运行,观察 $I-V$ 曲线。

如图 5-26-3 所示,在反向截止方向,电流应该非常小(μA 级别),且为负值。在正向导通方向,应该观察到在一个临界电压值以上,电流以指数形式上升到最大电流上限。

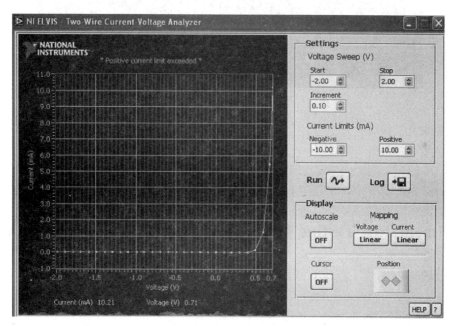

图 5-26-3　二极管特性曲线

(6) 改变显示按钮【线性/对数】,观察不同比例下的图像。

(7) 试着使用游标操作。当沿着轨线拖动游标时,它会显示对应的 (I, V) 坐标。

临界电压与二极管的半导体材料有关。对于硅二极管,临界电压约为 0.6V,而对于锗二极管则约为 0.3V。估计临界电压的一种方法是在正向导通区域最大电流值附近做一条切线(图 5-26-4),在切线的交点上,对应电压轴的值就是临界电压。观察发光二极管的 (I, V) 特性曲线。临界电压 V_T 为切线与电压轴的交点。

(8) 使用二端电流-电压分析器,确定红色、黄色和绿色发光二极管的临界电压,填写以下内容:

红色发光二极管＿＿＿＿＿＿＿＿＿＿＿＿V

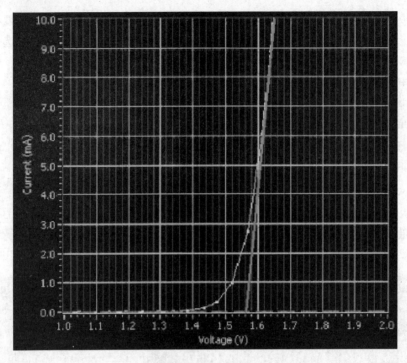

图 5 - 26 - 4 临界电压

黄色发光二极管_____V

绿色发光二极管_____V

(赵　喆　李百芳)

Experiment 26　Testing Diodes and Determining Their Polarities

【Objects】

(1) To test diodes and determine their polarities.

(2) To make the characteristic curve of a diode.

【Apparatus】

NI ELVIS software, a light emitting diode, computer (including OS).

【Theory】

A semiconductor junction diode is a polar device with one end often labeled with a band that is called cathode while the other lead is called anode. While there are many ways to indicate this polarity in the packaging of a diode, one thing is always the same. A positive voltage applied to the anode will result in the diode being forward biased so that current can flow. We can use NI ELVIS to figure out the diode polarity.

【Procedure】

1. Test diodes and determine their polarities

(1) Launch the NI ELVIS Instrument Launcher and select DMM.

(2) Click on the [▶┤] button.

(3) Connect one of the LEDs to the workstation leads DMM (current) HI and LO.

When the diode blocks the current, the display will read the same value as it does when no diode is connected (open circuit). When the diode allows current to flow, the LED will give off light and the display will read a voltage level less than the open circuit value. Try a red LED in both directions. When you see light, the diode lead connected to the LO or Black banana jack is the anode.

(4) You can use this simple test on other diodes to determine their polarity. For a silicon rectifying diode in the forward bias direction, the display will show a voltage less than 3.5 volts and display the word "GOOD", as shown in Fig. 5 – 26 – 1. In the reverse bias direction, the display will read the open circuit value (~3.5 V) and display the word "OPEN", as shown in Fig. 5 – 26 – 2.

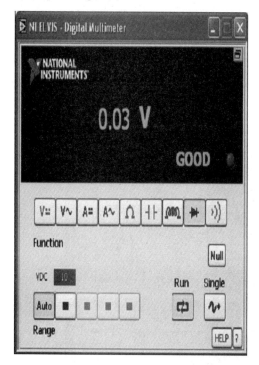

Fig. 5 – 26 – 1 The forward bias direction display of a silicon rectifying diode

Fig. 5 – 26 – 2 The reverse bias direction display of a silicon recitfying diode

(5) How does this work? The display shows the voltage required to generate a small current flow of about 1 mA. In the forward bias region, this voltage level is low and related to the materials used in the manufacture of the diode. In the reverse bias direction, no current flows and the tester displays the open circuit voltage, about 3.5 volts.

2. Draw the characteristic curve of a diode

The characteristic curve of a diode, which is a plot of the current flowing through the device as a function of the voltage across the diode, properly displays its electronic properties.

(1) Place the silicon diode across the DMM (current) leads. Make sure the anode is connected to the black banana input.

(2) Launch the NI ELVIS Instrument Launcher and select Two Wire Current - Voltage Analyzer. A new SFP will pop up which allows one to display the $(I-V)$ curve for the device under test. This SFP will apply a test voltage to the diode from a starting voltage level to an ending level in incremental voltage steps (selectable for all users).

(3) For a silicon diode, set the following parameters:

Start   -2 V

Stop   +2.0 V

Increment   0.1 V

(4) Notice the maximum current in either direction can be set to ensure the diode does not operate in a current region where damage may occur.

(5) Click on run and see the $I-V$ curve appear.

As shown in Fig. 5-26-3, in the reverse bias direction, the current should be very small (microamps) and negative. In the forward bias direction, you should see that above a threshold voltage the current rises exponentially to the maximum current limit.

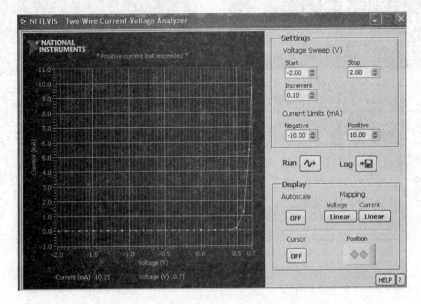

Fig. 5-26-3   The (I, V) characteristic curve for a light emitting diode

(6) Try changing the Display buttons [Linear/Log] to see the curve plotted on a different scale. Try the cursor operation. It gives the (I, V) coordinate values as you drag the cursor along the trace.

(7) It turns out that the threshold voltage is related to the semiconductor material of the diode. For silicon diodes, the threshold voltage is about 0.6 volts while for germanium diodes it is about 0.3 volts. One way to estimate the threshold voltage is to fit a tangent line in the

forward bias region near the maximum current (as shown in Fig. 5-26-4). The point where the tangent intersects, the voltage axis defines the threshold voltage.

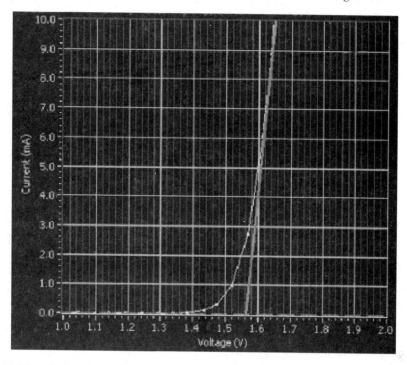

Fig. 5-26-4 The threshold voltage V_T is given by the intersection of the tangent line with the voltage

(8) Using the Two Wire Current – Voltage Analyzer to determine the threshold voltage by a red, a yellow and a green LED and fill in the following chart.

Red LED _____ V

Yellow LED _____ V

Green LED _____ V

Do you see any trend?

(Zhao Zhe)

实验二十七　电路仿真实验

【实验目的】

（1）使用 Multisim 软件包、NI ELVIS 3.0 教育平台软件和 LabVIEW Signal Express 软件对电路进行交互式仿真。

（2）使用交互式仿真来测量增益、带宽和带通滤波器的中心频率。在仿真过程中与元件进行交互，观察反馈电阻的改变对于整个电路的影响。

【实验器材】

NI Multisim 10.0 软件、NI ELVIS 和 NI ELVIS 3.0 软件、NI LabVIEW Signal Express 软件。

【实验原理】

Multisim 仿真软件可以充分利用 SPICE 的优势与特点而不必了解低级仿真的具体细节。使用 Multisim 构建的虚拟仪器，可以通过点击鼠标快捷交互式地进行 SPICE 的仿真。这些虚拟仪器是模拟的台式仪器，可以连接到电路图中，并且通过常见的接口（例如示波器或频谱分析仪）报告仿真结果。随着对仿真概念和细节的进一步掌握，可以容易地实现更多的高级应用。Multisim 软件具备强大的仿真能力，并能大幅度简化电路仿真的过程。

1. 仿真工具栏

如果你有 Multisim 的 MCU 模块，你还可以使用仿真工具栏实现高级调试功能，包括：step into、step out of、step over 以及断点。图 5 – 27 – 1 所示即为仿真工具栏。

2. 虚拟仪器

Mulitisim 软件提供了多于 25 个的虚拟仪器来测量电路的仿真性能。这些虚拟仪器（例如示波器、函数发生器和万用表）连接到电路上，可以像实际的仪器一样工作。它

图 5 – 27 – 1　仿真工具栏

们与实验室中使用的仪器看上去是一样的。使用虚拟仪器是检测电路性能和显示仿真结果的最简单的方法。

使用 Multisim 软件，你还可以在电路图中放置某种特定的定制 LabVIEW 虚拟仪器。这些虚拟仪器可以帮助实现 Multisim 仿真的用户化，并扩展其仿真能力。它们并不需要安装 LabVIEW，并可以从 NI 的程序员区（ni.com/devzone，搜索"Multisim"）上直接下载。

3. 仪器工具栏

使用如图 5 – 27 – 2 所示的仪器工具栏，即可以放置任何所需要的 Mulitisim 虚拟仪器。

图 5 – 27 – 2　仪器工具栏

4. 示波器

Multisim 中默认的双通道示波器显示电气信号的幅度和频率变化，见表 5 – 27 – 1。它能提供一个或者两个时间信号强度的图表，或者帮助用户进行两个波形之间的比较。

表 5 - 27 - 1　示波器

工具栏图标	原理图符号	仪器前面板
	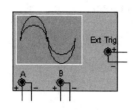	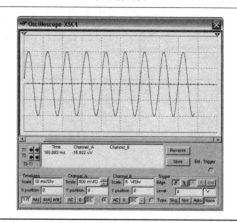

5. 波特绘图仪

波特绘图仪生成电路的频率响应的绘图，对分析滤波器电路是非常有用的。可以使用波特绘图仪来测量信号电压增益或者相移。将波特图附到电路中时，还可以进行频谱分析，如表 5 - 27 - 2 所示。

表 5 - 27 - 2　波特绘图仪

工具栏图标	原理图符号	仪器前端面板
	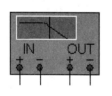	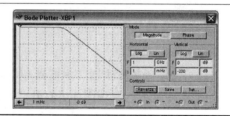

6. 函数发生器

函数发生器是一个可以提供正弦波、三角波和方波信号的电压源。它可以便利并实际地为电路提供激励信号。你可以改变波形并控制频率、幅度、占空比和直流偏移等。函数发生器的频率范围很大，足够产生常规的交流信号，以及音频和射频信号，见表 5 - 27 - 3。

表 5 - 27 - 3　函数发生器

工具栏图标	原理图符号	仪器前端面板

7. 测量探针

测量探针是测量电路中不同位置的电压、电流以及频率的一种快速而简便的方法。对电路进行仿真时,点击测量工具栏探针图标,则在电路中的任意节点上,光标将变成一个探针形状提示可以放置探针,见表 5 – 27 – 4。

表 5 – 27 – 4　测量探针

工具栏图标	原理图符号
	V: V(p-p): V(rms): V(dc): I: I(p-p): I(rms): I(dc): Freq.:

8. 分析功能

Multisim 提供了很多分析功能,这些功能都是使用仿真来产生所需分析功能的数据。这些分析功能,从基本的到极为复杂的,通常需要进行一项分析作为另一项分析的一部分。为了配置并开始一项分析,选择 Simulate/Analyses,并且选择需要的分析功能。对于每项分析而言,可以设定参数来告知 Multisim 如何准确地进行所需的分析。除了 Multisim 提供的分析功能外,还可以使用用户输入式 SPICE 命令来创建自定义的分析功能。

【实验步骤】

1. 练习

练习使用函数发生器、示波器和波特分析仪来测量带通滤波器的特性。直接利用 Exercise 1 中完成的原理图或者打开 "Exercise 1（Complete） – Bandpass Filter. ms10" 来进行练习。可用于仿真的原理图如图 5 – 27 – 3 所示。

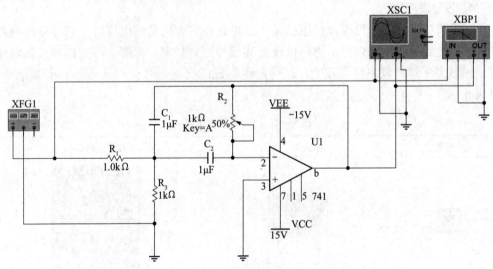

图 5 – 27 – 3　函数发生器高亮化后的仪器工具栏

2. 构建仿真电路

（1）放置并连接函数发生器

① 在仪器工具栏上首先点击函数发生器，它在图 5-27-4 中被圈出，并在原理图中再次点击以放置函数发生器。

图 5-27-4　函数发生器高亮化后的仪器工具栏

② 将函数发生器的正极端子连接到电路的输入上。

③ 将参考端子连接到地上。

你的电原理图看起来应该和图 5-27-5 一样。

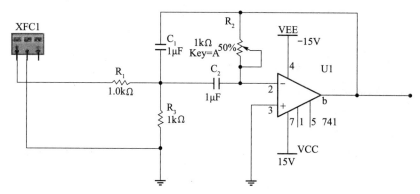

图 5-27-5　使用函数发生器的原理图

（2）放置并连接示波器

① 首先，在仪器工具栏上点击示波器，它在图 5-27-6 中被圈出。在原理图中再次点击以放置示波器。

图 5-27-6　示波器高亮化的仪器工具栏

② 将通道 A 和通道 B 的负极端子连接到地上（可以根据需要摆放地元件）。

③ 将通道 A 的正极端子连接到电路的输入端上。

④ 将通道 B 的正极端子连接到电路的输出端上。

现在改变连接到通道 B 的正极端子上的连线颜色。连接到示波器端子上的连线颜色要与示波器前面板上显示的曲线颜色相同。

⑤ 在连接到通道 B+上的红色连线上点击右键，并选择 Segment Color。在调色板对话框上选择暗蓝色。此时的原理图见图 5-27-7。

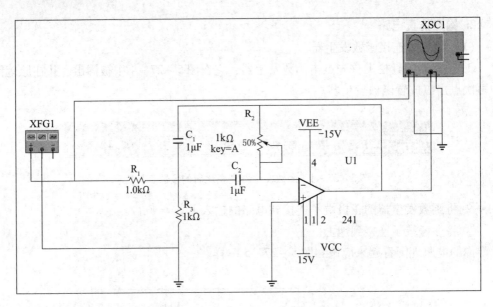

图 5-27-7 摆放了示波器的原理图

(3) 放置并连接波特分析仪

① 首先，在仪器工具栏上点击波特分析仪，它在图 5-27-8 中被圈出。在原理图中再次点击以放置波特分析仪。

图 5-27-8 波特分析仪被高亮化的仪器工具栏

② 将波特分析仪的 IN 和 OUT 的负极端子连接到地上。
③ 将 IN 正极端子连接到电路的输入端。
④ 将 OUT 正极端子连接到电路的输出端。

此时的电路可以用于仿真，如图 5-27-9 所示。

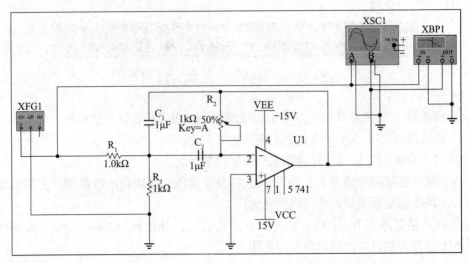

图 5-27-9 可进行仿真的完整原理图

图 5-27-10 仿真按钮被圈出的仿真工具栏

（4）使用函数发生器和示波器对电路进行仿真，并研究它的瞬态特性

① 在仿真工具栏上点击仿真按钮来启动仿真，此按钮在图 5-27-10 中被圈出。

② 在原理图上双击函数发生器和示波器的图标来打开它们的前面板。点击示波器上的 Reverse 按钮将背景颜色变成白色，这样可以更容易观察结果。

③ 按照图 5-27-11 和图 5-27-12 所示配置函数发生器和示波器，确保电路正常工作。

图 5-27-11 函数发生器配置

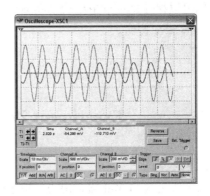

图 5-27-12 示波器配置

④ 在电路仿真过程中，研究电阻 R_2 的变化是如何影响电路的总增益和相位漂移的。为了改变电位计的阻值，将鼠标悬在元件上点击并拖动滑动条。

⑤ 将滑动条拉到 100% 时，表示元件具有 1kΩ 的完全阻值。

（5）使用波特分析仪测量电路的频率特性

① 从仿真工具栏上点击 Stop 按钮来停止仿真。

② 将电位器 R_2 的阻值调到 100%，重新开始仿真来获得电路的交流分析。

③ 双击波特分析仪，将前面板打开。点击 Reverse 按钮将背景变成白色。波特分析仪的前面板看起来应该类似于图 5-27-13。现在测量中心频率和增益。

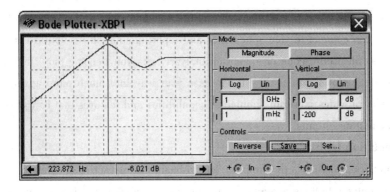

图 5-27-13 波特分析仪显示了中心频率的测量值

④ 在光标上点击右键，并且选择 Go To Next Y_ MAX => ，这就是中心频率。如

果中心频率不同，请确认电位计被设定为100%的阻值，停止并重新启动仿真。

期望的中心频率：223.9Hz

期望的总增益：-6.0dB

测量电路的-3dB点。-3dB点是增益在最大值下3dB处的频率。在这个例子中，-3dB点是电路增益为-9dB时的频率。

⑤ 在波特分析仪的前面板中移动光标来测量电路的-3dB点。

期望的下-3dB点：117Hz

期望的上-3dB点：433Hz

你可以保存这一章中生成的仿真结果，以与测量值进行比较，方法是在每个前面板中点击Save按钮。

（李百芳　亓　霞）

Experiment 27　Simulating Circuits

【Objects】

(1) To use NI Multisim 10.0, NI ELVIS and the NI ELVIS 3.0 software and NI LabVIEW Signal Express to simulate circuits.

(2) To use interactive simulation to measure the gain, bandwidth, and center frequency of the bandpass filter. Interact with components while simulation is running to see the effects of a changing feedback resistance on overall circuit operation.

【Apparatus】

NI Multisim10.0, NI ELVIS and the NI ELVIS 3.0 software, NI LabVIEW Signal Express.

【Theory】

With Multisim, you can gain the advantages of SPICE without needing knowledge of low-level simulation details. You can quickly and interactively conduct SPICE simulation with the click of a mouse using Multisim virtual instruments, which are simulated benchtop instruments that can connect to your schematics and report simulation results through familiar interfaces such as an oscilloscope or spectrum analyzer. As you become more familiar with the concepts and details of simulation, you can ease your way into much more advanced applications. Multisim provides powerful simulation capabilities and features that can greatly ease the process of simulating circuits.

1. Simulation toolbar

The Simulation toolbar is the simplest way to begin an interactive simulation of a circuit in Multisim. It contains buttons to start, stop, and pause simulation. If you have the Multisim MCU Module, you can also use the simulation toolbar for advanced debugging including stepping into, out of, and over, as well as breakpoints. Fig. 5-27-1 illustrates the simulation toolbar.

2. Virtual instruments

Multisim provides more than 25 virtual instruments to measure the simulated behavior of circuits. These instruments, such as oscilloscopes, function generators, and multimeters, connect to circuits, and you can use them just like their real-world counterparts. They look similar to instruments used in the lab. Using virtual instruments is the easiest way to examine circuit behavior and show the results of a simulation.

Fig. 5-27-1 Simulation toolbar

Using Multisim, you can also place certain custom LabVIEW virtual instruments on schematics. These instruments help you customize and extend Multisim simulation capabilities, do not require LabVIEW installation, and are downloadable for free from the NI Developer Zone (ni. com/devzone, search "Multisim").

3. Instruments toolbar

With the instruments toolbar, as shown in Fig. 5-27-2. You can place any of the Multisim virtual instruments.

图 5-27-2 Instruments toolbar

4. Oscilloscope

The default dual-channel oscilloscope in Multisim displays the magnitude and frequency variations of electronic signals, as shown in Tab. 5-27-1. It can provide a graph of the strength of one or two signals over time, or help you compare one waveform with another.

Tab. 5-27-1 Oscilloscope

toolbar icon	schematic symbol	instrument front panel

5. Bode plotter

The Bode plotter, which produces a graph of a circuit's frequency response, is most useful for analyzing filter circuits. You can use the bode plotter to measure a signal's voltage gain or phase shift. When you attach the bode plotter to a circuit, you can perform a spectrum analysis, as shown in Tab. 5 - 27 - 2.

Tab. 5 - 27 - 2 Bode plotter

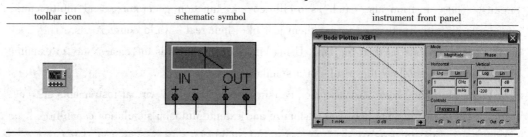

6. Function generator

The function generator is a voltage source that supplies sine, triangular, or square waves. It provides a convenient and realistic way to supply stimulus signals to a circuit. You can change the waveform and control its frequency, amplitude, duty cycle, and DC offset. The function generator's frequency range is great enough to produce conventional AC as well as audio - and radio - frequency signals, as shown in Tab. 5 - 27 - 3.

Tab. 5 - 27 - 3 Function generator

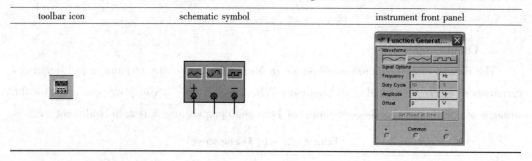

7. Measurement probes

The measurement probe is a fast and easy way to check voltage, current, and frequency readings at different points in the circuit. When simulating a circuit, click on the measurement probe toolbar icon, and the mouse cursor turns into a probe that informs you on any node in the circuit, as shown in Tab. 5 - 27 - 4.

Tab. 5 - 27 - 4 Measurement probe

toolbar icon	schematic symbol
	V: V(p-p): V(rms): V(dc): I: I(p-p): I(rms): I(dc): Freq.:

8. Analyses

Multisim offers many analyses, all of which use simulation to generate the data for the desired analysis. These analyses, which range from basic to extremely sophisticated, often require you to perform one analysis as part of another. To configure and begin an analysis, select Simulate/Analyses and choose the desired analysis. Fig. 5 – 27 – 1 lists all available Multisim analyses. For each analysis, you configure the settings that tell Multisim how to exactly perform the desired analysis. In addition to the analyses provided by Multisim, you can create user – defined analyses based on user – entered SPICE commands.

【Procedure】

1. Practise

This exercise uses the function generator, oscilloscope, and bode analyzer to measure the characteristics of the bandpass filter. You can work directly with the completed schematic from Exercise 1 or open "Exercise 1 (Complete) – Bandpass Filter. ms10" to continue the exercise. The schematic ready for simulation is shown in Fig. 5 – 27 – 3.

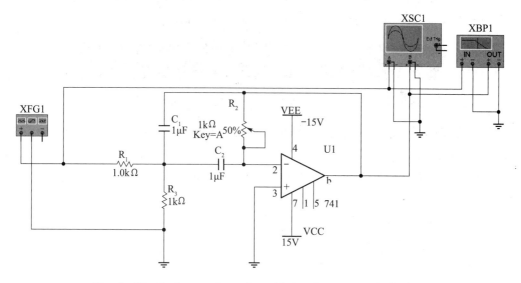

Fig. 5 – 27 – 3 Instruments toolbar with function generator highlighted

2. Conduct SPICE simulation

(1) Place and connect the function generator

① Click first on the function generator from the instruments toolbar circled in Fig. 5 – 27 – 4 and then click again on the schematic to place the instrument.

Fig. 5 – 27 – 4 Instruments toolbar with function generator highlighted

② Connect the positive terminal of the function generator to the input of the circuit.

③ Connect the center reference terminal to a ground terminal.

Your schematic should look like Fig. 5 - 27 - 5.

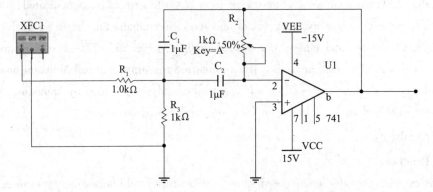

Fig. 5 - 27 - 5   Schematic with Function Generator

(2) Place and connect the oscilloscope

① Click first on the oscilloscope from the instruments toolbar circled in Fig. 5 - 27 - 6 and then click again on the schematic to place the instrument.

Fig. 5 - 27 - 6   Instruments toolbar with oscilloscope highlighted

② Connect the negative terminals of channels A and B to ground (place new ground components if needed).

③ Connect the positive trace of Channel A to the input of the circuit.

④ Connect the positive trace of Channel B to the output of the circuit.

Now change the color of the wire leading into the positive terminal of Channel B. The color of the wires connected to the terminals of this oscilloscope should match the color of the traces displayed on the oscilloscope's front panel.

⑤ Right - click on the red wire leading into channel B + and choose Segment Color. Pick a dark blue from the palette dialog.

Your schematic should look like Fig. 5 - 27 - 7.

(3) Place and connect the bode analyzer

① Click first on the bode analyzer from the instruments toolbar circled in Fig. 5 - 27 - 8 and then click again on the schematic to place the instrument.

② Connect the negative terminals IN and OUT of the bode analyzer to a ground terminal.

③ Connect the positive IN terminal to the input of the circuit.

④ Connect the positive OUT terminal to the output of the circuit.

Your circuit is now ready for simulation. It should resemble Fig. 5 - 27 - 9.

(4) Use the function generator and oscilloscope to simulate the circuit and explore its

transient characteristics

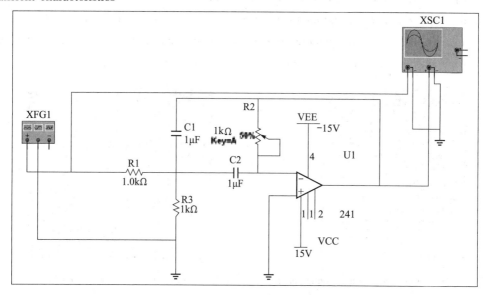

Fig. 5-27-7 Schematic with oscilloscope placed

Fig. 5-27-8 Instruments toolber with bode analyzer highlighted

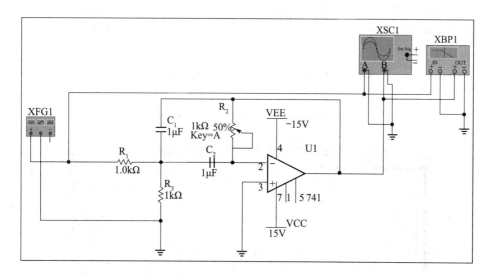

Fig. 5-27-9 Completed schematic ready for simulation

Fig. 5-27-10 Simulation toolbar with simulate brttom circled

① Start the simulation by clicking the simulate button from the simulation toolbar circled in Fig. 5-27-10.

② Open the function generator and oscillosco

pe front panels by double-clicking on their schematic symbols. Click the Reverse button on the oscilloscope to turn its background color white, making it easier to see the results.

③ Verify your circuit is operating as expected by configuring the function generator and oscilloscope as shown in Fig. 5-27-11 and Fig. 5-27-12.

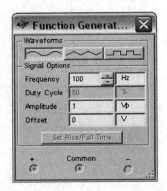

Fig. 5-27-11 Function generator configuration   Fig. 5-27-12 Oscilloscope configuration

④ While the circuit is simulating, investigate how to change the resistance of R_2 affects the overall gain and phase shift of the circuit. To change the resistance of the potentiometer, hover the mouse over the component and click and drag the slider.

⑤ Drag the slider to 100 percent, which represents the full 1 kΩ resistance of the component.

(5) Using the bode analyzer, measure the frequency characteristics of the circuit

① Stop simulation by clicking the Stop button from the simulation toolbar.

② Start the simulation to rerun the AC analysis to reflect the new value of 100 percent for the potentiometer R_2.

③ Open the bode analyzer front panel by double-clicking on it. Click the Reverse button to turn the background color white. The bode analyzer front panel should look similar to Fig. 5-27-13. Now measure the center frequency and gain.

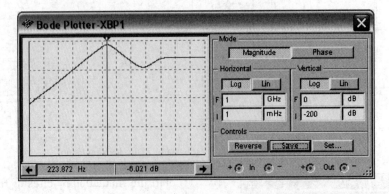

Fig. 5-27-13 Bode analyzer showing center frequency measuieme

④ Right-click on the cursor and choose Go To Next Y_ MAX = >. This is the center frequency. If your center frequency is different, verify that the potentiometer is set to 100 percent and stop and restart the simulation.

Expected center frequency: 　　　　　　223.9 Hz

Expected overall gain: 　　　　　　　　-6.0 dB

Measure the -3 dB points of the circuit. The -3 dB points are the frequencies at which the gain is 3 dB below its maximum value. In this case, the -3 dB points are the frequencies where the gain of the circuit is -9 dB.

⑤ Drag the cursor on the front panel of the Bode analyzer to measure the two -3 dB points of the circuit.

You can save the simulation results generated in this section for comparison with measurements by clicking the Save button from each front panel.

（Qi Xia, Wang Xiao-fei）

实验二十八　使用 myDAQ 捕获声音信号

【实验目的】

（1）使用 myDAQ 通过 LabVIEW 获取数据并把数据显示在用户界面上。

（2）编写一个 LabVIEW VI，实现从 myDAQ 上读取音频信号，并能够显示信号的频段。

【实验器材】

myDAQ、麦克风、LabVIEW 软件、计算机（含操作系统）。

【实验原理】

将一个麦克风连接到 myDAQ 上的音频输入，编写一个 LabVIEW VI，该 VI 将实现从 myDAQ 上读取音频信号，并显示信号的频段。

【实验步骤】

1. 从 myDAQ 上读取音频信号，并显示信号的频段

（1）启动 LabVIEW。

（2）在 LabVIEW 的开始界面中，选择打开现有的 VI，并找到 Audio Equalizer Starting Point. vi，如图 5-28-1 所示。

（3）可以看到 NI 已经创建了前面板的主要元素，一个波形图，一个停止按钮和一个选项卡控件。我们将在接下来的练习中加入更多的控件在前面板上。

（4）通过选择 Window Show Block Diagram 或者按 <Ctrl-E> 打开程序框图，你可以看到和前面板相应的三个接线端，如图 5-28-2 所示。

（5）放置一个 DAQ 助手 vi 到程序框图中。

①右键单击程序框图打开函数选板，如图 5-28-3 所示。

②通过 Express » Input » DAQ Assistant 选择 DAQ 助手。

③单击程序框图放置 DAQ 助手。

（6）配置 DAQ 助手

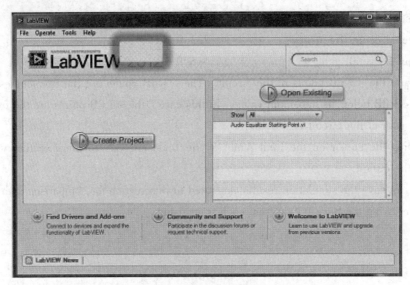

图 5-28-1　LabVIEW 的启动界面

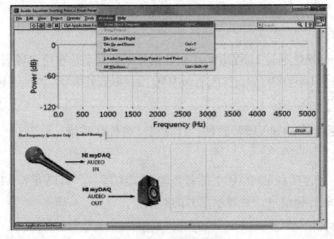

图 5-28-2　程序框图显示

①快速创建新任务窗口会弹出。如果没有弹出，双击 DAQ 助手图标。通过导航选择 Acquire Signals Analog Input Voltage。

②在下一个窗口中，展开 myDAQ 的输入并选择 audioInputLeft 和 audioInputRight（按住 Shift 键可以选择多个通道）。最后点击 Finish。

③接下来 DAQ 助手窗口出现。在信号输入范围区域内，修改 Max 和 Min 的值为 1.5V 和 -1.5V。确保两个通道（Voltage_0 和 Voltage_1））都进行了修改。在定时设置中，修改采样模式为连续采集，设置每通道采样数为 5k，并把采样率设置为 20k。你的窗口应该和图 5-28-4 所示一致。

图 5-28-3　函数选板显示

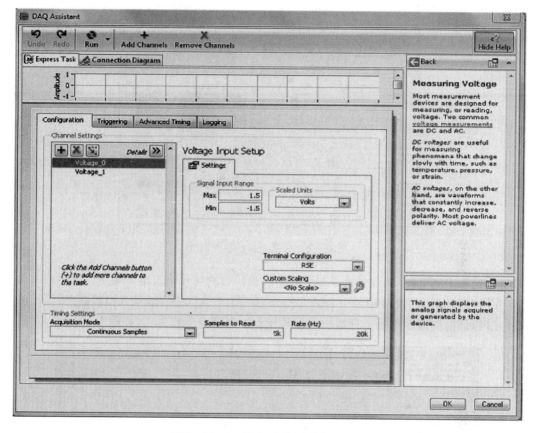

图 5-28-4　MyDAQ 的助手窗口显示

④配置结束后，点击 OK 并等待该函数创建。当创建完成后，你可能会看到确保自动创建循环的提示，如图 5-28-5 所示。如果你不希望该 VI 再次提示，选择 No。

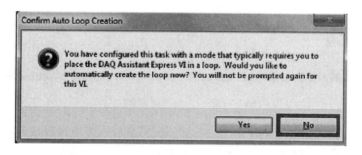

图 28-5　函数创建窗口显示

(7) 放置一个频谱测量函数到程序框图中，如图 5-28-6 所示。

①通过函数选板导航到 Express《Signal Analysis》Spectral，创建一个频谱测量函数。放置该函数到程序框图中。

②在配置频谱测量窗口中，修改选择测量为幅值（峰值）。其他设置保持默认值。并点击 OK。

(8) 连接函数和测试 VI。

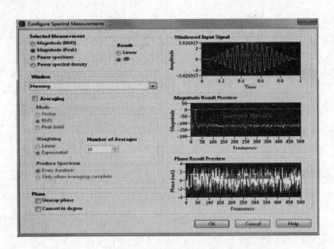

图 28-6 频谱测量函数的设置

①连接 DAQ 助手的数据接线端至频谱测量函数的信号输入接线端上。
②连接频谱测量的 FFT-（Peak）输出接线端至波形图表的输入接线端上。
③返回前面板并点击运行按钮（工具栏左上角的白色箭头），如图 5-28-7 所示。

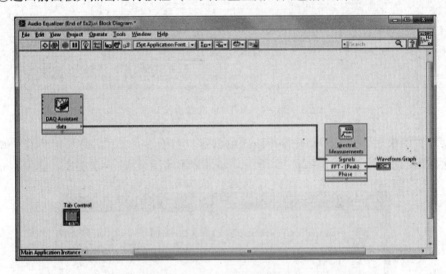

图 5-28-7 返回前面板运行

当你点下运行按钮后看到什么结果？让不同声音进入麦克风时试着多次运行该 VI。波形图表上的结果是否有改变？为什么在图表上有两条不同的线（蓝色和灰色）？

2. 连续采集数据

通常情况下，用户会希望能够连续采集数据。在 LabVIEW 中为了能够重复执行一个操作，我们使用 While 循环。这个结构能够重复执行包含在其中的代码直至满足用户定义的停止条件。在这种情况下，我们会使用一个前面板上的停止按钮作为停止条件。

（1）放置一个 While 循环在程序框图中。
①在函数选板中选择 Programming » Structures » While Loop。
②在程序框图中单击左上角，并拖拽鼠标至右下方来放置一个 While 循环。该循环

应该将你程序框图中的所有对象都包含在内。

③While 循环右下角的停止标志是条件接线端，你可以用它来停止循环。连接停止按钮至条件接线端。你的 VI 显示应该如图 5-28-8 所示。

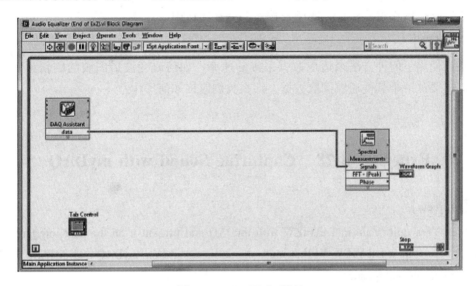

图 5-28-8　While 循环

（2）返回前面板。

（3）移动你的停止按钮（在 While 循环中创建的）到一个你可以方便点击的位置。你可以通过选中停止按钮并拖拽来实现该动作。前面板如图 5-28-9 所示。

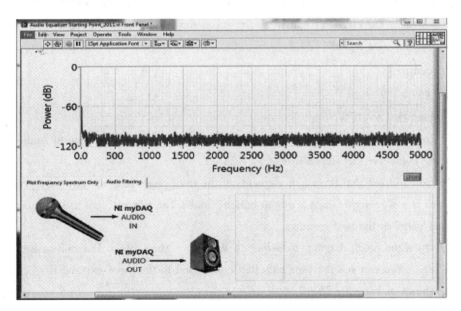

图 5-28-9　前面板显示

（4）运行该 VI，检验该 VI 是否实现了你预期的功能。

注意：

可以从函数选板中放置 DAQ 助手至程序框图中。右键点击程序框图打开函数选板，选择 Express Input 找到 DAQ 助手。当你打开函数选板后，你可以点击左上角的图钉按钮，这个操作可以保证你的函数选板一直在窗口中。

【思考题】

我们使用了何种方法来适应数据采集的速率？如何改变波形图表的刷新频率？如果我们想要从一个其他通道读取数据，那么我们应该如何操作？

（赵 喆 亓 霞）

Experiment 28　Capturing Sound with myDAQ

【Objects】

（1）To acquire data in LabVIEW with myDAQ and present it on the user interface.

（2）To write a LabVIEW VI that reads the audio signal from myDAQ and displays the frequency band of that signal.

【Apparatus】

myDAQ, a microphone, LabVIEW software, computer（including OS）.

【Theory】

We have a microphone connected to the audio input of the myDAQ. In this exercise, we're going to write a LabVIEW VI that reads the audio signal from the my DAQ and displays the frequency band of that signal.

【Procedure】

Part 1

（1）Launch LabVIEW.

（2）In the Getting Started window, select Open Existing, and navigate to Audio Equalizer Starting Point. vi, as shown in Fig. 5 - 28 - 1.

（3）You will see that the major elements of the front panel have already been created for you. There is a Waveform Graph, a Stop button, and a Tab Control, we will add more items to the front panel in the next exercise.

（4）Open the block diagram by selecting Window ≪ Show Block Diagram or by pressing ＜Ctrl－E＞. You can see the terminals that correspond to the three existing front panel objects, as shown in Fig. 5 - 28 - 2.

（5）Place a DAQ Assistant on the block diagram.

①Right - click on the block diagram to open the Functions Palette, as shown in Fig. 5 - 28 - 3.

②Select the DAQ Assistant by navigating to Express ≫ Input ≫ DAQ Assistant.

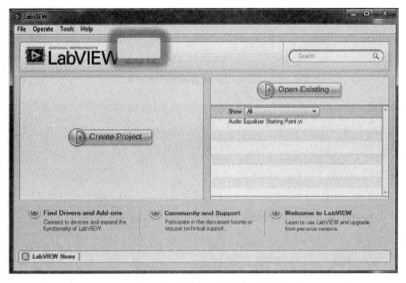

Fig. 5-28-1 The launch LabVIEW display

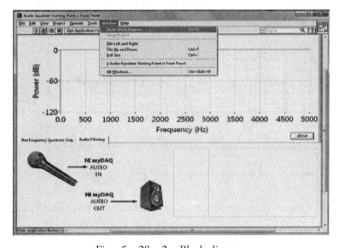

Fig. 5-28-2 Block diagram

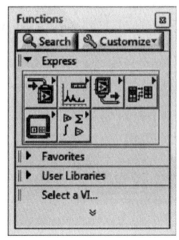

Fig. 5-28-3 The functions palette

③Click on the block diagram to place the DAQ Assistant.

(6) Configure the DAQ Assistant.

①The Create New Express Task window will appear. If it does not, double-click on the DAQ Assistant icon. Navigate to Acquire Signals » Analog Input » Voltage.

②In the next window, expand the myDAQ entry and select audio Input Left and audio Input Right (hold down the Shift key to select multiple channels). Click Finish.

③The DAQ Assistant window appears. In the Signal Input Range area, change the Max and Min to 1.5 and -1.5 Volts. Make sure to do this for both channels (Voltage_ 0 and Voltage_ 1). Under Timing Settings, change the Acquisition Mode to Continuous Samples, set Samples to Read to 5k, and Rate (Hz) to 20k. Your window should match the one shown in Fig. 5-28-4.

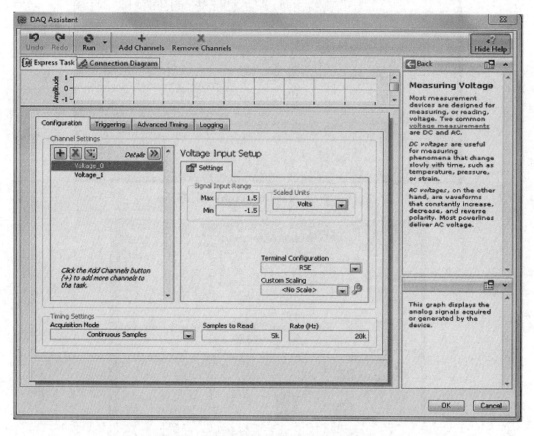

Fig. 5-28-4 The DAQ Assistant Window

④When finished, click OK and wait for the function to build. When the build is complete, you may see the Confirm Auto Loop Creation prompt. Should this occur, select No. See Fig. 5-28-5.

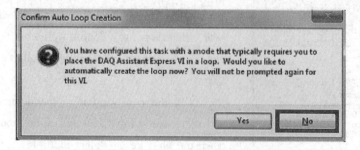

Fig. 5-28-5 Confirm Auto Loop Creation prompt

(7) Place a Spectral Measurements function on the block diagram, as shown

in Fig. 5 – 28 – 6.

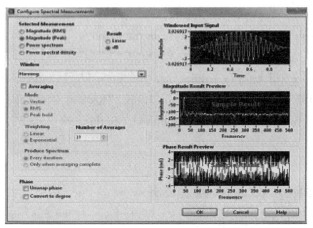

Fig. 5 – 28 – 6 Create a Spectral Measurements function

①Create a Spectral Measurements function by navigating to Express « Signal Analysis « Spectral in the Functions palette. Place the function on the block diagram.

②In the Configure Spectral Measurements window, change the Selected Measurement to Magnitude (Peak). Leave all other default settings unchanged. Click OK.

(8) Connect the functions and test the VI.

①Wire the data terminal of the DAQ Assistant to the Signals input of the Spectral Measurements.

②Wire the FFT – (Peak) output of the Spectral Measurements to the input terminal of the Waveform Graph.

③Return to the front panel and click the Run button (the white arrow in the upper left corner), as shown in Fig. 5 – 28 – 7.

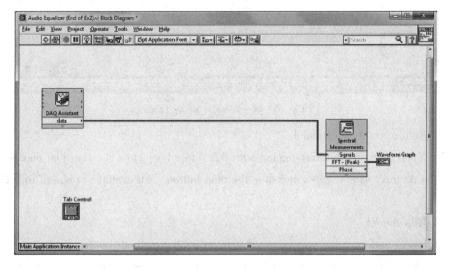

Fig. 5 – 28 – 7 Return to the front panel and Run

· 335 ·

What did you see when you clicked the Run button? Try running the VI multiple times while making different sounds into the microphone. Do the results on the graph change? Why are there two different lines (blue and gray) on the graph?

Part 2

Oftentimes a user will want to acquire data continuously. In order to repeat an operation indefinitely in LabVIEW, we can use a While Loop. This will repeat the code contained inside until a user-defined stop condition is achieved. In this case, we will use the Stop button on the front panel.

【Procedure】

(1) Place a While Loop on the block diagram.

①Navigate to Programming « Structures « While Loop on the Functions palette.

②Click in the upper left corner of the block diagram, and drag your mouse to the lower right corner to place the While Loop. The loop should surround all the objects on the diagram.

③The stop sign icon in the lower right corner of the loop is the conditional terminal, which is used to stop the loop. Wire the Stop button terminal into the conditional terminal. Your VI should look like the image on the next page. As shown in Fig. 5-28-8.

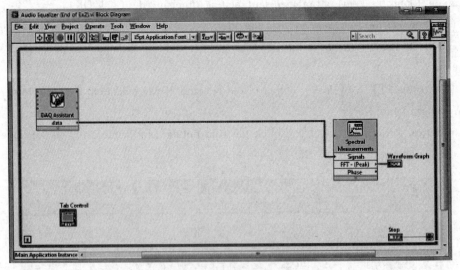

Fig. 5-28-8 The While Loop

(2) Return to the front panel.

(3) Move your Stop button (created with the While Loop) to a location that makes it easy to use. To do this, simply click and drag the Stop button. An example is shown in Fig. 5-28-9.

(4) Run the VI.

Did the VI perform what you expected it to do? What methods could we use to adjust the rate at which data is acquired? How could we change the frequency of updating the Waveform Graph? If we wanted to read in data from a different channel, what would we do?

Note

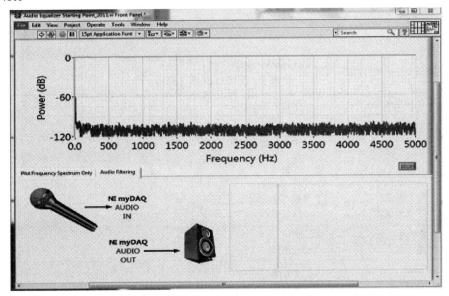

Fig. 5-28-9   The front panel

You can place the DAQ Assistant on your block diagram from the functions palette. Right-click the block diagram to open the functions palette and go to Express « Input to find it. When you bring up the functions palette, press the small pushpin in the upper left-hand corner of the palette. This tacks down the palette so that it doesn't disappear. This step is omitted in the following exercises, but you should repeat it.

(Zhao Zhe, Zhi Zhuang-zhi)

附　表

附表1　国际单位制（SI）

	物理量名称	单位名称	单位符号	用其他SI单位表示式
基本单位	长度	米	m	
	质量	千克（公斤）	kg	
	时间	秒	s	
	电流	安培	A	
	热力学温度	开［尔文］	K	
	物质的量	摩［尔］	mol	
	发光强度	坎［德拉］	cd	
辅助单位	［平面］角	弧度	rad	
	立体角	球面度	sr	
导出单位	面积	平方米	m^2	
	速度	米每秒	m/s	
	加速度	米每二次分秒	m/s^2	
	密度	千克每立方米	kg/m^3	
	频率	赫［兹］	Hz	s^{-1}
	力	牛［顿］	N	$kg \cdot m/s^2$
	压强、压力	帕［斯卡］	Pa	N/m^2
	功、能量、热量	焦［耳］	J	$N \cdot m$
	功率、辐射能量	瓦［特］	W	J/s
	电量、电荷	库［仑］	C	$A \cdot s$
	电势、电压、电动势	伏［特］	V	W/A
	电容	法［拉］	F	C/V
	电阻	欧［姆］	Ω	V/A
	磁通	韦［伯］	Wb	$V \cdot s$
	磁感强度	特［斯拉］	T	Wb/m^2
	电感	亨［利］	H	Wb/A
	光通量	流［明］	lm	
	光照度	勒［克斯］	lx	lm/m^2
	黏度	帕［斯卡］秒	$Pa \cdot S$	
	表面张力	牛［顿］每米	N/m	
	比热容	焦［耳］每千克开［尔文］	$J/(kg \cdot K)$	
	热导率	瓦［特］每米开［尔文］	$W/(m \cdot K)$	
	电容率	法［拉］每米	F/m	
	磁导率	亨［利］每米	H/m	

附表2　基本物理常量（1986年推荐值）

名　称	符　号	量　值
牛顿引力常量	G	$6.67259(85) \times 10^{-11} N \cdot m^2 \cdot kg^{-2}$
真空电容率	ε_0	$8.8542\cdots \times 10^{-12} F \cdot m^{-1}$
真空磁导率	μ_0	$12.566370614\cdots \times 10^{-7} H \cdot m^{-1}$
真空中的光速	c	$2.99792458 \times 10^8 m \cdot s^{-1}$
阿伏伽德罗常量	N_A	$6.0221367(36) \times 10^{23} mol^{-1}$
理想气体的摩尔体积	V_m	$0.0224136(19) m^3 \cdot mol^{-1}$
玻尔兹曼常量	k	$1.380658(12) \times 10^{-23} J \cdot K^{-1}$

续表

名　称	符　号	量　值
气体常量	R	8.314 510 (70) J·mol^{-1}·K^{-1}
质子质量	m_p	1.672 623 1 (10) ×10^{-27} kg
电子质量	m_e	9.109 389 7 (54) ×10^{-31} kg
元电荷	e	1.602 1892 ×10^{-19} C
普朗克常量	h	6.626 075 5 (40) ×10^{-34} J·s
里德伯常量	R_∞	1.097 373 12 ×10^7 m^{-1}

注：本表根据最小二乘法平差得出，括号内的数字是给定值最后几位数的一个标准偏差的不确定度。

附表3　水的密度（kg/m^{-3}）与温度的关系

温度（℃）	密度	温度（℃）	密度	温度（℃）	密度	温度（℃）	密度
0	999.840	14	999.244	28	996.231	42	991.432
1	999.898	15	999.099	29	995.943	43	991.031
2	999.940	16	998.943	30	995.646	44	990.623
3	999.964	17	998.774	31	995.339	45	990.208
4	999.972	18	998.595	32	995.024	46	989.786
5	999.965	19	998.404	33	994.700	47	987.358
6	999.940	20	998.203	34	994.369	48	988.922
7	999.901	21	998.991	35	994.029	49	988.479
8	999.848	22	997.769	36	993.681	50	988.040
9	999.781	23	997.537	37	993.325	60	983.191
10	999.699	24	997.295	38	992.962	70	977.770
11	999.605	25	997.044	39	992.591	80	971.790
12	999.497	26	996.782	40	992.212	90	965.304
13	999.377	27	996.511	41	991.826	100	958.360

附表4　20℃时常用固体和液体的密度（×10^3）

物质	密度 ρ(kg·m^{-3})	物质	密度 ρ(kg·m^{-3})
铝	2.70	醋酸	1.05
铜	8.90	丙酮	0.79
金	19.3	碳酸	1.07
铁	7.86	蓖麻油	0.96
铅	11.3	三氯甲烷	1.49
铂	21.5	汽油	0.68
银	10.5	甘油	1.26
钽	16.8	煤油	0.800
锡	7.30	润滑油	0.900
锌	6.92	汞	13.60
云母	2.90	硫酸	1.83
石英	2.60	硝酸	1.50
石膏	2.32	纯乙醇	0.80

附表5 水的黏滞系数（10^{-6}Pa·s）与温度的关系

温度（℃）	黏滞系数	温度（℃）	黏滞系数	温度（℃）	黏滞系数	温度（℃）	黏滞系数
0	1787.8	11	1271.3	22	957.0	33	752.3
1	1731.3	12	1236.3	23	935.8	34	737.1
2	1672.8	13	1202.8	24	914.2	35	722.5
3	1619.1	14	1170.9	25	893.7	36	708.5
4	1567.4	15	1140.4	26	873.7	37	694.7
5	1518.8	16	1111.1	27	854.8	38	681.4
6	1472.8	17	1082.8	28	836.0	39	668.5
7	1428.4	18	1055.6	29	818.0	40	656.0
8	1386.0	19	1029.9	30	800.7		
9	1346.2	20	1005.0	31	784.0		
10	1307.7	21	981.0	32	767.9		

附表6 几种液体的黏滞系数

液体	温度（℃）	黏度（10^{-6}Pa·s）	液体	温度（℃）	黏度（10^{-6}Pa·s）
汽油	0	1788	甘油	-20	134×10^6
	18	530		0	121×10^5
甲醇	0	817		20	1499×10^3
	20	584		100	12 945
乙醇	-20	2780	蜂蜜	20	650×10^4
	0	1780		80	100×10^3
	20	1190	鱼肝油	20	45600
乙醚	0	296		80	4600
	20	243	汞	-20	1855
变压器油	20	19800		0	1685
蓖麻油	10	242×10^4		20	1554
葵花籽油	20	50000		100	1224

附表7 在不同温度下与空气接触的水的表面张力系数 α

温度（℃）	α（10^{-3} N·m^{-1}）	温度（℃）	α（10^{-3} N·m^{-1}）	温度（℃）	α（10^{-3} N·m^{-1}）
0	75.62	16	73.34	30	71.15
5	74.90	17	73.20	40	69.55
6	74.76	18	73.05	50	67.90
8	74.48	19	72.89	60	66.17
10	74.20	20	72.75	70	64.41
11	74.07	21	72.60	80	62.60
12	73.92	22	72.44	90	60.74
13	73.78	23	72.28	100	58.84
14	73.64	24	72.12		
15	73.48	25	71.96		

附表8 几种液体的表面张力系数（20℃）

液体	$\alpha \times 10^{-3}$ (N/m)
水	73
甘油	65
乙醚	17
乙醇	22
汞	540

附表9 常见物质的折射率（20℃）

物质	n
水	1.333
乙醇	1.362
丙酮	1.359
三氯甲烷	1.447
玻璃	1.52～1.752
岩盐	1.544
金刚石	2.417
丁香油	1.530～1.535
薄荷油	1.460～1.417

附表10 常用药物的比旋光度 $[\alpha]_D^{25℃}$

药物名称	比旋光度	药物名称	比旋光度
葡萄糖	+52.5°～+53°	茴香油	+12°～+24°
麦芽糖	+132.5°	桂皮油	-1°～+1°
转化糖	-19.7°	葡甲胺	-15.5°～-17.5°
蔗糖	+66°	青霉素钠	+301°
可的松（乙醇溶剂）	+209°	硫酸链霉素	-84°
红霉素（乙醇溶剂）	-78°	吗啡（甲醇溶剂）	-132°
氯霉素（乙醇溶剂）	+18.6°	盐酸麻黄碱	-35.5°
土霉素	-27.1°	奎宁碱（乙醇溶剂）	-169°
乳酸	+3.8°	肾上腺素（0.5mol/dm^{-3}HCl溶剂）	-53.5°
薄荷油	-18°～-32°		

附表11　热电偶电动势的基本值

正端 负端	铜 康铜	铁 康铜	镍-铬 镍	铂铑 铂
测量温度（℃）	基　本　值			
	mV	mV	mV	mV
-200	-5.7	-8.15		
-100	-3.40	-4.75		
0	0	0	0	0
100	4.25	5.37	4.10	0.643
200	9.20	10.95	8.13	1.436
300	14.90	16.56	12.21	2.316
400	21.00	22.16	16.40	3.251
500	27.41	27.85	20.65	4.221
600	34.31	33.64	24.91	5.224
700		39.72	29.14	6.260
800		46.22		7.329
900		53.14	33.30	8.432
1000			37.36	9.570
1100			41.31	10.741
1200			45.16	11.935
1300			48.89	13.138
1400			52.46	14.337
1500				15.530
1600				16.716

注：在 0~400℃（对铂铑-铂热电偶是 0~600℃）范围内，允许偏差是 ±3℃，超过此范围时，允许偏差是 ±0.75%（铂铑-铂热电偶为 ±0.5%）。
表中台阶粗线（根据工作经验）表示在洁净空气中长时间使用热电偶时的极限温度。

附表12　常用金属的弹性模量

名称	弹性模量 E (10^9 N·m^{-2})	名称	弹性模量 E (10^9 N·m^{-2})
铝	68.7	铝合金 1100	68.7
铜	108	铝青铜合金	33.4
金	76.5	铍青铜合金	35.3
银	73.6	碳钢 AISI1020	207
锌	88.3	不锈钢	196
镍	206	合金铜	200
铬	240	钛合金	114

附表 13　部分材料的导热系数

名称	密度（kg·m^{-3}）	导热系数（J·m^{-2}·K^{-1}）
空气（0℃）		2.4×10^{-2}
氮气（0℃）		1.4×10^{-1}
铝		2.0×10^{2}
铜		3.9×10^{2}
钢		4.6×10^{1}
钢筋混凝土*	2400	1.55
碎石混凝土	2000	1.16
粉煤灰矿砂混凝土	1930	0.70
大理石、花岗石、玄武石	2800	3.49
砂石、石英石	2400	2.03
重石灰岩	2000	1.16
矿渣砖	1400	5.8×10^{-1}
砂（湿度<1%）	1600	8.1×10^{-1}
胶合板	600	1.7×10^{-1}
软木板	180	5.6×10^{-1}
沥青油毡	600	1.7×10^{-1}
石棉板	300	4.7×10^{-2}
聚氯乙烯（塑料泡沫）	18.9	3.0×10^{-2}
聚氨酯	324	2.0×10^{-2}

* 有关数据是在正常温度条件测定，否则将有较大差异。

（张红艳　支壮志）

参考文献

[1] 钱峰，潘人培．大学物理实验［M］．修订版．北京：高等教育出版社，2005．
[2] 吕斯骅 段家忯．新编基础物理实验［M］．北京：高等教育出版社，2006．
[3] 周殿清．基础物理实验［M］．北京：科学出版社，2009．
[4] 陈玉林，李传起．大学物理实验［M］．北京：科学出版社，2007．
[5] 张平．大学物理实验［M］南京：南京大学出版社，2008．
[6] 梁家惠，李朝荣，徐平，等．基础物理实验［M］．北京：北京航空航天大学出版社，2006．
[7] 邱菊，崔丽彬，孔爽，等．大学物理实验［M］．北京：北京工业大学出版社，2010．
[8] 李寿松．物理实验教程［M］．2版．高等教育出版社，2003．
[9] 邓金祥，刘国庆．大学物理实验［M］．北京：北京工业大学出版社，2005．
[10] 钟鼎．大学物理实验［M］．天津：天津大学出版社，2006．
[11] 曾照芳．临床检验仪器学［M］．2版．北京：人民卫生出版社，2011．
[12] 朱蕴璞，孔德仁，王芳．传感器原理及应用［M］．北京：国防工业出版社，2005．
[13] 俞阿龙，李正，孙红兵，等．传感器原理及其应用［M］．南京：南京大学出版社，2009．
[14] 周旭．现代传感器技术［M］．北京：国防工业出版社，2007．
[15] 王成．医疗仪器原理［M］．上海：上海交通大学出版社，2008．
[16] 武文君．多参数监护仪质量控制检测技术［M］．北京：中国计量出版社，2010．
[17] 吴建刚．现代医用电子仪器原理与维修［M］．北京：电子工业出版社，2005．
[18] 朱文玉．人体解剖生理学［M］．北京：北京大学医学出版社，2002．
[19] 吴建平，传感器原理及应用［M］．北京：机械工业出版社，2012．
[20] 刘克哲．物理学［M］．北京：高等教育出版社，2001．
[21] 胡新珉．医学物理学［M］．北京：人民卫生出版社，2001．
[22] 遐龄．医用物理学［M］．长春：吉林科学技术出版社，2000．
[23] 张三慧．大学基础物理学［M］．北京：清华大学出版社，2011．
[24] 世博强．LabVIEW 6.1编程技术实用教程［M］．北京：中国铁道出版社，2002．
[25] 汪敏生．LabVIEW基础教程［M］．北京：电子工业出版社，2002．
[26] 刘君华，贾惠芹，丁晖等．虚拟仪器图形化编程语言LabVIEW教程［M］．西安：西安电子科技大学出版社，2001．
[27] Robert H B. LabVIEW 6.1实用教程［M］．乔瑞萍，等．译．北京：电子工业出版社，2003．
[28] 赵会兵．虚拟仪器技术规范与系统集成［M］．北京：清华大学出版社/北方交通大学出版社，2003．